AF443342

Advances in Experimental Medicine and Biology

Volume 1041

More information about this series at http://www.springer.com/series/5584

Alexander Birbrair

Editor

Stem Cell Microenvironments and Beyond

 Springer

Editor
Alexander Birbrair
Department of Pathology
Federal University of Minas Gerais
Belo Horizonte, MG, Brazil

ISSN 0065-2598   ISSN 2214-8019   (electronic)
Advances in Experimental Medicine and Biology
ISBN 978-3-319-69193-0   ISBN 978-3-319-69194-7   (eBook)
DOI 10.1007/978-3-319-69194-7

Library of Congress Control Number: 2017959608

Printed on acid-free paper

This Springer imprint is published by Springer Nature
The registered company is Springer International Publishing AG
The registered company address is: Gewerbestrasse 11, 6330 Cham, Switzerland

Preface

This book *Stem Cell Microenvironments and Beyond* presents contributions by expert researchers and clinicians in the multidisciplinary areas of medical and biological research. The chapters provide timely detailed overviews of recent advances in the field. The texts are about stem cell microenvironments in different tissues and under distinct pathophysiological conditions. The authors focus on the modern methodologies and the leading-edge concepts in the field of stem cell biology. In recent years, remarkable progress has been made in the identification and characterization of the stem cell niches using state-of-the-art techniques. These advantages facilitated the identification of cellular components of the stem cell niche and the definition of the molecular basis of physical interaction between stem cells and their niches and revealed key niche signals involved in stem cell regulation. Just like the ecological niche of an organism, a stem cell niche is unique to the individual or small population and guides its dynamics. This book describes the major components of various stem cell microenvironments such as soluble factors, cell-cell interactions, extracellular matrix proteins, and physical forces. Thus, this book is an attempt to describe the most recent developments in the area of stem cell behavior regulation which is one of the emergent hot topics in the field of molecular and cellular biology today. Here, we present a selected collection of detailed chapters on what we know so far about the stem cell niches in various tissues and under distinct pathophysiological conditions. Twelve chapters written by experts in the field summarize the present knowledge about the physiological and pathophysiological roles of tissue microenvironments in stem cell regulation.

Daniel Lucas from the University of Michigan School of Medicine introduces our current understanding of the hematopoietic stem cell niche and discusses some of the open questions in the field for future research. Marina Konopleva and Yoko Tabe from the University of Texas MD Anderson Cancer Center describe recent research on several key components of specific niches that provide a sanctuary where leukemia stem cells evade chemotherapy-induced death and acquire a drug-resistant phenotype. Teresa V. Bowman and colleagues from Albert Einstein College of Medicine discuss knowledge that we gained from zebrafish about niche factors critical for early hemogenic endothelial induction as well as hematopoietic stem cell specification, migration, and expansion. Raúl E. Russo and colleagues from Instituto de Investigaciones Biológicas Clemente Estable focus on spinal cord ependymal neural stem cell niche regulation. Ilias Kazanis and colleagues from the University of Cambridge summarize the recent developments on the role of the microenvironment and how it affects neural stem cells in the brain. Akiva Mintz and his group from Columbia University Medical Center introduce the concept of glioblastoma stem cells and detail the latest findings within the microenvironment where these cells survive, proliferate, and differentiate. Christoph Handschin and colleagues from the University of Basel give an overview of the players in the skeletal muscle stem cell microenvironment and their mutual interactions with stem cells. Kiminori Sato from Kurume University School of Medicine addresses the importance of the maculae flavae of the human vocal fold as a stem cell microenvironment. Maria P. Alcolea from Wellcome Trust-Medical Research Council Cambridge Stem Cell Institute compiles recent observations on esophageal epithelial stem cell biology and how microenvironmental changes may lead to esophageal disease and cancer. Sujit K Bhutia and colleagues from the National Institute of Technology discuss the dynamic interplay between oral cancer stem cells and the tumor microenvironment in carcinogenesis. Maria Angelica Miglino and Phelipe Oliveira Favaron from the University of Sao Paulo describe the microenvironment and applications of yolk sac and amniotic membrane-derived stem cells for human and veterinary regenerative medicine. Finally, Carmine Gentile and colleagues from the University of Sydney update us with the latest technologies based on our knowledge of the stem cell niche and current approaches for engineering artificial stem cell microenvironments.

It is hoped that the articles published in this book will become a source of reference and inspiration for future research ideas. I would like to express my deep gratitude to my wife Veranika Ushakova and Mr. Sivachandran Ravanan from Springer, who helped at every step of the execution of this project.

Belo Horizonte, MG, Brazil   Alexander Birbrair

Contents

Contributors

Maria P. Alcolea Wellcome Trust-Medical Research Council Cambridge Stem Cell Institute, Cambridge, UK

Department of Oncology, University of Cambridge, Hutchison/MRC Research Centre, Cambridge, UK

Evangelia Andreopoulou Lab of Developmental Biology, Department of Biology, University of Patras, Patras, Greece

Asterios Arampatzis Wellcome Trust-MRC Cambridge Stem Cell Biology Institute, University of Cambridge, Cambridge, UK

School of Medicine, Aristotle University of Thessaloniki, Thessaloniki, Greece

Sujit K. Bhutia Department of Life Science, National Institute of Technology, Rourkela, Odisha, India

Alexander Birbrair Department of Pathology, Federal University of Minas Gerais, Belo Horizonte, MG, Brazil

Teresa V. Bowman Gottesman Institute for Stem Cell Biology and Regenerative Medicine, Albert Einstein College of Medicine, Bronx, NY, USA

Department of Developmental and Molecular Biology, Albert Einstein College of Medicine, Bronx, NY, USA

Departments of Molecular Biology and Medicine (Oncology), Albert Einstein College of Medicine, Bronx, NY, USA

Ivana Dinulovic Biozentrum, University of Basel, Basel, Switzerland

Phelipe Oliveira Favaron Surgery Department, School of Veterinary Medicine and Animal Science, University of Sao Paulo, Sao Paulo, SP, Brazil

Gemma Figtree Sydney Medical School, University of Sydney, Sydney, NSW, Australia

Regula Furrer Biozentrum, University of Basel, Basel, Switzerland

Carmine Gentile Sydney Medical School, University of Sydney, Sydney, NSW, Australia

Beth Israel Deaconess Medical Center, Harvard Medical School, Boston, MA, USA

Christoph Handschin Biozentrum, University of Basel, Basel, Switzerland

Ilias Kazanis Lab of Developmental Biology, Department of Biology, University of Patras, Patras, Greece

Wellcome Trust-MRC Cambridge Stem Cell Biology Institute, University of Cambridge, Cambridge, UK

Marina Konopleva Department of Leukemia and Stem Cell Transplantation and Cellular Therapy, The University of Texas MD Anderson Cancer Center, Houston, TX, USA

Daniel Lucas Department of Cell and Developmental Biology, University of Michigan School of Medicine, Ann Arbor, MI, USA

Center for Organogenesis, University of Michigan School of Medicine, Ann Arbor, MI, USA

The University of Michigan Comprehensive Cancer Center, University of Michigan, Ann Arbor, MI, USA

Nicolás Marichal Neurofisiología Celular y Molecular, Instituto de Investigaciones Biológicas Clemente Estable, Montevideo, Uruguay

Institute of Physiological Chemistry, University Medical Center, Johannes Gutenberg University Mainz, Mainz, Germany

Damia Mawad Faculty of Science, School of Materials Science and Engineering, University of New South Wales, Sydney, NSW, Australia

Maria Angelica Miglino Surgery Department, School of Veterinary Medicine and Animal Science, University of Sao Paulo, Sao Paulo, SP, Brazil

Akiva Mintz Department of Radiology, Columbia University College of Physicians and Surgeons, New York, NY, USA

Prajna Paramita Naik Department of Life Science, National Institute of Technology, Rourkela, Odisha, India

Sara Nik Gottesman Institute for Stem Cell Biology and Regenerative Medicine, Albert Einstein College of Medicine, Bronx, NY, USA

Department of Developmental and Molecular Biology, Albert Einstein College of Medicine, Bronx, NY, USA

Prashanta Kumar Panda Department of Life Science, National Institute of Technology, Rourkela, Odisha, India

Melina Patsoni Lab of Developmental Biology, Department of Biology, University of Patras, Patras, Greece

Cecilia Reali Neurofisiología Celular y Molecular, Instituto de Investigaciones Biológicas Clemente Estable, Montevideo, Uruguay

Raúl E. Russo Neurofisiología Celular y Molecular, Instituto de Investigaciones Biológicas Clemente Estable, Montevideo, Uruguay

Kiran Kumar Solingapuram Sai Department of Radiology, Columbia University College of Physicians and Surgeons, New York, NY, USA

Kiminori Sato Department of Otolaryngology—Head and Neck Surgery, Kurume University School of Medicine, Kurume, Japan

Anirudh Sattiraju Department of Radiology, Columbia University College of Physicians and Surgeons, New York, NY, USA

Yoko Tabe Department of Leukemia, The University of Texas MD Anderson Cancer Center, Houston, TX, USA

Department of Next Generation Hematology Laboratory Medicine, Juntendo University of Medicine, Tokyo, Japan

Omar Trujillo-Cenóz Neurofisiología Celular y Molecular, Instituto de Investigaciones Biológicas Clemente Estable, Montevideo, Uruguay

Joshua T. Weinreb Gottesman Institute for Stem Cell Biology and Regenerative Medicine, Albert Einstein College of Medicine, Bronx, NY, USA

Department of Developmental and Molecular Biology, Albert Einstein College of Medicine, Bronx, NY, USA

Chapter 1
Stem Cell Microenvironments and Beyond

Alexander Birbrair

Abstract Endogenous stem cells are indispensable to keep tissue homeostasis due to their unique ability to generate more specialized cell types in an organized way depending on the body needs. Precise control over stem cell differentiation is essential for organogenesis and tissue homeostasis. Stem cells reside in specialized microenvironments, also called niches, which maintain them in an undifferentiated and self-renewing state. The cellular and molecular mechanisms of stem cell maintenance are key to the regulation of homeostasis and likely contribute to several disorders when altered during adulthood. Extensive studies in a various tissues have shown the importance of the niche in modulating stem cell behavior, including bone marrow, skin, intestine, skeletal muscle, vocal cord, brain, spinal cord, stomach, esophagus, and others. In recent past, extraordinary advancement has been made in the identification and characterization of stem cell niches using modern state-of-art techniques. This progress lead to the definition of the main cellular components in the microenvironment where stem cells reside and the identification of molecular mechanisms by which stem cell behavior is controlled, revealing key niche signals involved in stem cell regulation. Similar to the ecological niche of an organism, a stem cell niche is exclusive to the specific type of stem cell and guides its dynamics. This book describes the major cellular and molecular components of various stem cells microenvironments in different organs and at distinct pathophysiological conditions, such as cell-cell interactions, extra-cellular matrix proteins, soluble factors, and physical forces. Although several advances have been made in our understanding of the signals that promote stem cell activation or quiescence, several components of the stem cells microenvironment remain unknown due to the complexity of niche composition and its dynamics. Further insights into these cellular and molecular mechanisms will have important implications for our understanding of organ homeostasis and disease. In this book, we present a selected collection of detailed chapters on what we know so far about the stem cell niches in various tissues and under distinct pathophysiological conditions. Twelve chapters written by experts in the field summarize the present knowledge about the physiological function and pathophysiological role of the stem cell regulation by the microenvironment.

A. Birbrair (✉)
Department of Pathology, Federal University of Minas Gerais, Belo Horizonte, MG, Brazil
e-mail: birbrair@icb.ufmg.br

© Springer International Publishing AG 2017
A. Birbrair (ed.), *Stem Cell Microenvironments and Beyond*,
Advances in Experimental Medicine and Biology 1041,
DOI 10.1007/978-3-319-69194-7_1

Keywords Stem cells • Niche • Microenvironment

1.1 Editorial

Endogenous stem cells are indispensable to keep tissue homeostasis due to their unique ability to generate more specialized cell types in an organized way depending on the body needs (Hall and Watt 1989). Precise control over stem cell differentiation is essential for organogenesis and tissue homeostasis (Watt and Hogan 2000). Stem cells reside in specialized microenvironments, also called niches (Schofield 1978), which maintain them in an undifferentiated and self-renewing state. The cellular and molecular mechanisms of stem cell maintenance are key to the regulation of homeostasis and likely contribute to several disorders when altered during adulthood. Extensive studies in a various tissues have shown the importance of the niche in modulating stem cell behavior, including bone marrow (Birbrair and Frenette 2016), skin (Fuchs 2009), intestine (Tan and Barker 2014), skeletal muscle (Yin et al. 2013), vocal cord (Kurita et al. 2015), brain (Koutsakis and Kazanis 2016), spinal cord (Marichal et al. 2016), stomach (Bartfeld and Koo 2017), esophagus (Alcolea et al. 2014), and others (Borges et al. 2017; Scadden 2014). In recent past, extraordinary advancement has been made in the identification and characterization of stem cell niches using modern state-of-art techniques. This progress lead to the definition of the main cellular components in the microenvironment where stem cells reside and the identification of molecular mechanisms by which stem cell behavior is controlled, revealing key niche signals involved in stem cell regulation. Similar to the ecological niche of an organism, a stem cell niche is exclusive to the specific type of stem cell and guides its dynamics. This book describes the major cellular and molecular components of various stem cells microenvironments in different organs and at distinct pathophysiological conditions, such as cell-cell interactions, extra-cellular matrix proteins, soluble factors, and physical forces. Although several advances have been made in our understanding of the signals that promote stem cell activation or quiescence, several components of the stem cells microenvironment remain unknown due to the complexity of niche composition and its dynamics. Further insights into these cellular and molecular mechanisms will have important implications for our understanding of organ homeostasis and disease.

In this book, we present a selected collection of detailed chapters on what we know so far about the stem cell niches in various tissues and under distinct pathophysiological conditions. Twelve chapters written by experts in the field summarize the present knowledge about the physiological function and pathophysiological role of the stem cell regulation by the microenvironment.

Alexander Birbrair

Editor

Acknowledgments   Alexander Birbrair is supported by a grant from Pró-reitoria de Pesquisa/ Universidade Federal de Minas Gerais (PRPq/UFMG) (Edital 05/2016).

References

Alcolea MP, Greulich P, Wabik A, Frede J, Simons BD, Jones PH (2014) Differentiation imbalance in single oesophageal progenitor cells causes clonal immortalization and field change. Nat Cell Biol 16:615–622

Bartfeld S, Koo BK (2017) Adult gastric stem cells and their niches. Wiley Interdiscip Rev Dev Biol 6

Birbrair A, Frenette PS (2016) Niche heterogeneity in the bone marrow. Ann N Y Acad Sci 1370:82–96

Borges IDT, Sena IFG, de Azevedo PO, Andreotti JP, de Almeida VM, de Paiva AE, Pinheiro Dos Santos GS, de Paula Guerra DA, Dias Moura Prazeres PH, Mesquita LL et al (2017) Lung as a niche for hematopoietic progenitors. Stem Cell Rev

Fuchs E (2009) Finding one's niche in the skin. Cell Stem Cell 4:499–502

Hall PA, Watt FM (1989) Stem cells: the generation and maintenance of cellular diversity. Development 106:619–633

Koutsakis C, Kazanis I (2016) How necessary is the vasculature in the life of neural stem and progenitor cells? Evidence from evolution, development and the adult nervous system. Front Cell Neurosci 10:35

Kurita T, Sato K, Chitose S, Fukahori M, Sueyoshi S, Umeno H (2015) Origin of vocal fold stellate cells in the human macula flava. Ann Otol Rhinol Laryngol 124:698–705

Marichal N, Fabbiani G, Trujillo-Cenoz O, Russo RE (2016) Purinergic signalling in a latent stem cell niche of the rat spinal cord. Purinergic Signal 12:331–341

Scadden DT (2014) Nice neighborhood: emerging concepts of the stem cell niche. Cell 157:41–50

Schofield R (1978) The relationship between the spleen colony-forming cell and the haemopoietic stem cell. Blood Cells 4:7–25

Tan DW, Barker N (2014) Intestinal stem cells and their defining niche. Curr Top Dev Biol 107:77–107

Watt FM, Hogan BL (2000) Out of Eden: stem cells and their niches. Science 287:1427–1430

Yin H, Price F, Rudnicki MA (2013) Satellite cells and the muscle stem cell niche. Physiol Rev 93:23–67

Chapter 2
The Bone Marrow Microenvironment for Hematopoietic Stem Cells

Daniel Lucas

Abstract The main function of the microenvironment in the bone marrow (BM) is to provide signals that regulate and support the production of the billions of blood cells necessary to maintain homeostasis. The best characterized BM microenvironment is the niche that regulates hematopoietic stem cells. Efforts from many different laboratories have revealed that the niche is mainly perivascular and that blood vessels and perivascular stromal cells are the key components. In addition numerous cell types have been shown to be components of the niche. Here we discuss our current understanding of the niche and the evidence supporting the role of different types of cells in regulating hematopoietic stem cell numbers and function in vivo.

Keywords Bone marrow • Hematopoiesis • Hematopoietic stem cell • Niche • Perivascular • Niche heterogeneity

2.1 Introduction

Hematopoietic stem cells (HSC) are multipotent cells capable of giving rise to all types of blood cells and regenerating a healthy hematopoietic system when transplanted into irradiated recipients. HSC reside in the bone marrow where they tightly associate with multicellular structures that provide a unique microenvironment that supports and regulate HSC. In the bone marrow these structures are called HSC niches as defined by Schofield in 1978 who was the first to propose the existence of niches capable of regulating HSC function and differentiation (Schofield 1978).

D. Lucas (✉)
Department of Cell and Developmental Biology, University of Michigan School of Medicine, Ann Arbor, MI, USA

Center for Organogenesis, University of Michigan School of Medicine, Ann Arbor, MI, USA

The University of Michigan Comprehensive Cancer Center, University of Michigan, Ann Arbor, MI, USA
e-mail: dlucasal@med.umich.edu

© Springer International Publishing AG 2017
A. Birbrair (ed.), *Stem Cell Microenvironments and Beyond*,
Advances in Experimental Medicine and Biology 1041,
DOI 10.1007/978-3-319-69194-7_2

Niche cells produce different molecules (e.g. CXCL12 and SCF) that regulate HSC numbers, quiescence, self-renewal and trafficking (Asada et al. 2017b; Birbrair and Frenette 2016; Crane et al. 2017; Ramalingam et al. 2017; Sanchez-Aguilera and Mendez-Ferrer 2017; Yu and Scadden 2016). Losses of niche cells, or niche-derived signals, irrevocably cause alterations in some or all of these functions (Asada et al. 2017b; Birbrair and Frenette 2016; Crane et al. 2017; Ramalingam et al. 2017; Sanchez-Aguilera and Mendez-Ferrer 2017; Yu and Scadden 2016). The purpose of this chapter is to describe our current understanding of the HSC niche and discuss some of the open questions in the field for future research.

2.2 Identification of Niche Cells

The cellular composition and function of the HSC niche is an area of intense research and new candidate niche cells and HSC regulators are reported every year. Different methods have been used to identify niche cells in vivo and it is necessary to understand the limitations of these approaches to correctly interpret the literature.

Manipulation of the number of candidate niche cells: although the existence of HSC niches was proposed in 1978 their existence was not formally proven until 2003 in two seminal studies from the Scadden and Li laboratories (Calvi et al. 2003; Zhang et al. 2003). These showed that genetic modifications that caused expansion of osteoblastic (bone-forming) cells and trabecular bone in the bone marrow also led to increases in HSC numbers (Calvi et al. 2003; Zhang et al. 2003). Many other studies have used genetic approaches to expand or ablate candidate niche cells in vivo which led to the identification of perivascular and periarteriolar cells, mega-karyocytes and several other cells as components of BM HSC niches (Asada et al. 2017a; Bruns et al. 2014; Kunisaki et al. 2013; Mendez-Ferrer et al. 2008; Nakamura-Ishizu et al. 2014; Zhao et al. 2014). The limitations of this approach are that (1) cell expansion/ablation frequently cannot distinguish whether the crosstalk between the niche cell and the HSC is direct or indirect (e.g. between offspring derived from the ablated cell and the HSC); and (2) that ablation of large numbers of cells in the BM might lead to non-specific activation of HSC.

Conditional deletion of HSC-supportive factors in candidate niche cells: In this method the gene encoding a factor known to regulate HSC (e.g. CXCL12 or SCF) is conditionally deleted via Cre-mediated recombination in candidate niche populations and the effect of this deletion on HSC (e.g. depletion, proliferation or mobilization) is then quantified (Ding and Morrison 2013; Ding et al. 2012). The big advantage of these methods is that, in contrast to cell ablation, it does not perturb the basic cellular architecture of the bone marrow. The main caveat for these approaches is that, to be successful, it is necessary to achieve almost complete Cre-mediated deletion of the targeted allele, exclusively, in the candidate cell but no other niche components. This is frequently not easy because many of the Cre-drivers (including some that were thought to be lineage-specific) used to target niche cells recombine in more than one type of stromal cell in the bone marrow (Zhang and Link 2016).

Imaging the interaction of HSC with candidate niche cells: The discovery that the SLAM family markers CD48 and CD150 could be used to image mouse HSC (as Lin⁻CD48⁻CD150⁺ cells) in long bones revolutionized the HSC niche field (Kiel et al. 2005). Experiments measuring the interaction of this "SLAM" HSC with different types of candidate niche cells and structures led to the discovery that most HSC are located in perivascular areas of the bone marrow and the discovery of peri-arteriolar cells, megakaryocytes and osteolineage cells, among others, as candidate niche cells (Asada et al. 2017a; Bruns et al. 2014; Kunisaki et al. 2013; Mendez-Ferrer et al. 2008; Nakamura-Ishizu et al. 2014; Nombela-Arrieta et al. 2013; Silberstein et al. 2016; Zhao et al. 2014). HSC could also be labeled as CD117⁺α-catulin-GFP⁺ cells using *α-catulin-GFP* reporter mice (Acar et al. 2015) or Hoxb5⁺ cells using *Hoxb5-cherry* reporter mice (Chen et al. 2016) although these models have not yet been widely used. The main limitation of imaging approaches is that they are correlative: they can determine whether HSC are in the proximity of a candidate niche cell but additional functional validation is required. Another concern is how to define proximity. Different groups have used different cut-offs to define HSC proximity to a candidate niche cell and a consensus has yet to emerge. The current state of the art to assess for specific interactions is based on testing whether the HSC distribution observed in vivo is statistically different from a computer-generated random distribution (Acar et al. 2015; Kunisaki et al. 2013).

2.3 Cellular Composition and Organization of the Murine Bone Marrow HSC Niche

Several excellent recent reviews have extensively described the role of each of the known niche cell types (Asada et al. 2017b; Birbrair and Frenette 2016; Crane et al. 2017; Ramalingam et al. 2017; Sanchez-Aguilera and Mendez-Ferrer 2017; Yu and Scadden 2016). We will thus focus on describing the cellular composition and organization of the HSC niche. Figure 2.1 summarizes discoveries from many different laboratories over the last 15 years and highlights the complexity of the HSC niche.

Niche cells associated with the vasculature: The bone marrow is a highly vascularized organ where arteries and arterioles enter the tissue and transform into a sinusoidal network that is drained by a central vein (Kunisaki et al. 2013). The vast majority of HSC are located adjacent to, or in close proximity (less than 5 μm) of, a blood vessel (Kiel et al. 2005). HSC arise from an hemogenic endothelium during development and remain associated with blood vessels through life (Ramalingam et al. 2017). Endothelial cells are an indispensable component of the HSC niche; they produce factors like CXCL12, SCF, E-SELECTIN and NOTCH ligands that regulate HSC self-renewal and trafficking HSC (reviewed in Ramalingam et al. (2017)). Endothelial-cell derived Notch ligands also regulate BM angiogenesis and

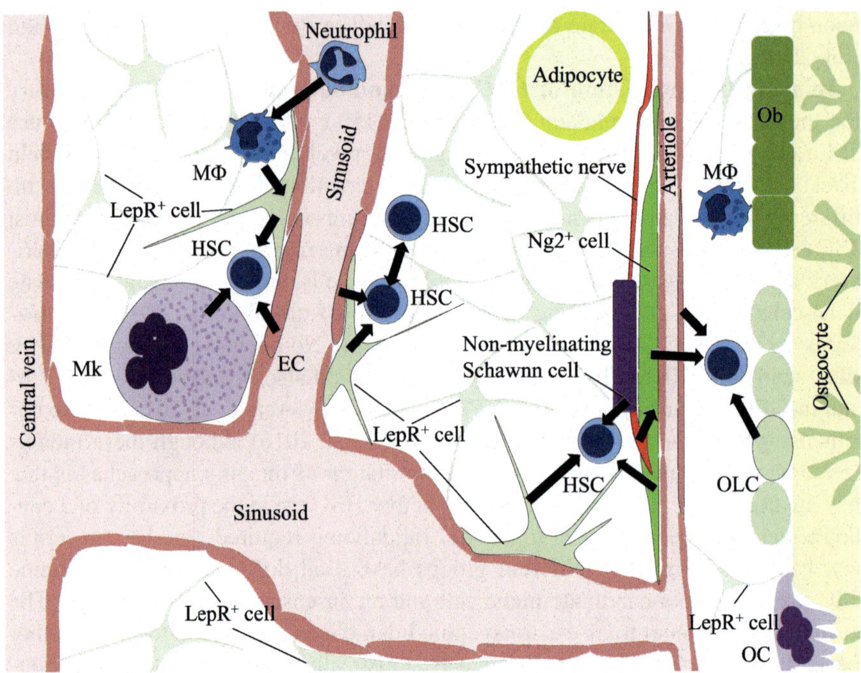

Fig. 2.1 Structure of the HSC niche. For simplicity hematopoietic cells are depicted with nuclei whereas stromal cells are shown without nucleus. Block arrows indicate direct regulation. Most hematopoietic stem cells (HSC) are located near perivascular areas (arterioles or sinusoids) where endothelial cells (EC) provide critical signals for HSC maintenance and function. LepR+ reticular cells form a tridimensional network intimately associated with the vasculature. They produce signals like CXCL12 and SCF that regulate HSC directly. Megakaryocytes (Mk) are associated with the sinusoids and restrict HSC proliferation via CXCL4, TGFβ and TPO signaling. Bone marrow macrophages (MΦ) are interspersed through the bone marrow but enriched in the endosteal surface close to osteoblasts. They promote HSC retention in the bone marrow by regulating CXCL12 production by perivascular stromal cells through an unknown mechanism. Macrophages are in turn regulated by aged neutrophils that return from the periphery. When macrophages phagocytose these aged neutrophils they became activated triggering CXCL12 downregulation in perivascular niche cells and thus HSC release. In the arterioles Ng2+ periarteriolar cells promote quiescence in a subset of HSC suggesting the existence of a periarteriolar niche. Sympathetic nerves enter the bone marrow associated with arteries and regulate HSC trafficking by controlling CXCL12 production by perivascular stromal cells. Associated with sympathetic fibers are non-myelinating Schwann cells that restrict HSC proliferation via TGFβ signaling. Osteolineage cells (OLC) in the endosteal surface of the bone regulate HSC maintenance via embigin and angiogenin signaling

abundance of other stromal components of the niche illustrating their central role in the niche (Kusumbe et al. 2016; Ramasamy et al. 2014). In close contact with sinusoids and arterioles, and connected to them and to each other via different adhesion molecules (including GAP junctions (Gonzalez-Nieto et al. 2012; Schajnovitz et al. 2011)), is a tridimensional network of LepR+ stromal cells that are also major sources of the CXCL12 and SCF that maintain HSC (Asada et al. 2017a; Ding and Morrison 2013; Ding et al. 2012; Greenbaum et al. 2013). These cells have been identified using different types of genetic reporters (e.g. CXCL12-reporter (Omatsu et al. 2010; Sugiyama et al. 2006), Nestin-GFP^{dim} (Kunisaki et al. 2013; Mendez-Ferrer et al. 2010) and LepR-cre-lineage traced cells (Ding and Morrison 2013; Ding et al. 2012)) and cell surface markers (Chan et al. 2009; Pinho et al. 2013) but these largely overlap and likely label the same cell (reviewed in Hanoun and Frenette (2013)). Importantly these cells are enriched in osteoprogenitors and are capable of giving rise to bone in vivo (Mendez-Ferrer et al. 2008; Omatsu et al. 2010; Zhou et al. 2014). Bone marrow arterioles are ensheathed by Ng2$^+$ stromal cells. Ng2$^+$ arterioles are enriched in quiescent HSC that also display low levels of reactive oxygen species (ROS) when compared to sinusoids. Ablation of Ng2$^+$ cells or conditional CXCL12 deletion in Ng2$^+$ cells causes loss of quiescent HSC (Asada et al. 2017a; Itkin et al. 2016; Kunisaki et al. 2013). Arterioles also provide physical support and associate with sympathetic nerves that enter the bone marrow from the periphery (Kunisaki et al. 2013). Both sympathetic nerves and associated non-myelinating Schwann cells are components of the HSC niche. Sympathetic nerves produce norepinephrine which acts via β3 adrenergic receptors in BM stromal cells to control CXCL12 release and thus HSC trafficking from the bone marrow to the blood (Katayama et al. 2006; Mendez-Ferrer et al. 2008, 2010). GFAP$^+$ non-myelinating Schwann cells ensheath BM sympathetic nerves and are major sources of active TGFβ that promotes HSC quiescence (Yamazaki et al. 2011). Another component of the niche associated with the vasculature are megakaryocytes. These are multinucleated hematopoietic cells that reside in the sinusoids where they produce platelets. Imaging studies showed that approximately 35% of HSC are in contact with a megakaryocyte (Bruns et al. 2014; Zhao et al. 2014). These cells function by restricting HSC proliferation via CXCL4, TGFβ and Thrombopoietin (Bruns et al. 2014; Nakamura-Ishizu et al. 2014, 2015; Zhao et al. 2014). Megakaryocyte ablation or loss of megakaryocyte-derived signals does not cause loss of HSC. Instead it causes HSC proliferation and relocation away from sinusoids, loss of quiescence and eventual exhaustion (Bruns et al. 2014; Nakamura-Ishizu et al. 2014, 2015; Zhao et al. 2014). Because megakaryocytes are hematopoietic cells this regulation suggest the existence of a feedback loop through which HSC are informed of their cellular output via regulation by their own progeny. In line with these results it has been shown that HSC and multipotent hematopoietic progenitors can also directly promote HSC proliferation via production of ESL which functions by limiting TGFβ availability (Leiva et al. 2016).

Niche cells not associated with the vasculature: The endosteal (inner) surface of the bone and trabecular areas are enriched in HSC in the steady-state and during regeneration suggesting that bone-lining cells might be niche components. Imaging analyses of HSC after transplantation revealed that many HSC where in contact with osteolineage (bone-forming) cells in the endosteal surface of the bone (Silberstein et al. 2016). Purification of these cells revealed that they were enriched in mRNAs for embigin and angiogenin. Deletion of these molecules in osteolineage cells led to loss of HSC quiescence and proliferation thus indicating that embigin and angiogenin were HSC-regulatory molecules and that bone-lining osteolineage cells were components of the HSC niche (Goncalves et al. 2016; Silberstein et al.

2016). Macrophages are another type of cell that does not associate with the vasculature but it regulates HSC trafficking (in this case indirectly) by targeting perivascular stromal cells. Many macrophages are found adjacent to the bone surface and are defined as "osteomacs". Macrophage ablation induces a reduction of CXCL12 production by perivascular niche cells thus triggering HSC mobilization from the bone marrow to the periphery (Chow et al. 2011; Christopher et al. 2011; Winkler et al. 2010). Another cell that indirectly regulates HSC trafficking are aged neutrophils that return from the periphery to the bone marrow where they are phagocytosed by macrophages activating them and ultimately triggering CXCL12 downregulation in perivascular niche cells and HSC release to the circulation (Casanova-Acebes et al. 2013). An open question is that while the crosstalk between macrophages and perivascular niche cells is well established the molecules and receptors involved in this crosstalk are not known.

Other candidate niche cell: In addition to the ones described above other cell types in the bone marrow have been proposed to regulate HSC function in some studies. However, other reports have found little to no effect of these cells on HSC and additional studies are needed to clarify their function. Bones (vertebrae) enriched in adipocytes contain fewer HSC and "fatless" mice, which have fewer BM adipocytes, showed faster HSC recovery after lethal irradiation and transplantation in long bones suggesting that adipocytes *negatively* regulate HSC during regeneration (Naveiras et al. 2009). In contrast, drug-induced adipocyte expansion in vivo had no effect on HSC suggesting that adipocytes *do not form part of the niche* during homeostasis (Spindler et al. 2014). A third, recent, report suggests that adipocytes are a *proregenerative* component of the niche: The authors found that BM adipocytes also produce SCF and that Adipoq-Cre/ER mice can be used to lineage-trace adipocytes and a subset of LepR$^+$ cells that contain the adipocyte progenitors. They also found that SCF derived from Adipoq-Cre/ER$^+$ lineage-traced cells had no effect in HSC numbers or function in the steady-state but it was required for HSC engraftment after transplantation in long bones (Zhou et al. 2017). In agreement "fatless" mice, which have a twofold reduction in BM adipocytes, had normal HSC numbers in the steady-state but impaired HSC recovery after BM transplantation in long-bones (Zhou et al. 2017). In contrast, the vertebrae of the fatless mice had increased HSC numbers and increased sinusoidal vessels in the steady-state suggesting that the vertebrae-specific effects are mediated via adipocyte-induced remodeling of the sinusoids in these bones (Zhou et al. 2017). More analyses are needed to dissect direct and indirect effects of adipocytes in HSC and niche cell regulation in different bones. It is also not clear what parts of the observed phenotypes are mediated by mature BM adipocytes or more immature Adipoq-Cre$^+$LepR$^+$ progenitors.

Another non-hematopoietic cell that has been proposed to be a component of the HSC niche is the osteoblast. These are cells located in the endosteal surface of the bone and that are actively producing new bone matrix. They were shown to support HSC maintenance in vitro (Taichman and Emerson 1994; Taichman et al. 1996). This together with the fact that the central marrow contains fewer HSC than areas closer to the bone (Nilsson et al. 2001); and that mice with constitutive expansion of the osteoblastic lineage have increased HSC numbers (Calvi et al. 2003; Zhang et al. 2003) strongly suggested that osteoblast were niche components. However, mice deficient in biglycan have a twofold reduction in bone marrow osteoblasts but no changes in HSC numbers or function (Kiel et al. 2007). Similarly, strontium treatment expands osteoblasts but has no effect on HSC numbers (Lymperi et al. 2008). These results suggest that the HSC expansion observed by Calvi et al., and Zhang et al., were mediated by expansion of osteoprogenitors (including LepR+ cells and angiogenin-producing osteolineage cells) and not mature osteoblasts which seem dispensable for HSC maintenance in the steady-state.

Osteocytes are bone cells that are completely embedded in the bone. Conditional deletion of Gsα in osteocytes using *DMP1-cre* mice led to expansion of BM myeloid cells but not HSC suggesting that osteocytes can control hematopoiesis (Fulzele et al. 2013). A different study showed that osteocyte depletion did not affect bone marrow hematopoietic stem and progenitor numbers in the steady-state but blocked G-CSF-induced mobilization (Asada et al. 2013). These studies suggest that osteocytes can regulate physiological and emergency hematopoiesis but do not regulate HSC during homeostasis.

Osteoclasts are multinucleated cells that arise via differentiation of myeloid progenitors. They reside in the surface of the bone where they digest the mineralized matrix to promote bone resorption. The first study to propose a role for osteoclasts in regulating HSC showed that treatment of mice with the cytokine RANKL increased osteoclasts numbers and mobilized HSC to the circulation. RANKL-induced mobilization was inhibited in PTPε-knockout females-which have a mild impairment in osteoclast function-suggesting that the observed phenotype was mediated by osteoclasts (Kollet et al. 2006). However, a different study found that acute osteoclast depletion via treatment with the bisphosphonate zoledronate did not mobilize HSC (Winkler et al. 2010). Lymperi et al., found that chronic osteoclast ablation with the bisphosphonate alendronate reduced HSC numbers (Lymperi et al. 2011). In contrast Miyamoto et al., examined three mouse models of osteoclasts deficiency (op/op, c-Fos-deficient and RANKL-deficient) and found increases in HSC numbers (Miyamoto et al. 2011). Paradoxically, the same authors also found that HSC mobilization was *reduced* in mice with a genetic mutation that increased osteoclasts numbers but *increased* in wild-type mice in which osteoclasts have been inhibited via alendronate treatment (Miyamoto et al. 2011). Thus the function of osteoclasts in the steady-state niche is far from clear and additional studies are needed to clarify the role of these cells in regulating HSC and their niches.

2.4 Heterogeneity of the HSC Niche

The niche cells described above are in different spatial locations in the BM and have distinct effects on regulation of HSC like maintenance (e.g. endothelial cells, LepR+ cells (Ding and Morrison 2013; Ding et al. 2012; Greenbaum et al. 2013)); inhibition of proliferation (e.g. megakaryocytes or non-myelinating Schwann cells (Bruns et al. 2014; Nakamura-Ishizu et al. 2014, 2015; Yamazaki et al. 2011; Zhao et al. 2014); or retention (e.g. macrophages (Chow et al. 2011; Christopher et al. 2011; Winkler et al. 2010). These raise the possibility that different spatial locations in the bone marrow regulate different HSC pools. In this section we discuss the evidence for, and against, spatially distinct niches.

Arteriolar and sinusoidal niches: A plethora of imaging and functional analyses have conclusively demonstrated that most HSC reside close to, and are regulated by, perivascular niches (Acar et al. 2015; Bruns et al. 2014; Chen et al. 2016; Kiel et al. 2005; Kunisaki et al. 2013; Mendez-Ferrer et al. 2010; Nombela-Arrieta et al. 2013). In addition several studies have proposed the existence of distinct sinusoidal and periarteriolar niches. Kunisaki *et al.,* showed that ~30–35% of all Lin$^-$CD48$^-$CD41$^-$CD150$^+$ HSC were located within 20 μm of CD31$^+$Sca1$^+$ BM arterioles (Kunisaki et al. 2013). These authors also showed that BM arterioles were ensheathed by Ng2$^+$/Nestin-GFP^{bright} cells and that these structures were enriched in quiescent HSC when compared to sinusoids. In agreement with these results depletion of Ng2$^+$ perivascular cells led to reduced BM HSC numbers, relocation of HSC away from arterioles and loss of HSC quiescence (Kunisaki et al. 2013). In a follow up manuscript the same group showed that conditional *Cxcl12* deletion in Ng2$^+$ perivascular cells using *Ng2-cre^{ER}* or *Myh11-Cre^{ERT2}* mice caused loss of BM HSC and relocation of HSC away from arterioles. In the same study Cxcl12 deletion in LepR$^+$ cells had no effect on BM HSC numbers (Asada et al. 2017a). In contrast *Scf* deletion in Ng2$^+$ cells had no effect on BM HSC numbers but *Scf* deletion in LepR$^+$ cells causes a dramatic HSC loss. These results suggest that Ng2$^+$ periarteriolar and LepR$^+$ (which are mainly perisinusoidal albeit some LepR$^+$ are close to arterioles (Ding and Morrison 2013; Ding et al. 2012; Kunisaki et al. 2013)) maintain HSC via the production of different cytokines (Asada et al. 2017a). It is important to note, however, that these results contrast with a previous study that showed that LepR$^+$ cells maintain HSC via CXCL12 (Ding and Morrison 2013). Megakaryocytes promote HSC quiescence ((Bruns et al. 2014; Nakamura-Ishizu et al. 2014, 2015; Zhao et al. 2014) and are located exclusively in the sinusoids. Megakaryocyte ablation induces HSC proliferation and HSC relocation away from sinusoids but does not disrupt HSC interaction with arterioles suggesting that sinusoidal and arteriolar HSC niches are functionally independent (Bruns et al. 2014). The fact that non-myelinating Schwann cells, that are associated with the sympathetic nerves in arterioles (Kunisaki et al. 2013), restrict HSC proliferation (Yamazaki et al. 2011) further supports the concept of a periarteriolar niche that promotes HSC quiescence. Additional data supporting the existence of functionally distinct arteriolar and sinusoidal niches was presented by Itkin et al. These authors quantified the percentage of Lin$^-$CD48$^-$CD150$^+$ hematopoietic stem and progenitor cells (HSPC) that stained positive for reactive oxygen species (ROS) in the bone marrow. They found that all HSPC adjacent to arteries were ROS$^-$ whereas ~35% of all HSPC adjacent to sinusoids where ROS$^+$. Increases in vascular leakiness using *Cdh5-cre^{ERT2}:Cxcr4$^{lox/lox}$* or *Cdh5-cre^{ERT2}:Fgfr1/2$^{lox/lox}$* mice revealed increased numbers of ROS$^+$ HSPC close to sinusoids indicating that vascular permeability controls HSPC metabolic state (Itkin et al. 2016). The same study showed that BM sinusoids are the exclusive trafficking site and that increased vascular permeability promotes BM HSPC release into the circulation (Itkin et al. 2016). Kusumbe et al., has also shown that arterioles support HSPC. Constitutive activation of the Notch pathway in endothelial cells led to a threefold increase in BM arterioles and associated perivascular cells and a ~ 50% increase in HSC numbers indicating that arteriole density controls HSC frequency (Kusumbe et al. 2016).

While the reports above support the idea that arterioles and periarteriolar cells maintain and regulate a subset of HSC in the bone marrow other reports have found no specific HSC association with arterioles. Acar et al., using *α-catulin-GFP* reporter mice to label c-kit⁺GFP⁺ HSC found that 10% of these HSC were close (less than 5 μm) to arterioles and they found no enrichment in quiescent HSC close to arterioles (Acar et al. 2015).

Endosteal niches: As discussed above it is clear that most HSC reside in perivascular niches. It is less clear whether the endosteal surface of the bone also provides a niche for HSC during homeostasis. A note of caution: some studies in the literature have used the term "endosteal" to describe cells located in inner surface of the bone while other studies have used this term to define cells with bone-forming potential. The term "osteoblastic" have been used in a similar way. Through this chapter we have used "endosteal" to describe a spatial location (proximity to the inner surface of the bone) and "osteoblastic" to define differentiation potential into bone cells. Studies in which different fractions of the bone marrow were purified and then tested for HSC numbers have shown that HSC are enriched in the BM that remains associated with the bone after fractionation supporting the idea of an endosteal niche (Nilsson et al. 2001). However, imaging studies have shown that less than 5% of HSC are in direct contact with the cells that line the endosteal surface of the bone (Acar et al. 2015). One possible explanation for this discrepancy is the fact that most arterioles are located within ~40 μm of the bone surface (Itkin et al. 2016; Kunisaki et al. 2013) and periarteriolar niches might thus account for HSC enrichment in endosteal areas. However, the fact that macrophages are enriched in areas close to the bone "osteomacs" (Christopher et al. 2011; Winkler et al. 2010) and that bone-lining osteolineage cells are major sources of the HSC regulatory factors embigin and angiogenin suggest that endosteal cells also support HSC (Goncalves et al. 2016; Silberstein et al. 2016). A second explanation for the lack of HSC in contact with the bone surface is that endosteal cells might be capable of exerting their HSC-regulatory function over longer distances than the ones used by endothelial and perivascular cells. Additional studies are needed to define whether endosteal signals regulate HSC during steady-state hematopoiesis.

2.5 Niches for Hematopoietic Progenitors Downstream of the HSC

Hematopoiesis occurs in a hierarchical manner with more immature, multipotent, stem and progenitor cells progressively committing into lineage-restricted progenitors before giving rise to mature blood cells (reviewed in Babovic and Eaves (2014)).

These hematopoietic progenitors also depend on signals from the microenvironment for maintenance and differentiation but they are much less characterized than HSC niches although significant progress has been made in identifying these cells. Osteoblasts do not seem to regulate HSC function during homeostasis but they provide a niche for common lymphoid progenitors (CLP) via CXCL12 production (Ding and Morrison 2013; Greenbaum et al. 2013). LepR$^+$ cells are also necessary for CLP maintenance via CXCL12 and IL7 production (Cordeiro Gomes et al. 2016). Bone marrow macrophages provide a niche for differentiating erythroid cells (Chow et al. 2013; Ramos et al. 2013). Differentiating erythroid cells regulate HSPC numbers via ACKR1 (Duchene et al. 2017). In the case of myeloid cells it is known that stromal cells regulate monocyte egress (Shi et al. 2011) and proliferation after TLR activation (Boettcher et al. 2012, 2014); *Crebbp* haploinsufficiency in stromal cells increased myelopoiesis (Zimmer et al. 2011) whereas IκBκ or DICER deletion in the stroma lead to myeloproliferative disease (Raaijmakers et al. 2010; Rupec et al. 2005). Also, osteocytes regulate myelopoiesis via GαS signaling (Fulzele et al. 2013). Identifying the specific niches for each hematopoietic progenitor is necessary in order to understand how normal and pathological hematopoiesis occurs. This has become more relevant with discoveries suggesting that HSC might not be the main source of blood cells in the steady-state. Two independent analyses of steady-state hematopoiesis suggest that more differentiated hematopoietic progenitors are responsible for most blood cell production with rare input from HSC (Busch et al. 2015; Sun et al. 2014) although these results have been challenged recently (Sawai et al. 2016). In addition mice lacking functional HSC or mice in which most HSC have been ablated are capable of maintaining almost normal hematopoiesis (Jones et al. 2015; Schoedel et al. 2016).

2.6 Concluding Remarks

The bone marrow niche is incredibly complex with multiple cells cooperating to integrate signals and provide input to regulate HSC, hematopoietic progenitors and blood cell output. Understanding how these niches function in vivo is essential to understand how hematopoiesis is regulated and might lead to the development of therapies that target specific niche components to modulate blood cell production as needed.

References

Acar M, Kocherlakota KS, Murphy MM, Peyer JG, Oguro H, Inra CN, Jaiyeola C, Zhao Z, Luby-Phelps K, Morrison SJ (2015) Deep imaging of bone marrow shows non-dividing stem cells are mainly perisinusoidal. Nature 526:126–130

Asada N, Katayama Y, Sato M, Minagawa K, Wakahashi K, Kawano H, Kawano Y, Sada A, Ikeda K, Matsui T et al (2013) Matrix-embedded osteocytes regulate mobilization of hematopoietic stem/progenitor cells. Cell Stem Cell 12:737–747

Asada N, Kunisaki Y, Pierce H, Wang Z, Fernandez NF, Birbrair A, Ma'ayan A, Frenette PS (2017a) Differential cytokine contributions of perivascular haematopoietic stem cell niches. Nat Cell Biol 19(3):214–223

Asada N, Takeishi S, Frenette PS (2017b) Complexity of bone marrow hematopoietic stem cell niche. Int J Hematol 106:45–54

Babovic S, Eaves CJ (2014) Hierarchical organization of fetal and adult hematopoietic stem cells. Exp Cell Res 329:185–191

Birbrair A, Frenette PS (2016) Niche heterogeneity in the bone marrow. Ann N Y Acad Sci 1370:82–96

Boettcher S, Ziegler P, Schmid MA, Takizawa H, van Rooijen N, Kopf M, Heikenwalder M, Manz MG (2012) Cutting edge: LPS-induced emergency myelopoiesis depends on TLR4-expressing nonhematopoietic cells. J Immunol 188:5824–5828

Boettcher S, Gerosa RC, Radpour R, Bauer J, Ampenberger F, Heikenwalder M, Kopf M, Manz MG (2014) Endothelial cells translate pathogen signals into G-CSF-driven emergency granulopoiesis. Blood 124:1393–1403

Bruns I, Lucas D, Pinho S, Ahmed J, Lambert MP, Kunisaki Y, Scheiermann C, Schiff L, Poncz M, Bergman A et al (2014) Megakaryocytes regulate hematopoietic stem cell quiescence through CXCL4 secretion. Nat Med 20:1315–1320

Busch K, Klapproth K, Barile M, Flossdorf M, Holland-Letz T, Schlenner SM, Reth M, Hofer T, Rodewald HR (2015) Fundamental properties of unperturbed haematopoiesis from stem cells in vivo. Nature 518:542–546

Calvi LM, Adams GB, Weibrecht KW, Weber JM, Olson DP, Knight MC, Martin RP, Schipani E, Divieti P, Bringhurst FR et al (2003) Osteoblastic cells regulate the haematopoietic stem cell niche. Nature 425:841–846

Casanova-Acebes M, Pitaval C, Weiss LA, Nombela-Arrieta C, Chevre R, A-González N, Kunisaki Y, Zhang D, van Rooijen N, Silberstein LE et al (2013) Rhythmic modulation of the hematopoietic niche through neutrophil clearance. Cell 153:1025–1035

Chan CK, Chen CC, Luppen CA, Kim JB, DeBoer AT, Wei K, Helms JA, Kuo CJ, Kraft DL, Weissman IL (2009) Endochondral ossification is required for haematopoietic stem-cell niche formation. Nature 457:490–494

Chen JY, Miyanishi M, Wang SK, Yamazaki S, Sinha R, Kao KS, Seita J, Sahoo D, Nakauchi H, Weissman IL (2016) Hoxb5 marks long-term haematopoietic stem cells and reveals a homogenous perivascular niche. Nature 530:223–227

Chow A, Lucas D, Hidalgo A, Mendez-Ferrer S, Hashimoto D, Scheiermann C, Battista M, Leboeuf M, Prophete C, van Rooijen N et al (2011) Bone marrow CD169(+) macrophages promote the retention of hematopoietic stem and progenitor cells in the mesenchymal stem cell niche. J Exp Med 208:261–271

Chow A, Huggins M, Ahmed J, Hashimoto D, Lucas D, Kunisaki Y, Pinho S, Leboeuf M, Noizat C, van Rooijen N et al (2013) CD169+ macrophages provide a niche promoting erythropoiesis under homeostasis and stress. Nat Med 19:429–436

Christopher MJ, Rao M, Liu F, Woloszynek JR, Link DC (2011) Expression of the G-CSF receptor in monocytic cells is sufficient to mediate hematopoietic progenitor mobilization by G-CSF in mice. J Exp Med 208:251–260

Cordeiro Gomes A, Hara T, Lim VY, Herndler-Brandstetter D, Nevius E, Sugiyama T, Tani-Ichi S, Schlenner S, Richie E, Rodewald HR et al (2016) Hematopoietic stem cell niches produce lineage-instructive signals to control multipotent progenitor differentiation. Immunity 45:1219–1231

Crane GM, Jeffery E, Morrison SJ (2017) Adult haematopoietic stem cell niches. Nat Rev Immunol 17(9):573–590

Ding L, Morrison SJ (2013) Haematopoietic stem cells and early lymphoid progenitors occupy distinct bone marrow niches. Nature 495:231–235

Ding L, Saunders TL, Enikolopov G, Morrison SJ (2012) Endothelial and perivascular cells maintain haematopoietic stem cells. Nature 481:457–462

Duchene J, Novitzky-Basso I, Thiriot A, Casanova-Acebes M, Bianchini M, Etheridge SL, Hub E, Nitz K, Artinger K, Eller K et al (2017) Atypical chemokine receptor 1 on nucleated erythroid cells regulates hematopoiesis. Nat Immunol 18:753–761

Fulzele K, Krause DS, Panaroni C, Saini V, Barry KJ, Liu X, Lotinun S, Baron R, Bonewald L, Feng JQ et al (2013) Myelopoiesis is regulated by osteocytes through Gsalpha-dependent signaling. Blood 121:930–939

Goncalves KA, Silberstein L, Li S, Severe N, Hu MG, Yang H, Scadden DT, Hu GF (2016) Angiogenin promotes hematopoietic regeneration by dichotomously regulating quiescence of stem and progenitor cells. Cell 166:894–906

Gonzalez-Nieto D, Li L, Kohler A, Ghiaur G, Ishikawa E, Sengupta A, Madhu M, Arnett JL, Santho RA, Dunn SK et al (2012) Connexin-43 in the osteogenic BM niche regulates its cellular composition and the bidirectional traffic of hematopoietic stem cells and progenitors. Blood 119:5144–5154

Greenbaum A, Hsu YM, Day RB, Schuettpelz LG, Christopher MJ, Borgerding JN, Nagasawa T, Link DC (2013) CXCL12 in early mesenchymal progenitors is required for haematopoietic stem-cell maintenance. Nature 495:227–230

Hanoun M, Frenette PS (2013) This niche is a maze; an amazing niche. Cell Stem Cell 12:391–392

Itkin T, Gur-Cohen S, Spencer JA, Schajnovitz A, Ramasamy SK, Kusumbe AP, Ledergor G, Jung Y, Milo I, Poulos MG et al (2016) Distinct bone marrow blood vessels differentially regulate haematopoiesis. Nature 532:323–328

Jones M, Chase J, Brinkmeier M, Xu J, Weinberg DN, Schira J, Friedman A, Malek S, Grembecka J, Cierpicki T et al (2015) Ash1l controls quiescence and self-renewal potential in hematopoietic stem cells. J Clin Invest 125:2007–2020

Katayama Y, Battista M, Kao WM, Hidalgo A, Peired AJ, Thomas SA, Frenette PS (2006) Signals from the sympathetic nervous system regulate hematopoietic stem cell egress from bone marrow. Cell 124:407–421

Kiel MJ, Yilmaz OH, Iwashita T, Yilmaz OH, Terhorst C, Morrison SJ (2005) SLAM family receptors distinguish hematopoietic stem and progenitor cells and reveal endothelial niches for stem cells. Cell 121:1109–1121

Kiel MJ, Radice GL, Morrison SJ (2007) Lack of evidence that hematopoietic stem cells depend on N-cadherin-mediated adhesion to osteoblasts for their maintenance. Cell Stem Cell 1:204–217

Kollet O, Dar A, Shivtiel S, Kalinkovich A, Lapid K, Sztainberg Y, Tesio M, Samstein RM, Goichberg P, Spiegel A et al (2006) Osteoclasts degrade endosteal components and promote mobilization of hematopoietic progenitor cells. Nat Med 12:657–664

Kunisaki Y, Bruns I, Scheiermann C, Ahmed J, Pinho S, Zhang D, Mizoguchi T, Wei Q, Lucas D, Ito K et al (2013) Arteriolar niches maintain haematopoietic stem cell quiescence. Nature 502:637–643

Kusumbe AP, Ramasamy SK, Itkin T, Mae MA, Langen UH, Betsholtz C, Lapidot T, Adams RH (2016) Age-dependent modulation of vascular niches for haematopoietic stem cells. Nature 532:380–384

Leiva M, Quintana JA, Ligos JM, Hidalgo A (2016) Haematopoietic ESL-1 enables stem cell proliferation in the bone marrow by limiting TGFbeta availability. Nat Commun 7:10222

Lymperi S, Horwood N, Marley S, Gordon MY, Cope AP, Dazzi F (2008) Strontium can increase some osteoblasts without increasing hematopoietic stem cells. Blood 111:1173–1181

Lymperi S, Ersek A, Ferraro F, Dazzi F, Horwood NJ (2011) Inhibition of osteoclast function reduces hematopoietic stem cell numbers in vivo. Blood 117:1540–1549

Mendez-Ferrer S, Lucas D, Battista M, Frenette PS (2008) Haematopoietic stem cell release is regulated by circadian oscillations. Nature 452:442–447

Mendez-Ferrer S, Michurina TV, Ferraro F, Mazloom AR, Macarthur BD, Lira SA, Scadden DT, Ma'ayan A, Enikolopov GN, Frenette PS (2010) Mesenchymal and haematopoietic stem cells form a unique bone marrow niche. Nature 466:829–834

Miyamoto K, Yoshida S, Kawasumi M, Hashimoto K, Kimura T, Sato Y, Kobayashi T, Miyauchi Y, Hoshi H, Iwasaki R et al (2011) Osteoclasts are dispensable for hematopoietic stem cell maintenance and mobilization. J Exp Med 208:2175–2181

Nakamura-Ishizu A, Takubo K, Fujioka M, Suda T (2014) Megakaryocytes are essential for HSC quiescence through the production of thrombopoietin. Biochem Biophys Res Commun 454:353–357

Nakamura-Ishizu A, Takubo K, Kobayashi H, Suzuki-Inoue K, Suda T (2015) CLEC-2 in mega-karyocytes is critical for maintenance of hematopoietic stem cells in the bone marrow. J Exp Med 212:2133–2146

Naveiras O, Nardi V, Wenzel PL, Hauschka PV, Fahey F, Daley GQ (2009) Bone-marrow adi-pocytes as negative regulators of the haematopoietic microenvironment. Nature 460:259–263

Nilsson SK, Johnston HM, Coverdale JA (2001) Spatial localization of transplanted hemopoietic stem cells: inferences for the localization of stem cell niches. Blood 97:2293–2299

Nombela-Arrieta C, Pivarnik G, Winkel B, Canty KJ, Harley B, Mahoney JE, Park SY, Lu J, Protopopov A, Silberstein LE (2013) Quantitative imaging of haematopoietic stem and pro-genitor cell localization and hypoxic status in the bone marrow microenvironment. Nat Cell Biol 15:533–543

Omatsu Y, Sugiyama T, Kohara H, Kondoh G, Fujii N, Kohno K, Nagasawa T (2010) The essential functions of adipo-osteogenic progenitors as the hematopoietic stem and progenitor cell niche. Immunity 33:387–399

Pinho S, Lacombe J, Hanoun M, Mizoguchi T, Bruns I, Kunisaki Y, Frenette PS (2013) PDGFR alpha and CD51 mark human Nestin(+) sphere-forming mesenchymal stem cells capable of hematopoietic progenitor cell expansion. J Exp Med 210:1351–1367

Raaijmakers MH, Mukherjee S, Guo S, Zhang S, Kobayashi T, Schoonmaker JA, Ebert BL, Al-Shahrour F, Hasserjian RP, Scadden EO et al (2010) Bone progenitor dysfunction induces myelodysplasia and secondary leukaemia. Nature 464:852–857

Ramalingam P, Poulos MG, Butler JM (2017) Regulation of the hematopoietic stem cell lifecycle by the endothelial niche. Curr Opin Hematol 24:289–299

Ramasamy SK, Kusumbe AP, Wang L, Adams RH (2014) Endothelial Notch activity promotes angiogenesis and osteogenesis in bone. Nature 507:376–380

Ramos P, Casu C, Gardenghi S, Breda L, Crielaard BJ, Guy E, Marongiu MF, Gupta R, Levine RL, Abdel-Wahab O et al (2013) Macrophages support pathological erythropoiesis in polycythemia vera and beta-thalassemia. Nat Med 19:437–445

Rupec RA, Jundt F, Rebholz B, Eckelt B, Weindl G, Herzinger T, Flaig MJ, Moosmann S, Plewig G, Dorken B et al (2005) Stroma-mediated dysregulation of myelopoiesis in mice lacking I kappa B alpha. Immunity 22:479–491

Sanchez-Aguilera A, Mendez-Ferrer S (2017) The hematopoietic stem-cell niche in health and leukemia. Cell Mol Life Sci 74:579–590

Sawai CM, Babovic S, Upadhaya S, Knapp DJ, Lavin Y, Lau CM, Goloborodko A, Feng J, Fujisaki J, Ding L et al (2016) Hematopoietic stem cells are the major source of multilineage hemato-poiesis in adult animals. Immunity 45:597–609

Schajnovitz A, Itkin T, D'Uva G, Kalinkovich A, Golan K, Ludin A, Cohen D, Shulman Z, Avigdor A, Nagler A et al (2011) CXCL12 secretion by bone marrow stromal cells is dependent on cell contact and mediated by connexin-43 and connexin-45 gap junctions. Nat Immunol 12:391–398

Schoedel KB, Morcos MN, Zerjatke T, Roeder I, Grinenko T, Voehringer D, Gothert JR, Waskow C, Roers A, Gerbaulet A (2016) The bulk of the hematopoietic stem cell population is dispens-able for murine steady-state and stress hematopoiesis. Blood 128(19):2285–2296

Schofield R (1978) The relationship between the spleen colony-forming cell and the haemopoietic stem cell. Blood Cells 4:7–25

Shi C, Jia T, Mendez-Ferrer S, Hohl TM, Serbina NV, Lipuma L, Leiner I, Li MO, Frenette PS, Pamer EG (2011) Bone marrow mesenchymal stem and progenitor cells induce monocyte emi-gration in response to circulating toll-like receptor ligands. Immunity 34:590–601

Silberstein L, Goncalves KA, Kharchenko PV, Turcotte R, Kfoury Y, Mercier F, Baryawno N, Severe N, Bachand J, Spencer JA et al (2016) Proximity-based differential single-cell analysis of the niche to identify stem/progenitor cell regulators. Cell Stem Cell 19:530–543

Spindler TJ, Tseng AW, Zhou X, Adams GB (2014) Adipocytic cells augment the support of primitive hematopoietic cells in vitro but have no effect in the bone marrow niche under homeostatic conditions. Stem Cells Dev 23:434–441

Sugiyama T, Kohara H, Noda M, Nagasawa T (2006) Maintenance of the hematopoietic stem cell pool by CXCL12-CXCR4 chemokine signaling in bone marrow stromal cell niches. Immunity 25:977–988

Sun J, Ramos A, Chapman B, Johnnidis JB, Le L, Ho YJ, Klein A, Hofmann O, Camargo FD (2014) Clonal dynamics of native haematopoiesis. Nature 514:322–327

Taichman RS, Emerson SG (1994) Human osteoblasts support hematopoiesis through the production of granulocyte colony-stimulating factor. J Exp Med 179:1677–1682

Taichman RS, Reilly MJ, Emerson SG (1996) Human osteoblasts support human hematopoietic progenitor cells in vitro bone marrow cultures. Blood 87:518–524

Winkler IG, Sims NA, Pettit AR, Barbier V, Nowlan B, Helwani F, Poulton IJ, van Rooijen N, Alexander KA, Raggatt LJ et al (2010) Bone marrow macrophages maintain hematopoietic stem cell (HSC) niches and their depletion mobilizes HSCs. Blood 116:4815–4828

Yamazaki S, Ema H, Karlsson G, Yamaguchi T, Miyoshi H, Shioda S, Taketo MM, Karlsson S, Iwama A, Nakauchi H (2011) Nonmyelinating Schwann cells maintain hematopoietic stem cell hibernation in the bone marrow niche. Cell 147:1146–1158

Yu VW, Scadden DT (2016) Heterogeneity of the bone marrow niche. Curr Opin Hematol 23:331–338

Zhang J, Link DC (2016) Targeting of mesenchymal stromal cells by Cre-recombinase transgenes commonly used to target osteoblast lineage cells. J Bone Miner Res 31:2001–2007

Zhang R, Xu Y, Ekman N, Wu Z, Wu J, Alitalo K, Min W (2003) Etk/Bmx transactivates vascular endothelial growth factor 2 and recruits phosphatidylinositol 3-kinase to mediate the tumor necrosis factor-induced angiogenic pathway. J Biol Chem 278:51267–51276

Zhao M, Perry JM, Marshall H, Venkatraman A, Qian P, He XC, Ahamed J, Li L (2014) Megakaryocytes maintain homeostatic quiescence and promote post-injury regeneration of hematopoietic stem cells. Nat Med 20:1321–1326

Zhou BO, Yue R, Murphy MM, Peyer JG, Morrison SJ (2014) Leptin-receptor-expressing mesenchymal stromal cells represent the main source of bone formed by adult bone marrow. Cell Stem Cell 15:154–168

Zhou BO, Yu H, Yue R, Zhao Z, Rios JJ, Naveiras O, Morrison SJ (2017) Bone marrow adipocytes promote the regeneration of stem cells and haematopoiesis by secreting SCF. Nat Cell Biol 19(8):891–903

Zimmer SN, Zhou Q, Zhou T, Cheng Z, Abboud-Werner SL, Horn D, Lecocke M, White R, Krivtsov AV, Armstrong SA et al (2011) Crebbp haploinsufficiency in mice alters the bone marrow microenvironment, leading to loss of stem cells and excessive myelopoiesis. Blood 118:69–79

Chapter 3
Leukemia Stem Cells Microenvironment

Yoko Tabe and Marina Konopleva

Abstract The dynamic interactions between leukemic cells and bone marrow (BM) cells in the leukemia BM microenvironment regulate leukemia stem cell (LSC) properties including localization, self-renewal, differentiation, and proliferation. Recent research of normal and leukemia BM microenvironments has revealed several key components of specific niches that provide a sanctuary where subpopulations of leukemia cells evade chemotherapy-induced death and acquire a drug-resistant phenotype, as well as the molecular pathways critical for microenvironment/leukemia interactions. Although the biology of LSCs shares many similarities with that of normal hematopoietic stem cells (HSCs), LSCs are able to outcompete HSCs and hijack BM niches. Increasing evidence indicates that these niches fuel the growth of leukemia cells and contribute to therapeutic resistance and the metastatic potential of leukemia cells by shielding LSCs. Not only "microenvironment-induced oncogenesis," but also a "malignancy-induced microenvironment" have been proposed. In this chapter, the key components and regulation of BM niches in leukemic BM is described. In addition, metabolic changes in LSCs, which are currently a subject of intense investigation, will also be discussed to understand LSC survival.

Keywords Microenvironment • Stem cell • Niche

Y. Tabe
Department of Leukemia, The University of Texas MD Anderson Cancer Center, Houston, TX, USA

Department of Next Generation Hematology Laboratory Medicine, Juntendo University Faculty of Medicine, Tokyo, Japan
e-mail: tabe@juntendo.ac.jp

M. Konopleva (✉)
Department of Leukemia, The University of Texas MD Anderson Cancer Center, Houston, TX, USA
e-mail: mkonople@mdanderson.org

© Springer International Publishing AG 2017
A. Birbrair (ed.), *Stem Cell Microenvironments and Beyond*,
Advances in Experimental Medicine and Biology 1041,
DOI 10.1007/978-3-319-69194-7_3

3.1 Introduction

The dynamic interactions between leukemic cells and bone marrow (BM) cells in the leukemia BM microenvironment regulate leukemia stem cell (LSC) properties including localization, self-renewal, differentiation, and proliferation (Tabe and Konopleva 2014). Despite the significant progress that has been achieved in chemotherapy-based and targeted treatments of several leukemia subsets, relapse remains common after an initial response (Dores et al. 2012; Sant et al. 2010), indicating the persistence of chemoresistant LSCs in the BM. Recent research of normal and leukemia BM microenvironments has revealed several key components of specific niches that provide a sanctuary where subpopulations of leukemia cells evade chemotherapy-induced death and acquire a drug-resistant phenotype, as well as the molecular pathways critical for microenvironment/leukemia interactions.

Although the biology of LSCs shares many similarities with that of normal hematopoietic stem cells (HSCs), LSCs are able to outcompete HSCs and hijack BM niches. Increasing evidence indicates that these niches fuel the growth of leukemia cells and contribute to therapeutic resistance and the metastatic potential of leukemia cells by shielding LSCs (Hanahan and Coussens 2012; Tabe et al. 2017). Not only "microenvironment-induced oncogenesis," but also a "malignancy-induced microenvironment" have been proposed (Shiozawa and Taichman 2010). BM niches are part of a complex of BM cells including bone-lining cells (osteoblasts and osteoclasts), mesenchymal stem cells (MSCs), sinusoidal endothelium and perivascular stromal cells, nonmyelinating Schwann cells and immune cells, which play distinct roles in the regulation of hematopoiesis (Ding et al. 2012; Kunisaki et al. 2013) Chemokine receptors (Rombouts et al. 2004; Zeng et al. 2009; Tabe et al. 2013; Nervi et al. 2009), adhesion molecules (Jin et al. 2006; Williams et al. 2013; Redondo-Muñoz et al. 2008; Jacamo et al. 2014), the sympathetic nervous system (Katayama et al. 2006), hypoxia-related proteins (Wellmann et al. 2004; Suda et al. 2011), and genetic and epigenetic abnormalities of leukemia-associated stroma cells have been proposed as key emerging therapeutic targets (Walkley et al. 2007a). In this chapter, the key components and regulation of BM niches in leukemic BM is described. In addition, metabolic changes in LSCs, which are currently a subject of intense investigation, will also be discussed to understand LSC survival.

3.2 Leukemic Microenvironment

3.2.1 Components of Microenvironmental Niches

Endosteal and vascular niches are anatomically closely related to distinct vascular structures, arterioles, and sinusoids, which work in concert (Ding et al. 2012; Kunisaki et al. 2013; Calvi et al. 2003; Kiel and Morrison 2008; Adams et al. 2006;

Mendelson and Frenette 2014). Recent advances in microscopy and transgenic animal models have revolutionized the understanding of these niches (Morrison and Scadden 2014). Quiescent HSCs associated with periarteriolar niches are found within the endosteal BM (Kunisaki et al. 2013), where arterioles run in proximity to the endosteal surface, accompanied by sympathetic nerve fibers ensheathed by non-myelinating Schwann cells. In turn, sinusoids, fenestrated and lined by reticular-shaped sinusoidal cells, are associated with less-quiescent HSCs re-localizing to perisinusoidal niches (Kunisaki et al. 2013; Kfoury et al. 2014). The surface of the endosteum in the endosteal niche is lined by osteoblasts and osteoclasts. Osteoblasts are progenitor bone-forming cells derived from pluripotent MSCs (Adams et al. 2006). Interactions between angiopoietin-1 (ANG1) in osteoblasts with its receptor Tyrosine kinase with immunoglobulin-like and EGF-like domains 2 (TIE2) on HSCs result in activation of β1-integrin and N-cadherin and enhanced adhesion between niche cells and HSCs, which contributes to the maintenance of stem cell quiescence (Arai et al. 2004). Notably, AML cells induce a preosteoblast-rich niche in the BM that in turn facilitates AML expansion (Battula et al. 2017). The C-X-C motif chemokine 12 (CXCL12), which is produced by osteoblasts, is the major chemoattractant of HSCs (Christopher et al. 2009).

Bone-resorbing osteoclasts participate in the initial formation and maintenance of cavities that constitute the endosteal niche (Adams et al. 2006; Mendelson and Frenette 2014). Osteoclastic bone resorption produces abundant active transforming growth factor beta (TGF-β) from bone, which is the largest latent reservoir of TGF-β (Morrison and Scadden 2014). The sympathetic nervous system is responsible for regulating HSCs residing in the periarteriolar position via norepinephrine signaling (Katayama et al. 2006). A mechanistic analysis showed that nonmyelinating Schwann cells activate latent TGF-β, and the neoplastic niche is altered by leukemic cells through sympathetic neuropathy (Hanoun et al. 2014). For example, MLL-AF9 acute myeloid leukemia (AML) cells transform the HSC niche, reducing the numbers of arteriole-associated niche cells and the density of their sympathetic nerve network (Hanoun et al. 2014; Price and Sipkins 2014). Sympathetic neuropathy by myeloproliferative neoplasia (MPN) alter the HSC niche and progression of the disease; MPN cells produce interleukin-1β (IL-1β) that destroys Schwann cells supporting sympathetic nerve fibers, followed by the apoptotic loss of Nestin-positive (Nestin+) cells and reduced HSC maintenance factors in the microenvironment, such as CXCL12, resulting in peripheral mobilization of HSCs and accelerated MPN cell expansion in the BM (Price and Sipkins 2014) The perivascular cells in the vascular niche include CXCL12-abundant reticular (CAR) cells (Sugiyama et al. 2006), Nestin+ MSCs (Méndez-Ferrer et al. 2010) and leptin receptor-positive (LepR+) MSCs (Ding et al. 2012) exhibiting significant overlap and expressing multiple soluble and membrane-bound factors that regulate stem cell self-renewal and retention (Doan and Chute 2012). The conditional deletion of stem cell factor from LepR+ perivascular stromal cells, including Nestin+ MSCs and CAR cells, significantly reduces the number of HSCs (Adams et al. 2006).

Multiple mature hematopoietic cells including T-regulatory (Treg) cells, macrophages, and megakaryocytes also participate in regulation of the BM microenvironment

(Jin et al. 2006). Treg cells provide a relative immune sanctuary for stem cells on the endosteal surface and participate in creating the BM niche and supporting stem cell function (Kfoury et al. 2014; Fujisaki et al. 2011). Macrophages promote retention of HSCs by regulating CXCL12 production in the BM (Chow et al. 2011). Megakaryocytes, localized with a subset of HSCs, promote stem cell quiescence through the production of CXCL4 (Bruns et al. 2014) and TGF-β (Zhao et al. 2014).

3.2.2 Modulation of the LSC Niche

The behavior of LSC is modulated by interactions and signals received within their BM microenvironment (Calvi et al. 2003; Kiel and Morrison 2008; Arai et al. 2004; Nilsson et al. 2005). Although LSCs share certain self-renewal and differentiation features with HSCs, they differ in their dysregulated activation of key pathways controlling proliferation, survival and invasion (Lane 2012). One study reported that BM niche components contribute to determining the leukemia phenotype by providing the necessary cytokines and cell contact-mediated signals to LSCs (Raaijmakers et al. 2010). LSCs, in turn, contribute to deregulation of the BM niche by their dominant proliferation-promoting signals, and MSCs participate in this process. MSCs in the BM constitute a heterogeneous population (Raaijmakers 2014), and heterotypic signaling from diseased "reprogrammed" MSCs may affect other cells in the BM. For example, MSCs are required to drive the initiation and progression of myelodysplastic syndrome (MDS), which is characterized by BM failure and predisposition for evolution into AML. Transcriptional profiling revealed the aberrant gene expression implicated in intercellular crosstalk, osteo/adipogenesis, inflammation, and fibrosis in MDS MSCs (Raaijmakers 2014). Deficiency of phosphatase and tensin homolog, a tumor suppressor and antagonist of the phosphoinositide 3-kinase pathway in hematopoietic cells and BM cells results in myeloproliferation that progresses to overt leukemia/lymphoma (Yilmaz et al. 2006). These findings support the concept of niche-induced oncogenesis; primary stromal dysfunction can result in secondary neoplastic disease. As a frontier study, Walkley et al. reported that dysfunction of the retinoblastoma protein, a central regulator of the cell cycle and a tumor suppressor, or of retinoic acid receptor-γ in the BM microenvironment contributes to the development of preleukemic myeloproliferative disease from originally nonmutated hematopoietic cells (Walkley et al. 2007a, b). In another study, conditional knockout of *DICER1*, a gene that regulates microRNA processing, in osteoblastic precursors resulted in BM failure and a predisposition for leukemia. Deleting *DICER1* causes reduced expression of *SBDS*, a gene mutated in Schwachman–Bodian–Diamond syndrome. Deletion of *SBDS* in mouse osteoprogenitors induces myelodysplasia and the development of AML (Raaijmakers et al. 2010). On the other hand, LSCs themselves create a "foster home," inducing reversible changes in BM stromal cell function or composition that result in survival of leukemic cells (Dührsen and Hossfeld 1996). Suppression of normal hematopoiesis in patients with leukemia and a relatively low tumor burden may reflect disruption of normal hematopoietic progenitor cell niches and creation of leukemia niches by leukemic cells (Colmone et al. 2008).

3.2.3 *CXCR4–CXCL12 Interactions and LSC Migration to the BM*

Interactions between LSCs and BM niches are recognized as the major cause of leukemia relapse. Leukemia cells highjack normal BM vascular niches dependent upon CXCL12 and E-selectin (Colmone et al. 2008). The chemokine CXCL12, produced by CAR cells, Nestin+ MSCs, and osteoblasts, is a key factor mediating homing and engraftment of LSCs into the BM niche. Levels of the CXCL12 receptor CXCR4 are significantly elevated in leukemia cells (Raaijmakers 2014), and the association between CXCR4 expression with the poor outcome of patients with leukemia has been reported (Rombouts et al. 2004). Of note, chemotherapy for patients with AML and imatinib treatment in patients with CML upregulates CXCR4 expression, which results in increased CXCL12/CXCR4 signaling and lodging into BM niches, fostering chemoresistance (Zeng et al. 2009; Sison et al. 2013; Tabe et al. 2012). Inhibiting CXCL12–CXCR4 interactions results in abolishment of CXCL12-induced chemotaxis, inactivation of prosurvival signaling pathways including phosphorylation of p44/42 mitogen-activated protein kinase and signal tranducer and activator of transcription 3 (STAT3), and decreases in BM stromal protective effects on chemotherapy-induced apoptosis in CLL and AML cells (Zeng et al. 2009; Zeng et al. 2006; Cho et al. 2015). Recruitment of CXCR4 and its downstream mediator Lyn into lipid rafts in CML cells contributes to imatinib resistance (Tabe et al. 2012). BM stromal cell-derived TGF-β1 is also known as a mediator of resistance during cytarabine treatment of AML (Tabe et al. 2013). Combined treatment with the CXCR4 inhibitor plerixafor, the TGFβ-neutralizing antibody 1D11, and cytarabine decreases the leukemia burden and prolongs survival in a leukemia mouse model, demonstrating that TGFβ and CXCL12 play a role in AML chemoresistance (Tabe et al. 2013). Overall, CXCL12–CXCR4 interactions in the BM microenvironment contribute to the chemoresistance of leukemic cells, and disruption of these interactions by CXCR4 inhibitors represents a rational strategy for blocking LSC homing to the BM niche and/or sensitizing AML cells to chemotherapy or kinase inhibitors. Clinical trials exploring this concept are underway (Uy et al. 2012); NCT02652871.

3.2.4 *Adhesion of LSCs to the BM Niche*

Adhesion to the stromal niche is crucial for LSCs because it directly supports self-renewal and protects cells from damage by chemotherapy. The transmembrane glycoprotein CD44 of LSCs, existing as a standard isoform (CD44s) and a range of variant isoforms (CD44v), is a key regulator of LSC homing to BM niches and maintenance of their primitive state (Jin et al. 2006). CD44 modulates interactions between LSCs and extracellular matrix components and growth factor ligands to promote CD44/ligand/receptor tyrosine kinase (RTK) complex formation and signal transduction (Nervi et al. 2009). CD44–hyaluronan interactions contribute to self-renewal, proliferation, differentiation, homing to the BM, and preservation of the integrity of the stem cell genome by decreasing DNA damage and enhancing DNA repair (Williams et al. 2013). CD44/ligand/RTK signaling may participate in reprogramming of leukemia cells to exhibit a more stem cell-like LSC phenotype by modulating microRNA expression to regulate promoter methylation status and gene expression (Williams et al. 2013). Although activities of CD44s or CD44v have been reported to be similar in hyaluronan-mediated regulation of HSC differentiation and MSC homing to BM (Avigdor et al. 2004), CD44v expression preferentially enforces maturation of self-renewing LSCs (Herrlich et al. 2000). Cells in the BM niche express integrins as cell adhesion receptors that link extracellular adhesion molecules with the intracellular actin cytoskeleton (Redondo-Muñoz et al. 2008) and are required for lodging of LSCs in the BM niche (Redondo-Muñoz et al. 2008). Integrin heterodimers, composed of an 18α subunits and an 8β subunit, regulate cell–cell adhesion, growth factor receptor signaling, cell lineage specification, differentiation, survival, proliferation, and migration (Prowse et al. 2011). Many of these functions parallel CD44 expression, suggesting integrin–CD44 interactions (Williams et al. 2013). Homing of HSCs to the BM requires a coordinated sequence of four steps, including E-selectin receptor/ligand interaction and engagement of CXCL12–CXCR4 signaling, resulting in activation of very late antigen–4 (VLA-4; integrin $\alpha4\beta1$); VLA-4 adherence to vascular cell adhesion molecule 1 (VCAM-1); and transmigration on endothelium (Sackstein 2011). VLA-4 binds to CD44v to form a docking complex for pro-matrix metalloproteinase-9, which is associated with transendothelial migration and invasion of chronic lymphocytic leukemia (CLL) B-cells (Redondo-Muñoz et al. 2008). Conditional deletion of $\alpha4$ sensitized BCR-ABL(+) leukemia to nilotinib and pharmacological VLA4 blockade with the antibody natalizumab prolong survival of NOD/SCID recipients of primary ALL when combined with chemotherapy, indicating a role for this integrin in chemoresistance of lymphoid malignancies (Hsieh et al. 2013). Mechanistic in vitro studies revealed that the interaction between vascular cell adhesion molecule 1 (VCAM-1) and VLA-4 play an integral role in the activation of NF-κB in both, the stromal and the tumor cell compartments (Jacamo et al. 2014). Nuclear factor-kappa beta activation via the VCAM-1/VLA-4 interaction causes increased numbers of dysplastic hematopoietic cells and progression into secondary AML (Rupec et al. 2005). Abnormal ANG1/TIE signaling has been detected both in endothelial cells and leukemia cells (Watarai et al. 2002; Wakabayashi et al. 2004). In leukemia cells, autocrine ANG1/TIE2 signaling activates STAT1/3/5/6 and extracellular regulated kinase pathways that increase proliferation, and TIE2/IP-3 kinase signaling supports AML cell survival (Watarai et al. 2002; Wakabayashi et al. 2004). Secretion of the proinflammatory cytokines TNF-α and IL-1β by LSCs upregulates endothelial adhesion receptors such as selectins, VCAM-1, and intercellular adhesion molecule-1, to support vascular adhesion (Stucki et al. 2001). It has been further demonstrated that *INTEGRIN β3* knockdown impairs homing, downregulates LSC transcriptional programs, and induces differentiation via the intracellular kinase Syk without affecting normal HSCs (Miller et al. 2013).

3.2.5 Hypoxia/HIF-1α Signaling and BM Vasculature

Overexpression of the hypoxia-regulated component hypoxia-inducible factor 1α (HIF-1α) has been shown in clusters of leukemia cells in BM specimens (Wellmann et al. 2004). Hypoxia affects LSC cycling, quiescence, metabolism and chemotherapy resistance, and HIF-1α could serve as a putative prognostic marker for high-risk leukemia and potential therapeutic target (Griessinger et al. 2014). HIF-1α-induced quiescence supports chemoresistance of AML cells (Griessinger et al. 2014; Drolle et al. 2015). The suppression of reactive oxygen species (ROS) and endoplasmic reticulum stress are the main proposed mechanisms of HIF-1α-induced anti-apoptosis in LSCs (Zhang et al. 2012). It has also been demonstrated that HIF-1α upregulates *CXCL12* gene expression in endothelial cells (Ceradini et al. 2004) and CXCR4 expression in AML cells (Fiegl et al. 2009), which increases migration and homing of circulating CXCR4-positive AML cells into ischemic tissue. Hypoxia could favor LSC niche metabolism, too (Korn and Méndez-Ferrer 2017). However, some studies have demonstrated that deleting HIF-1α induces AML and MPN progression (Velasco-Hernandez et al. 2014, 2015). These controversial results reflect the complexity of BM environmental control of leukemia cells. One of the key functions of hypoxia and HIF-1α is upregulation of growth factor vascular endothelial growth factor (VEGF) and stimulation of angiogenesis. Myeloid leukemia including AML, MPN, and MDS is correlated with increased BM angiogenesis (Hussong et al. 2000; Aguayo et al. 2000; Lundberg et al. 2000; Medinger et al. 2009; Pruneri et al. 1999; Ferrara et al. 2003), and disorganized BM vascularization is a common niche change in myeloid malignancies. The microvasculature) is an active component of the BM microenvironment, supplying appropriate oxygen and nutrients. VEGF secreted by leukemic cells activates receptors on both leukemic and endothelial cells and plays a vital role in the growth of leukemia cells (Ferrara et al. 2003). The direct HIF-1α inhibitor PX-478 decreases hypoxia-mediated VEGF expression in tumor xenografts, resulting in antitumor activity (Koh et al. 2008). Angiogenesis is stimulated in the LSC microenvironment, despite the lower oxygen tension in the BM than in arterial blood (Benito et al. 2011). However, it remains difficult to define the BM stem cell niche as severely hypoxic (Bonig and Papayannopoulou 2013). A quantitative imaging study demonstrated that HSCs in the endosteal BM niche closely interact with BM microvessels and yet highly express HIF1α, indicating that the hypoxic profile is at least, in part, through a cell-intrinsic mechanism rather than a lack of blood supply (Nombela-Arrieta et al. 2013). In fact, cytokines), hormones, and genetic modifications can stimulate HIF-1α signaling (Kuschel et al. 2012).

3.2.6 LSC Metabolism

Leukemia cells have two major metabolic challenges: how to meet the bioenergetic and biosynthetic demands of increased cell proliferation and how to survive fluctuations in external nutrient and oxygen availability in the BM environment. Some studies have indicated a link between the microenvironment and leukemia cell metabolism. The cholesterol-lowering drug lovastatin induces cell-autonomous inhibition of LSCs in a co-culture system with MSCs but does not inhibit AML cells cultured alone. Lovastatin pretreatment of LSC–stromal co-cultures also prolongs survival of mice injected with these cells (Seton-Rogers 2013; Hartwell et al. 2013). Adipocytes are the prevalent type of stromal cell in the adult BM, and fatty acids produced by adipocytes modulate the activity of signaling molecules (Carracedo et al. 2013; Behan et al. 2009). The finding that the rate of relapse after chemotherapy in mice injected with syngeneic leukemia cells was higher in obese mice than in normal-weight mice (Behan et al. 2009) suggests the possibility that the increased adipocyte content of adult BM promotes leukemia growth and negatively affects sensitivity to chemotherapy. BM stromal cells promote survival of AML cells via a metabolic shift from pyruvate oxidation to fatty acid β-oxidation (FAO), which causes mitochondrial uncoupling that diminishes mitochondrial ROS formation and decreases intracellular oxidative stress linked to the Bcl-2 anti-apoptotic machinery (Samudio et al. 2010). FAO is required for HSC self-renewal and quiescence (Ito et al. 2012). Another study demonstrated that AML stem cells rely on oxidative phosphorylation and are unable to utilize glycolysis when mitochondrial respiration is inhibited (Skrtić et al. 2011), showing that maintenance of mitochondrial function is essential for their survival (Lagadinou et al. 2013). These findings suggest that acetyl-CoA produced by FAO fuels the Krebs cycle and oxidative phosphorylation. In turn, more recent evidence suggests that metabolic signals play critical roles determining transcriptional regulation; metabolic enzymes are often present in transcriptional complexes, and thus provide a local supply of substrates/cofactors (Suzuki et al. 2009). Interestingly, AML cells alter the immune microenvironment via release of high concentrations of arginase II, which suppresses T cell proliferation, polarizes surrounding monocytes into a suppressive M2-like phenotype, and inhibits proliferation and differentiation of murine granulocyte-monocyte progenitors and human CD34$^+$ progenitors (Mussai et al. 2013). These findings suggest that metabolic features supporting the AML BM niche may yield novel therapeutic targets.

3.2.7 A Commentary on Likely Future Directions

Circulating leukemia cells are effectively eliminated by traditional or targeted therapies, whereas leukemia cells residing in the BM are chemoresistant and are responsible for relapse. The BM microenvironment contributes to increase leukemic cell

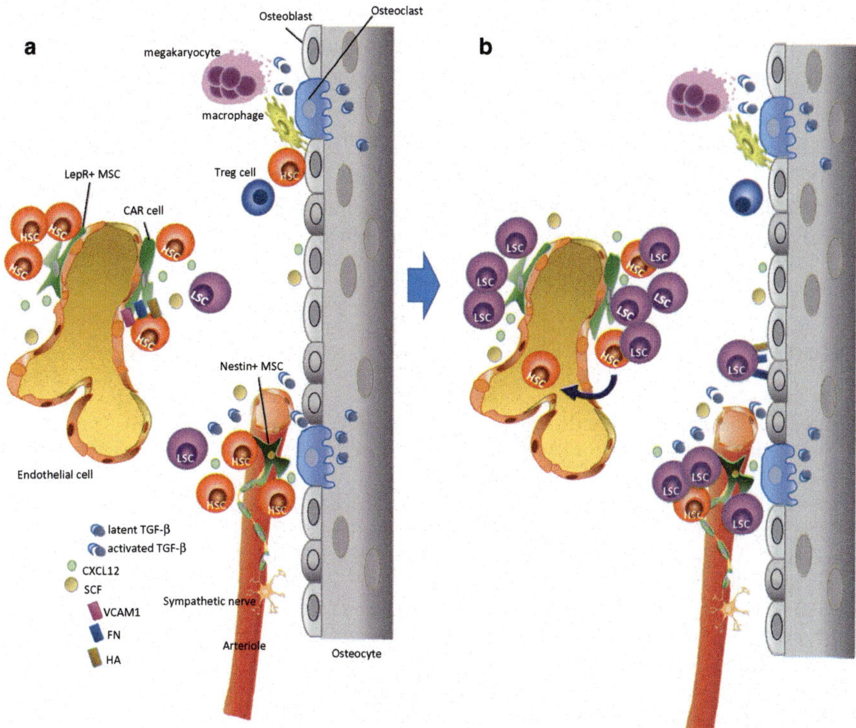

Fig. 3.1 Key components of the leukemic BM microenvironment and invasion of LSCs. (**a**) Components of normal hematopoietic stem cell (HSC) niches consist of multiple cell types including osteoblasts, Cxcl12-abundant reticular (CAR) cells, nestin-positive mesenchymal stem cells (MSCs), LepR-expressing perivascular cells, endothelial cells, immune cells (macrophages and T-regs) and Schwann cells wrapping sympathetic nerve fibers. (**b**) Leukemia stem cells (LSCs) hijack HSC marrow spaces including perivascular and endosteal niches. The BM stromal cells and osteoblasts produce a complex extracellular matrix (ECM) such as vascular cell adhesion molecule-1 (VCAM-1), fibronectin (FN) and hyaluronic acid (HA), which facilitate engraftment and adhesion of LSCs. Osteoblasts within endosteal niches produce transforming growth factor-β (TGF-β), which, in turn, promotes leukemia cell dormancy and decreases their chemosensitivity. CAR cells, nestin-positive MSCs, leptin receptor-positive perivascular cells, and endothelial cells may play a role in migration of leukemia cells into the BM microenvironment via cytokines, chemokines, and adhesion molecules

adhesion, provides growth factors, and promotes immunosuppression (Tabe and Konopleva 2014; Nwajei and Konopleva 2013). At the same time, leukemia cells are constantly adjusting their metabolic state in response to extracellular signaling and/or nutrient availability by making "decisions" such as quiescence, proliferation or differentiation in the constantly changing environment. By elucidating the role of the BM microenvironment in the pathogenesis of hematological tumors, recent studies have provided insight into the molecular mechanisms involved in stem cell activation and homing to the BM niche (Fig. 3.1). Long-term quiescence and self-renewal may be crucial for stress resistance of LSCs, and survival and proliferation of leukemic cells is critically regulated by the transition mechanisms from stem to progenitor cell, corresponding to metabolic changes, oxygen concentration, cytokine stimulation, or cell contact regulation (Katayama et al. 2006). The BM microenvironment presents an attractive target for novel therapeutic strategies. In particular, increasing insight into LSC development in their specific BM microenvironment may ultimately result in novel therapeutic strategies within a framework for targeting niche cells to attenuate leukemic progression or targeting LSCs without adversely affecting normal stem cell self-renewal.

Reference

Adams GB, Chabner KT, Alley IR, Olson DP, Szczepiorkowski ZM, Poznansky MC et al (2006) Stem cell engraftment at the endosteal niche is specified by the calcium-sensing receptor. Nature 439:599–603

Aguayo A, Kantarjian H, Manshouri T, Gidel C, Estey E, Thomas D et al (2000) Angiogenesis in acute and chronic leukemias and myelodysplastic syndromes. Blood 96:2240–2245

Arai F, Hirao A, Ohmura M, Sato H, Matsuoka S, Takubo K et al (2004) Tie2/angiopoietin-1 signaling regulates hematopoietic stem cell quiescence in the bone marrow niche. Cell 118:149–161

Avigdor A, Goichberg P, Shivtiel S, Dar A, Peled A, Samira S et al (2004) CD44 and hyaluronic acid cooperate with SDF-1 in the trafficking of human CD34+ stem/progenitor cells to bone marrow. Blood 103:2981–2989

Battula VL, Le PM, Sun JC, Nguyen K, Yuan B, Zhou X et al (2017) AML-induced osteogenic differentiation in mesenchymal stromal cells supports leukemia growth. JCI Insight 2:e90036

Behan JW, Yun JP, Proektor MP, Ehsanipour EA, Arutyunyan A, Moses AS et al (2009) Adipocytes impair leukemia treatment in mice. Cancer Res 69:7867–7874

Benito J, Shi Y, Szymanska B, Carol H, Boehm I, Lu H et al (2011) Pronounced hypoxia in models of murine and human leukemia: high efficacy of hypoxia-activated prodrug PR-104. PLoS One 6:e23108

Bonig H, Papayannopoulou T (2013) Hematopoietic stem cell mobilization: updated conceptual renditions. Leukemia 27:24–31

Bruns I, Lucas D, Pinho S, Ahmed J, Lambert MP, Kunisaki Y et al (2014) Megakaryocytes regulate hematopoietic stem cell quiescence through CXCL4 secretion. Nat Med 20:1315–1320

Calvi LM, Adams GB, Weibrecht KW, Weber JM, Olson DP, Knight MC et al (2003) Osteoblastic cells regulate the haematopoietic stem cell niche. Nature 425:841–846

Carracedo A, Cantley LC, Pandolfi PP (2013) Cancer metabolism: fatty acid oxidation in the limelight. Nat Rev Cancer 13:227–232

Ceradini DJ, Kulkarni AR, Callaghan MJ, Tepper OM, Bastidas N, Kleinman ME et al (2004) Progenitor cell trafficking is regulated by hypoxic gradients through HIF-1 induction of SDF-1. Nat Med 10:858–864

Cho BS, Zeng Z, Mu H, Wang Z, Konoplev S, McQueen T et al (2015) Antileukemia activity of the novel peptidic CXCR4 antagonist LY2510924 as monotherapy and in combination with chemotherapy. Blood 126:222–232

Chow A, Lucas D, Hidalgo A, Méndez-Ferrer S, Hashimoto D, Scheiermann C et al (2011) Bone marrow CD169+ macrophages promote the retention of hematopoietic stem and progenitor cells in the mesenchymal stem cell niche. J Exp Med 208:261–271

Christopher MJ, Liu F, Hilton MJ, Long F, Link DC (2009) Suppression of CXCL12 production by bone marrow osteoblasts is a common and critical pathway for cytokine-induced mobilization. Blood 114:1331–1339

Colmone A, Amorim M, Pontier AL, Wang S, Jablonski E, Sipkins DA (2008) Leukemic cells create bone marrow niches that disrupt the behavior of normal hematopoietic progenitor cells. Science 322:1861–1865

Ding L, Saunders TL, Enikolopov G, Morrison SJ (2012) Endothelial and perivascular cells maintain haematopoietic stem cells. Nature 481:457–462

Doan PL, Chute JP (2012) The vascular niche: home for normal and malignant hematopoietic stem cells. Leukemia 26:54–62

Dores GM, Devesa SS, Curtis RE, Linet MS, Morton LM (2012) Acute leukemia incidence and patient survival among children and adults in the United States, 2001-2007. Blood 119:34–43

Drolle H, Wagner M, Vasold J, Kütt A, Deniffel C, Sotlar K et al (2015) Hypoxia regulates proliferation of acute myeloid leukemia and sensitivity against chemotherapy. Leuk Res 39:779–785

Dührsen U, Hossfeld DK (1996) Stromal abnormalities in neoplastic bone marrow diseases. Ann Hematol 73:53–70

Ferrara N, Gerber HP, LeCouter J (2003) The biology of VEGF and its receptors. Nat Med 9:669–676

Fiegl M, Samudio I, Clise-Dwyer K, Burks JK, Mnjoyan Z, Andreeff M (2009) CXCR4 expression and biologic activity in acute myeloid leukemia are dependent on oxygen partial pressure. Blood 113:1504–1512

Fujisaki J, Wu J, Carlson AL, Silberstein L, Putheti P, Larocca R et al (2011) In vivo imaging of Treg cells providing immune privilege to the haematopoietic stem-cell niche. Nature 474:216–219

Griessinger E, Anjos-Afonso F, Pizzitola I, Rouault-Pierre K, Vargaftig J, Taussig D et al (2014) A niche-like culture system allowing the maintenance of primary human acute myeloid leukemia-initiating cells: a new tool to decipher their chemoresistance and self-renewal mechanisms. Stem Cells Transl Med 3:520–529

Hanahan D, Coussens LM (2012) Accessories to the crime: functions of cells recruited to the tumor microenvironment. Cancer Cell 21:309–322

Hanoun M, Zhang D, Mizoguchi T, Pinho S, Pierce H, Kunisaki Y et al (2014) Acute myelogenous leukemia-induced sympathetic neuropathy promotes malignancy in an altered hematopoietic stem cell niche. Cell Stem Cell 15:365–375

Hartwell KA, Miller PG, Mukherjee S, Kahn AR, Stewart AL, Logan DJ et al (2013) Niche-based screening identifies small-molecule inhibitors of leukemia stem cells. Nat Chem Biol 9:840–848

Herrlich P, Morrison H, Sleeman J, Orian-Rousseau V, König H, Weg-Remers S et al (2000) CD44 acts both as a growth- and invasiveness-promoting molecule and as a tumor-suppressing cofactor. Ann N Y Acad Sci 910:106–118. discussion 18-20

Hsieh YT, Gang EJ, Geng H, Park E, Huantes S, Chudziak D et al (2013) Integrin alpha4 blockade sensitizes drug resistant pre-B acute lymphoblastic leukemia to chemotherapy. Blood 121:1814–1818

Hussong JW, Rodgers GM, Shami PJ (2000) Evidence of increased angiogenesis in patients with acute myeloid leukemia. Blood 95:309–313

Ito K, Carracedo A, Weiss D, Arai F, Ala U, Avigan DE et al (2012) A PML–PPAR-δ pathway for fatty acid oxidation regulates hematopoietic stem cell maintenance. Nat Med 18:1350–1358

Jacamo R, Chen Y, Wang Z, Ma W, Zhang M, Spaeth EL et al (2014) Reciprocal leukemia-stroma VCAM-1/VLA-4-dependent activation of NF-κB mediates chemoresistance. Blood 123:2691–2702

Jin L, Hope KJ, Zhai Q, Smadja-Joffe F, Dick JE (2006) Targeting of CD44 eradicates human acute myeloid leukemic stem cells. Nat Med 12:1167–1174

Katayama Y, Battista M, Kao WM, Hidalgo A, Peired AJ, Thomas SA et al (2006) Signals from the sympathetic nervous system regulate hematopoietic stem cell egress from bone marrow. Cell 124:407–421

Kfoury Y, Mercier F, Scadden DT (2014) SnapShot: The hematopoietic stem cell niche. Cell 158:228–2e1

Kiel MJ, Morrison SJ (2008) Uncertainty in the niches that maintain haematopoietic stem cells. Nat Rev Immunol 8:290–301

Koh MY, Spivak-Kroizman T, Venturini S, Welsh S, Williams RR, Kirkpatrick DL et al (2008) Molecular mechanisms for the activity of PX-478, an antitumor inhibitor of the hypoxia-inducible factor-1alpha. Mol Cancer Ther 7:90–100

Korn C, Méndez-Ferrer S (2017) Myeloid malignancies and the microenvironment. Blood 129:811–822

Kunisaki Y, Bruns I, Scheiermann C, Ahmed J, Pinho S, Zhang D et al (2013) Arteriolar niches maintain haematopoietic stem cell quiescence. Nature 502:637–643

Kuschel A, Simon P, Tug S (2012) Functional regulation of HIF-1α under normoxia--is there more than post-translational regulation? J Cell Physiol 227:514–524

Lagadinou ED, Sach A, Callahan K, Rossi RM, Neering SJ, Minhajuddin M et al (2013) BCL-2 inhibition targets oxidative phosphorylation and selectively eradicates quiescent human leukemia stem cells. Cell Stem Cell 12:329–341

Lane SW (2012) Bad to the bone. Blood 119:323–325

Lundberg LG, Lerner R, Sundelin P, Rogers R, Folkman J, Palmblad J (2000) Bone marrow in polycythemia vera, chronic myelocytic leukemia, and myelofibrosis has an increased vascularity. Am J Pathol 157:15–19

Medinger M, Skoda R, Gratwohl A, Theocharides A, Buser A, Heim D et al (2009) Angiogenesis and vascular endothelial growth factor-/receptor expression in myeloproliferative neoplasms: correlation with clinical parameters and JAK2-V617F mutational status. Br J Haematol 146:150–157

Mendelson A, Frenette PS (2014) Hematopoietic stem cell niche maintenance during homeostasis and regeneration. Nat Med 20:833–846

Méndez-Ferrer S, Michurina TV, Ferraro F, Mazloom AR, Macarthur BD, Lira SA et al (2010) Mesenchymal and haematopoietic stem cells form a unique bone marrow niche. Nature 466:829–834

Miller PG, Al-Shahrour F, Hartwell KA, Chu LP, Järås M, Puram RV et al (2013) In Vivo RNAi screening identifies a leukemia-specific dependence on integrin beta 3 signaling. Cancer Cell 24:45–58

Morrison SJ, Scadden DT (2014) The bone marrow niche for haematopoietic stem cells. Nature 505:327–334

Mussai F, De Santo C, Abu-Dayyeh I, Booth S, Quek L, McEwen-Smith RM et al (2013) Acute myeloid leukemia creates an arginase-dependent immunosuppressive microenvironment. Blood 122:749–758

Nervi B, Ramirez P, Rettig MP, Uy GL, Holt MS, Ritchey JK et al (2009) Chemosensitization of acute myeloid leukemia (AML) following mobilization by the CXCR4 antagonist AMD3100. Blood 113:6206–6214

Nilsson SK, Johnston HM, Whitty GA, Williams B, Webb RJ, Denhardt DT et al (2005) Osteopontin, a key component of the hematopoietic stem cell niche and regulator of primitive hematopoietic progenitor cells. Blood 106:1232–1239

Nombela-Arrieta C, Pivarnik G, Winkel B, Canty KJ, Harley B, Mahoney JE et al (2013) Quantitative imaging of haematopoietic stem and progenitor cell localization and hypoxic status in the bone marrow microenvironment. Nat Cell Biol 15:533–543

Nwajei F, Konopleva M (2013) The bone marrow microenvironment as niche retreats for hematopoietic and leukemic stem cells. Adv Hematol 2013:953982

Price T, Sipkins DA (2014) Rewiring the niche: sympathetic neuropathy drives malignant niche transformation. Cell Stem Cell 15:261–262

Prowse AB, Chong F, Gray PP, Munro TP (2011) Stem cell integrins: implications for ex-vivo culture and cellular therapies. Stem Cell Res 6:1–12

Pruneri G, Bertolini F, Soligo D, Carboni N, Cortelezzi A, Ferrucci PF et al (1999) Angiogenesis in myelodysplastic syndromes. Br J Cancer 81:1398–1401

Raaijmakers MH (2014) Disease progression in myelodysplastic syndromes: do mesenchymal cells pave the way? Cell Stem Cell 14:695–697

Raaijmakers MH, Mukherjee S, Guo S, Zhang S, Kobayashi T, Schoonmaker JA et al (2010) Bone progenitor dysfunction induces myelodysplasia and secondary leukaemia. Nature 464:852–857

Redondo-Muñoz J, Ugarte-Berzal E, García-Marco JA, del Cerro MH, Van den Steen PE, Opdenakker G et al (2008) Alpha4beta1 integrin and 190-kDa CD44v constitute a cell surface docking complex for gelatinase B/MMP-9 in chronic leukemic but not in normal B cells. Blood 112:169–178

Rombouts EJ, Pavic B, Löwenberg B, Ploemacher RE (2004) Relation between CXCR-4 expression, Flt3 mutations, and unfavorable prognosis of adult acute myeloid leukemia. Blood 104:550–557

Rupec RA, Jundt F, Rebholz B, Eckelt B, Weindl G, Herzinger T et al (2005) Stroma-mediated dysregulation of myelopoiesis in mice lacking I kappa B alpha. Immunity 22:479–491

Sackstein R (2011) The biology of CD44 and HCELL in hematopoiesis: the 'step 2-bypass pathway' and other emerging perspectives. Curr Opin Hematol 18:239–248

Samudio I, Harmancey R, Fiegl M, Kantarjian H, Konopleva M, Korchin B et al (2010) Pharmacologic inhibition of fatty acid oxidation sensitizes human leukemia cells to apoptosis induction. J Clin Invest 120:142–156

Sant M, Allemani C, Tereanu C, De Angelis R, Capocaccia R, Visser O et al (2010) Incidence of hematologic malignancies in Europe by morphologic subtype: results of the HAEMACARE project. Blood 116:3724–3734

Seton-Rogers S (2013) Tumour microenvironment: Destroying leukaemia stem cell habitats. Nat Rev Cancer 13:821

Shiozawa Y, Taichman RS (2010) Dysfunctional niches as a root of hematopoietic malignancy. Cell Stem Cell 6:399–400

Sison EA, McIntyre E, Magoon D, Brown P (2013) Dynamic chemotherapy-induced upregulation of CXCR4 expression: a mechanism of therapeutic resistance in pediatric AML. Mol Cancer Res 11:1004–1016

Skrtić M, Sriskanthadevan S, Jhas B, Gebbia M, Wang X, Wang Z et al (2011) Inhibition of mitochondrial translation as a therapeutic strategy for human acute myeloid leukemia. Cancer Cell 20:674–688

Stucki A, Rivier AS, Gikic M, Monai N, Schapira M, Spertini O (2001) Endothelial cell activation by myeloblasts: molecular mechanisms of leukostasis and leukemic cell dissemination. Blood 97:2121–2129

Suda T, Takubo K, Semenza GL (2011) Metabolic regulation of hematopoietic stem cells in the hypoxic niche. Cell Stem Cell 9:298–310

Sugiyama T, Kohara H, Noda M, Nagasawa T (2006) Maintenance of the hematopoietic stem cell pool by CXCL12-CXCR4 chemokine signaling in bone marrow stromal cell niches. Immunity 25:977–988

Suzuki H, Forrest AR, van Nimwegen E, Daub CO, Balwierz PJ, Irvine KM et al (2009) The transcriptional network that controls growth arrest and differentiation in a human myeloid leukemia cell line. Nat Genet 41:553–562

Tabe Y, Konopleva M (2014) Advances in understanding the leukaemia microenvironment. Br J Haematol 164:767–778

Tabe Y, Jin L, Iwabuchi K, Wang RY, Ichikawa N, Miida T et al (2012) Role of stromal microenvironment in nonpharmacological resistance of CML to. Leukemia 26:883–892

Tabe Y, Shi YX, Zeng Z, Jin L, Shikami M, Hatanaka Y et al (2013) TGF-β-Neutralizing Antibody 1D11 Enhances Cytarabine-Induced Apoptosis in AML Cells in the Bone Marrow Microenvironment. PLoS One 8:e62785

Tabe Y, Yamamoto S, Saitoh K, Sekihara K, Monma N, Ikeo K et al (2017) Bone Marrow Adipocytes Facilitate Fatty Acid Oxidation Activating AMPK and a Transcriptional Network Supporting Survival of Acute Monocytic Leukemia Cells. Cancer Res 77:1453–1464

Uy GL, Rettig MP, Motabi IH, McFarland K, Trinkaus KM, Hladnik LM et al (2012) A phase 1/2 study of chemosensitization with the CXCR4 antagonist plerixafor in relapsed or refractory acute myeloid leukemia. Blood 119:3917–3924

Velasco-Hernandez T, Hyrenius-Wittsten A, Rehn M, Bryder D, Cammenga J (2014) HIF-1α can act as a tumor suppressor gene in murine acute myeloid leukemia. Blood 124:3597–3607

Velasco-Hernandez T, Tornero D, Cammenga J (2015) Loss of HIF-1α accelerates murine FLT-3(ITD)-induced myeloproliferative neoplasia. Leukemia 29:2366–2374

Wakabayashi M, Miwa H, Shikami M, Hiramatsu A, Ikai T, Tajima E et al (2004) Autocrine pathway of angiopoietins-Tie2 system in AML cells: association with phosphatidyl-inositol 3 kinase. Hematol J 5:353–360

Walkley CR, Olsen GH, Dworkin S, Fabb SA, Swann J, McArthur GA et al (2007a) A microenvironment-induced myeloproliferative syndrome caused by retinoic acid receptor gamma deficiency. Cell 129:1097–1110

Walkley CR, Shea JM, Sims NA, Purton LE, Orkin SH (2007b) Rb regulates interactions between hematopoietic stem cells and their bone marrow microenvironment. Cell 129:1081–1095

Watarai M, Miwa H, Shikami M, Sugamura K, Wakabayashi M, Satoh A et al (2002) Expression of endothelial cell-associated molecules in AML cells. Leukemia 16:112–119

Wellmann S, Guschmann M, Griethe W, Eckert C, von Stackelberg A, Lottaz C et al (2004) Activation of the HIF pathway in childhood ALL, prognostic implications of VEGF. Leukemia 18:926–933

Williams K, Motiani K, Giridhar PV, Kasper S (2013) CD44 integrates signaling in normal stem cell, cancer stem cell and (pre)metastatic niches. Exp Biol Med (Maywood) 238:324–338

Yilmaz OH, Valdez R, Theisen BK, Guo W, Ferguson DO, Wu H et al (2006) Pten dependence distinguishes haematopoietic stem cells from leukaemia-initiating cells. Nature 441:475–482

Zeng Z, Samudio IJ, Munsell M, An J, Huang Z, Estey E et al (2006) Inhibition of CXCR4 with the novel RCP168 peptide overcomes stroma-mediated chemoresistance in chronic and acute leukemias. Mol Cancer Ther 5:3113–3121

Zeng Z, Shi YX, Samudio IJ, Wang RY, Ling X, Frolova O et al (2009) Targeting the leukemia microenvironment by CXCR4 inhibition overcomes resistance to kinase inhibitors and chemotherapy in AML. Blood 113:6215–6224

Zhang H, Li H, Xi HS, Li S (2012) HIF1α is required for survival maintenance of chronic myeloid leukemia stem cells. Blood 119:2595–2607

Zhao M, Perry JM, Marshall H, Venkatraman A, Qian P, He XC et al (2014) Megakaryocytes maintain homeostatic quiescence and promote post-injury regeneration of hematopoietic stem cells. Nat Med 20:1321–1326

Chapter 4
Developmental HSC Microenvironments: Lessons from Zebrafish

Sara Nik, Joshua T. Weinreb, and Teresa V. Bowman

Abstract Hematopoietic stem cells (HSCs) posses the ability to maintain the blood system of an organism from birth to adulthood. The behavior of HSCs is modulated by its microenvironment. During development, HSCs acquire the instructions to self-renew and differentiate into all blood cell fates by passing through several developmental microenvironments. In this chapter, we discuss the signals and cell types that inform HSC decisions throughout ontogeny with a focus on HSC specification, mobilization, migration, and engraftment.

Keywords Zebrafish • Hematopoietic niche • Blood development • Hematopoietic stem cell

---

Sara Nik and Joshua T. Weinreb contributed equally.

S. Nik • J.T. Weinreb
Gottesman Institute for Stem Cell Biology and Regenerative Medicine, Albert Einstein College of Medicine, Bronx, NY, USA

Department of Developmental and Molecular Biology, Albert Einstein College of Medicine, Bronx, NY, USA

T.V. Bowman (✉)
Gottesman Institute for Stem Cell Biology and Regenerative Medicine, Albert Einstein College of Medicine, Bronx, NY, USA

Department of Developmental and Molecular Biology, Albert Einstein College of Medicine, Bronx, NY, USA

Departments of Molecular Biology and Medicine (Oncology), Albert Einstein College of Medicine, Bronx, NY, USA
e-mail: teresa.bowman@einstein.yu.edu

© Springer International Publishing AG 2017
A. Birbrair (ed.), *Stem Cell Microenvironments and Beyond*,
Advances in Experimental Medicine and Biology 1041,
DOI 10.1007/978-3-319-69194-7_4

4.1 Introduction

Hematopoietic stem cells (HSCs) are self-renewing, multi-potent cells with the capacity to give rise to all mature blood lineages. HSCs first acquire these skills during embryonic development as they move among a variety of niches, each providing different signals that help educate embryonic HSCs. Much of the earlier work on HSC development focused on the intrinsic factors dictating fate choices (reviewed in Orkin and Zon (2008)), but studies over the last decade have uncovered a wealth of information on the cell types and signals of the microenvironment that inform HSC decisions during ontogeny.

The use of the zebrafish (*Danio rerio*) to understand early patterning and organogenesis has exploded since the seminal genetic screens initiated by Christiane Haffter (1996). Since that time, over 600 papers have been published studying zebrafish hematopoiesis. Zebrafish possess many advantages that make it an excellent model to study developmental hematopoiesis. Hematopoietic cell types and gene programs are highly conserved from zebrafish to humans. The small size and extensive fertility of zebrafish are ideal attributes for performing unbiased forward genetic and chemical screens in a vertebrate model. Combined with the natural optical clarity of embryos, the use of transgenic zebrafish that express fluorescent proteins in a cell-type or tissue-type specific manner greatly facilitate live imaging of cells within their native microenvironment (reviewed in Zhang and Liu (2011)). Additionally, with the advent of CRISPR/Cas9 genome editing (Hwang et al. 2013) and tol2-based transgenesis (Kawakami et al. 2004), reverse genetic approaches are now also commonly used in zebrafish. These advantages have made zebrafish a rapidly emerging model system for the study of hematopoiesis.

In all vertebrates including zebrafish, hematopoiesis occurs in sequential waves (Fig. 4.1a). The earliest hematopoietic cells emerge during the primitive wave, which gives rise to mostly erythroid and myeloid cells arising from the lateral mesoderm during the first 24 h post fertilization (hpf) (equivalent to blood formation in the mammalian yolk sac) (reviewed in Robertson et al. (2016)). The final wave gives rise to definitive HSCs that persist into adulthood to maintain hematopoiesis throughout the organism's life. HSCs originate from the posterior lateral mesoderm (PLM). The cells emerge from specialized endothelial cells termed hemogenic endothelium found within the ventral wall of the dorsal aorta (the equivalent to the aorta-gonad-mesonephros (AGM) in mammals, Fig. 4.1b). Induction of HSCs is first detectable by expression of the transcription factors *gata2b* and *runx1* in the hemogenic endothelium around 24 h post fertilization (hpf) (Burns et al. 2002; Butko et al. 2015; Kalev-Zylinska et al. 2002). The process of HSC conversion from endothelium is termed the endothelial-to-hematopoietic transition (EHT) and involves the budding of HSC from the aortic endothelium (Bertrand et al. 2010; Boisset et al. 2010; Kissa and Herbomel 2010). Around 48–72 hpf, nascent HSCs migrate from the dorsal aorta (DA) via the circulation to an intermediate hematopoietic organ known as the caudal hematopoietic tissue (CHT) (the fetal liver equivalent in mammals) (Kissa et al. 2008; Murayama et al. 2006) (Fig. 4.1b). The CHT is the first site where HSC expand and differentiate into mature blood cells. The majority

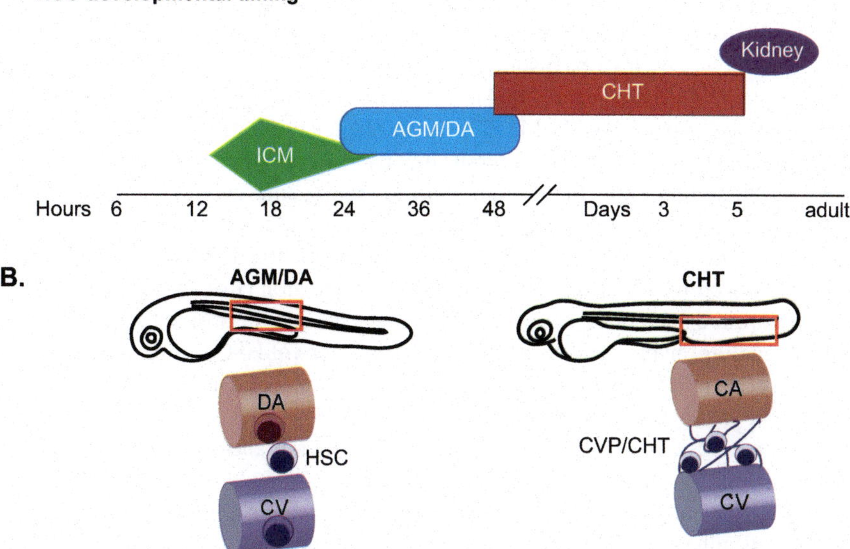

Fig. 4.1 The developmental timing and location of HSC development. (**a**) Timeline showing when and where primitive and definitive hematopoietic induction occurs in zebrafish. (**b**) Schema showing the location of the AGM/DA (*left*) and CHT (*right*) within zebrafish. Larger diagrams of the boxed regions are shown on the bottom. *ICM*, inner cell mass, *AGM* aorta-gonad-mesonephros, *DA* dorsal aorta, *CHT* caudal hematopoietic tissue, *CA* caudal artery, *CV* caudal vein, *CVP* caudal vein plexus

of HSCs then re-enter circulation and seed their final destination in the kidney marrow (equivalent to the bone marrow in mammals).

Each stage of HSC development is regulated by extrinsic cues from the local and systemic microenvironment. In this chapter, we will review the newest findings on the niche factors critical for early hemogenic endothelial induction as well as HSC specification, migration, and expansion. Understanding the key signals during ontogeny is not only important to developmental biologists, but could also have great clinical significance. Many of the players in embryonic niches are also important in adult HSC biology, thus new discoveries from development could enlighten the microenvironmental requirements necessary for maintaining adult HSC homeostasis. Moreover, uncovering how HSCs are normally produced in the embryo will help improve attempts to generate patient-specific HSCs from pluripotent stem cells in vitro (reviewed in Kyba and Daley (2003)).

4.2   Somite-Derived Niche Signals Promoting HSC Production

HSCs arise from specific mesoderm positioned in the posterior aspect of the embryo and lateral to somitic mesoderm termed the posterior lateral mesoderm (PLM) (Ho and Kimmel 1993). Recent work has demonstrated that this juxtaposition is critical for the early events of hemogenic endothelial and HSC specification. Specifically, several groups demonstrated that signals and cells emanating from the somite are required for proper HSC formation within the zebrafish embryo (Clements et al. 2011; Kim et al. 2014; Kobayashi et al. 2014; Lee et al. 2014; Nguyen et al. 2014; Pillay et al. 2016; Pouget et al. 2014).

During embryogenesis, PLM cells begin as bilateral strips along the lateral aspect of the embryo and then migrate medially (reviewed in Davidson and Zon (2004)). After migration, the endothelial and hemogenic endothelial progenitors within the PLM will form the dorsal aorta. During this journey, cells in the PLM make direct physical contact with the somites, a connection that Kobayashi and colleagues showed were essential for proper HSC formation (Kobayashi et al. 2014). The appropriate interaction between these cells is needed for proper transmission of Notch signaling, an important pathway for several steps of HSC formation (reviewed in Butko et al. (2016)). The Notch signaling pathway is well known to play a fundamental role in regulating cell fate decisions among adjacent cells through signaling between a transmembrane Notch receptor and membrane-spanning ligands on neighboring cells (Artavanis-Tsakonas et al. 1999). Thus, direct cell contact is the main modality for transmission of Notch signaling. In zebrafish, PLM cells expressing the cell-adhesion factor Jam1a interact with somite cells expressing Jam2a en route to the DA (Kobayashi et al. 2014). Knockdown of *jam1a* led to a decrease in Notch signaling and a decrease in HSC formation, but upon forced activation of Notch, specifically in endothelial precursors, HSC levels could be rescued.

Several additional studies have implicated Notch signaling in the early somitic niche. The non-canonical Wnt ligand, Wnt16, is highly expressed in somites and promotes HSC formation in a non-cell autonomous manner (Clements et al. 2011) (Fig. 4.2). Mechanistically, Wnt16 regulates the expression of two Notch ligands, *dlc (delta-c)* and *dld (delta-d),* which promotes Notch signaling for HSC induction. The relevant Notch receptor was later shown to be Notch3, which is expressed within the dorsal aorta but also earlier in the somites (Kim et al. 2014). Three out of four Notch receptors (Notch1a, Notch1b, and Notch3) are important for HSC formation, but only Notch3 function is needed during the stage when the PLM receives somite-derived signals. Epistasis analysis between Notch3 and Wnt16 demonstrated that Wnt16 lies upstream of Notch3 function within the somite (Kim et al. 2014).

Fibroblast growth factor (FGF) signaling provides a bridge between Wnt16 and Notch function during HSC emergence (Lee et al. 2014). Specifically, FGF signaling is required in the developing zebrafish somite for HSC formation during mid-somitogenesis (14–17 hpf), but not in the pre-endothelial PLM (Fig. 4.2). During this timeframe, FGF signaling informs HSC specification by relaying signals between Wnt16 and Dlc via the activity of its receptor, Fgfr4. Slightly later in development at the 23 somite-stage (~20.5 hpf), FGF signaling is a crucial player in establishing the HSC microenvironment around the dorsal aorta by regulating BMP pathway activity in the sub-aortic mesenchyme (Pouget et al. 2014). By modulating BMP pathway activity via transcriptional inhibition of *bmp4* and activation of the BMP antagonists, *noggin2* and *gremlin1a*, FGF provides a carefully regulated axis which functions as a developmental switch. Collectively, these data indicate that

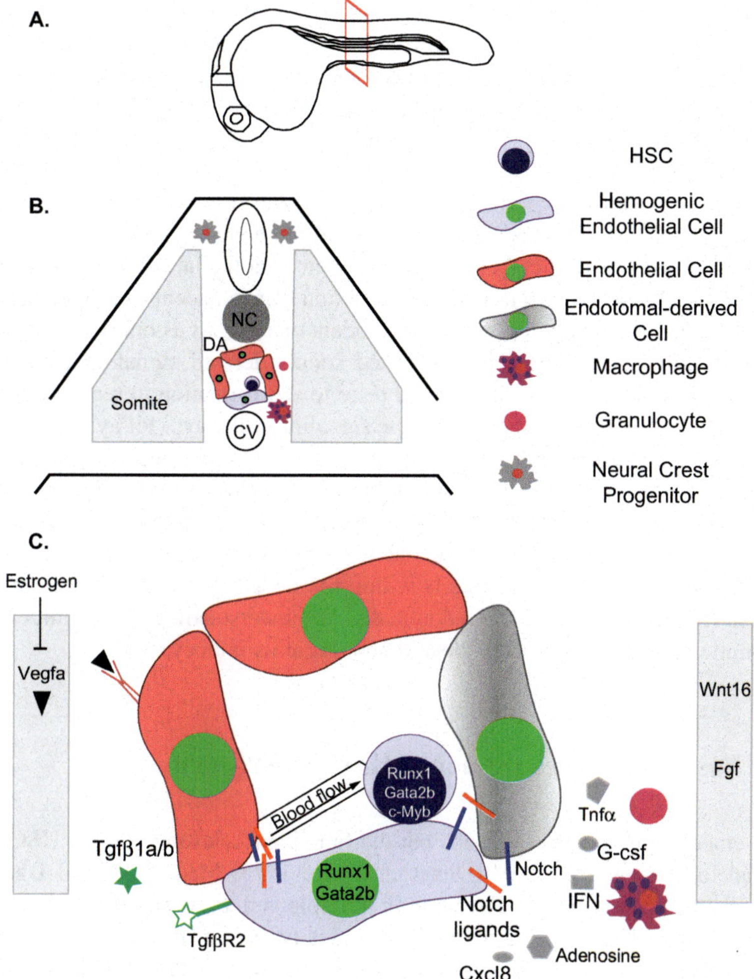

Fig. 4.2 The HSC specification niche. (**a**) Diagram showing the position within the zebrafish shown in the panel (**b**) cross-section. (**b**) Diagram of a cross-section within the dorsal aorta HSC specification niche in a zebrafish embryo. The legend is shown to the right. (**c**) The bottom panel highlights the most recent findings on the signaling pathways involved in the HSC specification niche. *HSC* hematopoietic stem cell, *NC* notochord, *DA* dorsal aorta, *CV* caudal vein, *vegfa* vascular endothelial growth factor a, *FGF* fibroblast growth factor, *TGFβ1* a/b-transforming growth factor β 1 a/b, *Tnfα* tumor necrosis factor α, *IFN* interferon, *Cxcl8-CXC* chemokine ligand 8, *Gcsf* granulocyte-colony stimulating factor

FGF provides a precise temporal mechanism to both activate and inhibit its own signaling activity to modulate HSC formation.

Work from the Currie laboratory found that not only does the somite provide signals as the PLM migrates, but cells derived from the somite can migrate and form a supportive niche within the dorsal aorta (Nguyen et al. 2014). Zebrafish *choker* mutants are loss-of-function for the *meox1* gene, which encodes a homeobox factor normally expressed in the early somite. Interestingly, loss of *meox1* resulted in an increase in HSC levels within the dorsal aorta. Nguyen and colleagues demonstrated that *choker* mutants also have an expansion of endotomal cells—a specialized subset of somitic cells that they showed give rise to endothelial cells that line the DA (Fig. 4.2). Although the somite-derived cells are known to contribute to the DA in other organisms (Jaffredo et al. 2013), findings from this study were the first evidence of its existence in zebrafish. Important to the hematopoietic field is the discovery that the endotomal cells are not hemogenic endothelial cells, but rather form an endothelial niche that promotes HSC induction from adjacent endothelial cells in a Cxcl12 (CXC chemokine ligand 12)-dependent manner. As a complement to these findings, Pillay et al. further characterized somite-derived signaling factors and found that retinoic acid (RA) is required prior to DA formation to regulate components of the Notch and Cxcl12 chemokine signaling pathways (Pillay et al. 2016).

The above studies demonstrate that the somite is a dynamic and integral niche for the earliest events of HSC formation in zebrafish. As HSCs originate in similar anatomical locations in mammals, it is highly probable that the influence of the somite on HSC development will be conserved from zebrafish to mammals. The recent identification of somite-derived cells within the developing dorsal aorta supports this idea (Jaffredo et al. 2013). Undoubtedly, our understanding of the influence of the somitic environment on HSC development is at its infancy.

4.3 Vascular Forces Influence HSC Emergence

HSCs emerge within blood vessels, but the functional relevance of this HSC birth site had long been a mystery (Dieterlen-Lievre 1975; Medvinsky and Dzierzak 1996; Walmsley et al. 2002). Research in multiple vertebrate systems has revealed that the environment of the vasculature, specifically the dorsal aorta, is a vital niche for emerging HSCs. Signals important for vessel growth and those that are key for vascular tone are not only critical for proper vascular development, but in many instances, play an additional role in HSC induction and survival.

VEGF (Vascular Endothelial Growth Factor) is a pro-angiogenic signal that promotes sprouting of intersegmental blood vessels and HSC formation (Connolly et al. 1989; Habeck et al. 2002). During development, there are a variety of inputs that regulate VEGF, keeping it at the appropriate level. Moreover, VEGF signaling feeds into additional downstream pathways important for HSC production. Carroll et al. uncovered estrogen as a titrator of Vegf signaling within the dorsal aorta endothelial niche in zebrafish (Carroll et al. 2014) (Fig. 4.2). In mammals, endogenous estrogen levels are generally low during early pregnancy, but increase as gestation goes on (Tulchinsky et al. 1972). There are several pieces of evidence suggesting that there are mechanisms intentionally in place to limit the amount of estrogen exposure on the fetus, including the expression of 17β-hydroxysteroid dehydrogenase type 2, which degrades estrogen (E2) in the umbilical arteries and veins

(Simard et al. 2011). During the first trimester of pregnancy maternal use of estrogen hormone supplementation is associated with an increased risk of infant acute leukemia, indicating that in utero exposure to estrogenic compounds may negatively impact fetal hematopoiesis (Pombo-de-Oliveira et al. 2006). Carroll and colleagues uncovered one mechanism to explain the negative impact of E2 on embryonic hematopoiesis (Carroll et al. 2014). They showed that excessive exposure to E2 from early somitogenesis (~12 hpf) until 24 hpf, the window of hemogenic endothelial specification, significantly decreased the formation of *runx1+* AGM HSCs. Activation of Notch signaling via overexpression of *vegfa* rescued hemogenic endothelium specification and HSC formation defects from excess E2. Overall, they show that maternally deposited E2 antagonizes the ventral limit of Vegf and downstream Notch signaling, allowing for the correct assignment of hemogenic endothelium and subsequent HSC formation (Carroll et al. 2014).

Prior work showed that TGFβ (Transforming Growth Factor β) signaling also regulates VEGF levels (reviewed in Holderfield and Hughes (2008) and Massague and Gomis (2006)). New work from the Patient lab showed that Vegf signaling could also regulate the TGFβ pathway (Monteiro et al. 2016) (Fig. 4.2). During zebrafish development, Vegfa signaling through its receptors Kdr (Kinase-insert domain receptor) and Kdr-like promote the expression of the TGFβ ligands *tgfβ1a* and *tgfβ1b* in endothelial cells. These ligands bind Tgfβ Receptor 2 (TgfβR2) to induce the signaling cascade. Monteiro and colleagues demonstrate that decreasing levels of TgfβR2 resulted in a severe decrease in expression of the HSC markers *runx1* and *gata2b* at 26–28 hpf but had no effect on endothelial or arterial markers, suggesting that HSC emergence is impaired at a step after dorsal aorta specification (Monteiro et al. 2016). Expression of the Notch ligand *jagged1a (jag1a)* downstream of TgfβR2 activation was shown as a critical mediator needed for expression of the HSC transcription factors *runx1* and *cmyb*.

These data suggest that TGFβ signaling is a positive regulator of HSC induction, but another group found that excessive TGFβ could suppress HSC formation. In a study of the transcriptional elongation mutant *spt5^{m806} (suppressor of Ty-5 homolog)*, Yang and colleagues showed that loss of transcriptional pausing lead to elevated TGFβ signaling, which was detrimental to HSC formation (Yang et al. 2016). Treatment of *spt5^{m806}* mutants with the TGFβ inhibitors SB505124 or LY2157299 restored HSCs in these embryos. Additionally, they showed that elevating TGFβ signaling in wild-type embryos via expression of a constitutively-active TGFβR1/ALK5 actually diminished HSC levels (Yang et al. 2016). The differences between these studies could be the differences between TGFβR1 and TGFβR2 signaling or could demonstrate that HSC emergence requires a balanced level of signaling. In adult HSCs, there are noted differences in how distinct HSC subtypes respond to different levels of TGFβ ligands (reviewed in Blank and Karlsson (2015)). It is also known that excess TGFβRI signaling can lead to hematologic defects including cytopenias as observed in myelodysplastic syndromes (Zhou et al. 2008). TGFβ receptors can act as homodimers or heterodimers (reviewed in Blank and Karlsson (2015)). The different responses to TGFβ receptor perturbation could therefore represent differences in receptor dimerization, downstream signaling, levels, or a combination of all three.

In addition to VEGF-regulated signaling, the biomechanical forces present in blood vessels were discovered to control embryonic HSC formation. Seminal work from the Zon, Daley, and Garcia-Cardena labs showed that blood flow-induced physical forces are critical to promote HSC production from endothelial cells both in murine and zebrafish models (Adamo et al. 2009; North et al. 2009) (Fig. 4.2). Specifically, they found that shear stress stimulated endothelial production of nitric oxide (NO), which induced HSC emergence. Through studies in zebrafish mutants devoid of a robust heartbeat and therefore possessing poor blood flow, Wang et al. later went on to show that the Kruppel-like transcription factor Klf2a directly regulates the NO signaling pathway allowing for HSC induction (Wang et al. 2011). KLF2 (or the zebrafish paralog Klf2a) is a zinc-finger transcription factor expressed in endothelial cells that is a known mediator of hemodynamic forces created by blood flow (Dekker et al. 2002; Lee et al. 2006). Several previous studies showed that KLF2 activates the expression of endothelial NO synthase (eNOS), which is a fundamental determinant of cardiovascular homeostasis, including systemic blood pressure, vascular remodeling, and angiogenesis (Dimmeler et al. 1999; Groenendijk et al. 2007; Lin et al. 2005; Parmar et al. 2006). In zebrafish development, Liu and colleagues showed that *klf2a* expression was induced by blood flow, and that Klf2a indeed regulated expression of the eNOS genes *nos1* and *nos2b* in vivo. They demonstrated that activation of the NO signaling pathway was an important downstream mediator of Klf2a as treatment with the NO donor SNAP (S-Nitroso-*N*-Acetyl-D,L-Penicillamine) partially rescues artery maturation and HSC production in *klf2a*-deficient embryos (Wang et al. 2011).

HSCs remain adjacent to blood vessels in the adult niche (reviewed in Crane et al. (2017)). In particular HSCs in the bone marrow are in close proximity to arterioles, which have different blood flow properties compared to venous or sinusoidal vessels (Kunisaki et al. 2013). Of note, HSC localization in the murine liver corresponds to the change in blood flow that occurs in the portal system after birth (Khan et al. 2016). These findings indicate that biomechanical forces are not only influencing HSC emergence, but likely play a function later in the life of a HSC.

4.4 Inflammatory Signaling Regulates HSC Emergence

Adult HSCs can proliferate in response to inflammatory cues from systemic infection or myeloablation and differentiate to replace lost effector immune cells (Baldridge et al. 2010; Essers et al. 2009; Feng et al. 2008; Kobayashi et al. 2015; Takizawa et al. 2011, 2012). This response of HSCs is not a secondary outcome to the loss of immune cells, but rather a direct response to inflammatory cytokines (Baldridge et al. 2010; Essers et al. 2009). More recently, similar cytokine signaling pathways have been found to play a critical role during embryonic HSC production independent of infection, a process termed sterile inflammation (Espin-Palazon et al. 2014; He et al. 2015; Li et al. 2014; Lim et al. 2017; Sawamiphak et al. 2014; Stachura et al. 2013). Specifically, roles for cytokines including tumor necrosis factors (TNFs) like TNFα and TNFβ, interferons (IFN) like IFNα and IFNγ, interleukins (IL) like IL-6 and growth factors like granulocyte colony stimulating factor (G-CSF) appear to be essential for definitive HSCs (Fig. 4.2).

NFκB (Nuclear factor κ-light-chain enhancer of activated B core) is a transcription factor stimulated by a variety of pro-inflammatory cytokines such as TNFs and toll-like receptor (TLR) agonists such as the bacterial pathogen-associated molecular pattern (PAMPs) lipopolysaccharide (LPS). TNFα binds to its receptors TNFR1 or TNFR2 to activate NF-κB (reviewed in Aggarwal et al. (2012) and Faustman and Davis (2010)). In zebrafish, Espin-Palazon et al. showed that Tnfα signaling specifically through Tnfr2 is required for definitive hematopoiesis (Espin-Palazon et al. 2014). Prior work showed Tnfr2 is also important for vascular development, but the hematopoietic requirement of Tnfr2 could be uncoupled from the vascular role by titrating the amount of knockdown of *tnfr2* (Espin et al. 2013). Tnfα-induced expression of *jag1a* activated Notch1a signaling required within endothelial cells to promote HSC formation. He et al. showed that NFκB-stimulation through Tlr4 is also utilized in both zebrafish and mice during definitive hematopoietic induction (He et al. 2015). Like Tnfr2, Notch signaling is a major downstream effector of Tlr4-regulated HSC production.

Type I and II Interferons are released in response to infection and stress and signal through the Interferon α or Interferon γ receptors, respectively (reviewed in Baldridge et al. (2011)). Work from the Speck, North, and Stainier labs demonstrated that both Type I and II Interferons are employed during zebrafish and murine development to promote HSC formation (Li et al. 2014; Sawamiphak et al. 2014). Expression of the ligand *ifnγ* and the Interferon γ receptor *crfb17 (cytokine receptor family b17)* are positively regulated by Notch signaling and blood flow (Sawamiphak et al. 2014). These data place the Interferon γ pathway downstream from early endothelial niche signals, and likely downstream of the effects of Tnfα- and Tlr4-mediated NFκB signaling.

Myeloid effector cells are some of the main producers of pro-inflammatory cytokines, thus the researchers examining the role of inflammatory signaling in embryonic HSC formation also assessed the contribution of myeloid cells on HSC development (Espin-Palazon et al. 2014; He et al. 2015; Li et al. 2014). Primitive myelopoiesis precedes HSC emergence, thus early neutrophils and macrophages are present to interact with precursors to HSCs. Ablation of all primitive myelopoiesis via decreasing the levels of the master myeloid transcription factor *pu.1* resulted in lower HSC formation, suggesting these cells could be early HSC niche cells and a key source of pro-inflammatory cytokines during development (Espin-Palazon et al. 2014; He et al. 2015; Li et al. 2014). Expansion of primitive neutrophils increased HSC numbers in a Tnfα-dependent manner, while the effects of Interferon γ could be more dependent on primitive macrophages (Espin-Palazon et al. 2014; Li et al. 2014).

In addition to classic myeloid effector cytokines, other inflammatory pathways were also recently identified to modulate HSC formation. Through a chemical screen, the Zon lab found that adenosine signaling could increase HSC numbers (Tamplin et al. 2015) (Fig. 4.2). The elevation in HSC levels was not due to expansion of existing HSCs, but rather an increased production of HSCs from hemogenic endothelium (Jing et al. 2015). Extracellular adenosine binds to the transmembrane adenosine receptors A_1, A_{2A}, A_{2B}, or A_3 (reviewed in Sousa and Diniz (2017)). The A_{2b} adenosine receptor was shown to be the main receptor important for the effect on HSC formation. They showed that endothelial-expressed A_{2b} activates the cAMP (cyclic Adenosine monophosphate)–PKA (Protein Kinase A) pathway, which up-regulates expression of the chemokine *cxcl8* (*il-8*), which in turn promotes the emergence of HSCs from the endothelium (Jing et al. 2015). CXCL8, also known as neutrophil chemotactic factor, induces chemotaxis and phagocytosis in neutrophils at sites of infections, and plays a major role in immunity and inflammation (reviewed in Kobayashi (2008)). Later in the chapter we will discuss an additional role for Cxcl88 signaling in HSC migration.

Adult HSCs must respond to an ever-changing milieu in animals exposed to life stressors, such as hematologic injuries and infections. It was long thought that this ability arose later in life when animals first encounter these stressful situations, but new data from studies of zebrafish embryogenesis indicate that HSCs are exposed to and utilize pro-inflammatory signals early in life. In addition to the utilization to promote HSC formation, it is possible that the early exposure to pro-inflammatory signaling is part of a HSC's "education" and that perturbation of these pathways in early life could have long-lasting impacts on adult HSC function.

4.5 Niche Signals from the Nervous System and Neural Crest

Groundbreaking work from the Frenette lab was among the first to demonstrate that the nervous system provided regulatory signals to the bone marrow HSC niche (Mendez-Ferrer et al. 2008). This early work showed a function specifically for nerves from the peripheral nervous system (PNS), which are derived from the neural crest (NC) (reviewed in Bronner and Simoes-Costa (2016)). NC cells are a migratory neural-ectoderm-derived cell population that, depending on their location within the developing embryo, gives rise to neurons, pigment cells, glia and endocrine cells of the parasympathetic and sympathetic nervous system. New work in the zebrafish uncovered a role for trunk NC in HSC formation (Damm and Clements 2017) (Fig. 4.2). Using time-lapse confocal microscopy in transgenic fluorescence reporter lines, Damm and Clements were able to demonstrate that NC precursors, in particular those arising from the trunk NC, migrate to and physically associate with the DA just prior to the initiation of HSC production. This ability of the NC to directly migrate from the neural tube to the DA is dependent on platelet-derived growth factor (PDGF) signaling. Perturbing the signaling cascade or the migratory path of the trunk NC removed these cells from the HSC microenvironment and negatively impacted HSC emergence. Future studies will be informative in discerning the role of the embryonic NC on HSC specification signals, but the close proximity of NC to emerging HSCs in the developing zebrafish suggest that short-range signaling through PDGF contributes to the HSC niche. Recently, Lim and colleagues found that PDGF acts downstream of Hif1α and that it induces a pro-proliferative effect on HSCs via IL-6 signaling (Lim et al. 2017). Together these studies suggest NC precursors could crosstalk with pro-inflammatory signals during HSC emergence.

In addition to neural crest inputs, recent studies suggest that the central nervous system (CNS) also provides environmental cues to HSCs (Kwan et al. 2016; Pierce et al. 2017)}. Kwan et al. uncovered an early role for the hypothalamic-pituitary-adrenal (HPA) axis in HSC emergence (Kwan et al. 2016). Through a prior chemical screen for regulators of HSC development, they identified serotonin as a positive regulator of stem cell formation (North et al. 2007). Serotonin is produced both in the CNS and the periphery and acts as a neurotransmitter to both CNS and PNS neurons (reviewed in Barnes and Sharp (1999)). Within the CNS, serotonin stimulates neurons in the hypothalamus activating the HPA cascade. This axis is a relay system among three endocrine glands that results in the balanced production of many hormones regulating diverse body processes, including digestion, the immune system, and energy storage and expenditure (reviewed in Del Rey et al. (2008)). One of the main hormones released by the adrenal gland in response to HPA stimulation is glucocorticoid. In zebrafish developmental hematopoiesis, GCs are the main downstream effectors mediating the positive effects of serotonin on HSC formation. Studies in the murine system demonstrate that the HPA axis and GC production can also modulate HSC mobilization, indicating a conserved role for the CNS in HSC regulation (Pierce et al. 2017).

4.6 Cellular Constituents Involved in HSC Engraftment of the CHT Niche

After their emergence, HSCs migrate to different locations at discrete developmental time points where they receive necessary inputs as part of their "education" to become fully functioning adult HSCs. Movement to these different niches is necessary as conditions that hinder HSC seeding of developmental microenvironments result in hematopoietic defects (reviewed in Karpova and Bonig (2015)). The stages involved in HSC migrations during development are similar to the steps in adult HSC mobilization and engraftment: (1) escape the current niche, (2) travel to and seed the next niche, and (3) grow and differentiate within the new niche. Studies of developmental HSC migrations and niche interactions thus hold great potential to inform human HSC transplant biology.

Visualization of HSC movements within and in between their native microenvironments is greatly enhanced in transparent zebrafish. Murayama and colleagues were among the first to image the migration of HSCs from the DA to the CHT, and established this anatomical location as the secondary niche for HSCs (Murayama et al. 2006). Since then, many groups have uncovered cellular constituents and vital signals regulating the stages of CHT engraftment. To understand the first step, how HSCs escape the DA niche, Travnickova et al. used in vivo real time imaging to examine the dynamic interactions between emerging HSCs and macrophages derived from the primitive wave of hematopoiesis and demonstrated the importance of this niche component in facilitating HSC mobilization (Travnickova et al. 2015) (Fig. 4.3a). By secreting matrix metalloproteinases (Mmps), these primitive macrophages are able to remodel the extracellular matrix, allowing HSCs to migrate through the AGM stroma, and begin to mobilize towards the CHT. Coupled with the studies on inflammatory signaling, these studies demonstrate that depending on developmental time point, primitive macrophages could therefore promote HSC production and expansion as well as migration.

To gain insight into the cellular events critical for CHT seeding, Tamplin et al. used high-speed, time-lapse imaging to monitor the dynamic interplay between HSCs and the CHT niche elements (Tamplin et al. 2015). They observed extensive vascular remodeling once HSCs enter into the CHT microenvironment. In a process they termed "endothelial cuddling," a cluster of ECs surrounds a single HSC (Fig. 4.3b). Within this "cuddle" niche, HSC are also in contact with a single stromal cell, and this interaction appeared to regulate the plane of cell division of the HSC. To uncover pathways regulating HSC-CHT interplay, they also performed a chemical screen and identified a novel drug lycorine, which is characterized as a putative anti-inflammatory molecule. Treatment of zebrafish with lycorine promoted HSC-CHT niche interactions by inducing significant gene expression changes that alter the adhesive properties of ECs and ultimately lead to a long-term increase in HSC number from early development to adulthood.

The CHT niche is comprised of two main cell types: the venous endothelial cells of the caudal vein (CV) plexus and stromal fibroblast reticular cells (FRCs). The origin of the FRCs was unknown until a recent study by Murayama et al., where they discovered that they in fact originate from the ventral border of the caudal somites, through an epithelial-to-mesenchymal transition (EMT) mechanism (Murayama et al. 2015) (Fig. 4.3b). In zebrafish *olaca* mutants, definitive hematopoiesis is greatly diminished due to a drastically compromised ability of HSCs to remain and differentiate within the CHT. However, this was not due to a defect within the HSCs, but rather in the CHT niche and its defective maturation of FRCs. The defective gene in *olaca* mutants is the nascent polypeptide-associated complex alpha subunit (NACA) gene whose function is to bind to nascent polypeptides on the ribosome and act as a chaperone system to ensure proper folding of emerging proteins (Kirstein-Miles et al. 2013; Lauring et al. 1995). Its deficiency in mammals has been associated with ER stress and the UPR, which can ultimately trigger apoptosis (Hotokezaka et al. 2009). Given this role, the authors investigated the role of ER stress-induced UPR/apoptosis during CHT niche formation in *olaca* mutants and observed that chemical treatment with chaperones rescued HSC levels and reduced cell death within the CHT niche (namely the stromal cells). Along with the work demonstrating the role of somites in the emergence of HSCs (Kobayashi et al. 2014; Nguyen et al. 2014), this work demonstrates a second contribution of somitic

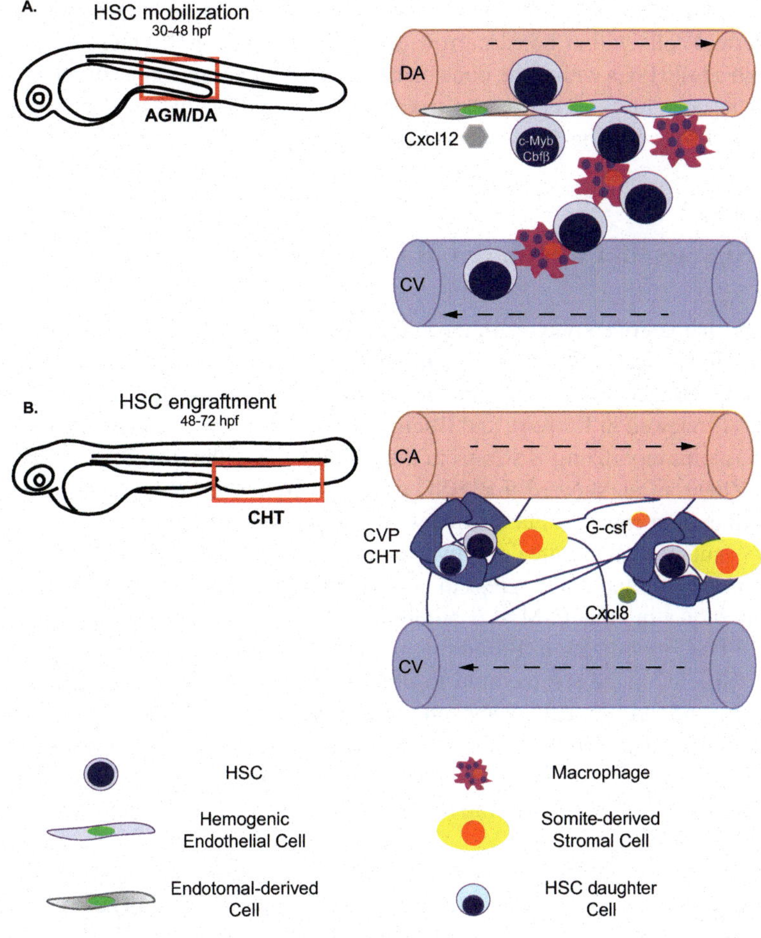

Fig. 4.3 Signals and cell types regulating developmental HSC mobilization and engraftment. (**a**) The left diagram shows the position within the zebrafish shown in the cross-section (*right*). The right panel highlights the most recent findings on the niche cells and signaling pathways involved in HSC mobilization/migration from the DA. Expression of the transcription factors c-Myb and Cbfβ in emerging HSCs is important to mediate egress from the DA. Appropriate levels of Cxcl12 are required for this process, and are regulated by endotomal-derived endothelial cells and HSC expression of c-Myb. Macrophages serve as escorts that facilitate HSC migration into the CV. (**b**) The left diagram shows the position within the zebrafish shown in the cross-section (*right*). The right panel highlights the most recent findings on the niche cells and signaling pathways involved in HSC engraftment in the CHT niche. Within the CHT, HSCs are surrounded by endothelial cells and contact a stromal cell, which provides cues on the orientation of the division plane. The stromal cells are derived from the somite and are critical for HSC maintenance in the CHT niche. Secreted factors, such as Cxcl8 and G-csf modify HSC interactions and proliferation within the niche. *HSC* hematopoietic stem cell, *NC* notochord, *DA* dorsal aorta, *CV* caudal vein, *Cxcl12-CXC* chemokine ligand 12, *Cbfβ* core binding protein beta, *CA* caudal artery, *CVP* caudal vein plexus, *CHT* caudal hematopoietic tissue, *Cxcl8-CXC* chemokine ligand 8, *Gcsf* granulocyte-colony stimulating factor

mesoderm-derived stromal cells within the CHT niche. Many non-hematopoietic mesodermal cells within bone marrow are derived from somitic mesoderm. Thus, although adult HSCs within the bone marrow are no longer in contact with developing somites or muscle, a conserved crosstalk between these two mesodermal compartments likely remains important into adulthood.

4.7 Signals that Regulate CHT Engraftment

The best characterized players in the adult HSC niche are the chemokine CXCL12 and the growth factor Stem Cell Factor (SCF) (reviewed in Crane et al. (2017)). CXCL12-mediated signaling is one of the main targets for therapeutic mobilization of HSCs (reviewed in Karpova and Bonig (2015)). Chemokine downregulation also plays a role in mobilizing HSCs from their initial site within the DA (Zhang et al. 2011). Zhang et al. showed that HSC egression from the DA was impaired in a zebrafish mutant *cmyb^{hkz3}*, due to upregulated expression of Cxcl12 (Fig. 4.3a). The *cmyb^{hkz3}* mutants can initiate definitive hematopoiesis normally, but the HSCs accumulate in the DA and don't egress to begin the migration and seeding process, collectively indicating that C-Myb is required to down-regulate Cxcl12 signaling and promote HSC mobilization. Based on this model, high levels of *cxcl12* expression within the HSC niche are required to prevent nascent HSCs from mobilizing too early from the DA. As HSCs develop and mature, *cxcl12* expression decreases to facilitate their release and migration to the CHT. Interestingly, high levels of *cxcl12* and *cmyb* are observed in the *meox1/choker* mutants that have expanded endotomal-contribution to the DA, suggesting the somite-derived cells could play a role in the crosstalk of C-Myb and Cxcl12.

Cbfβ, a non-DNA-binding subunit of the core-binding factor (Cbf), is also required for the release of HSCs from the AGM into circulation (Bresciani et al. 2014) (Fig. 4.3a). Similar to the *cmyb^{hkz3}* mutant, *cbfb* knockout zebrafish mutants show defects in HSC escape from the DA niche. In *cbfb* mutants, nascent HSCs emerge unaffected, but do not seed the CHT. This uncouples the role of CBFB from its interacting partner RUNX1, a key transcription factor that regulates the emergence of HSCs from the DA, and provides it an independent and temporally regulated function in HSC development. The downstream mediator of this egress defect in *cbfb* mutants is unknown, but given the importance of chemokine signaling in directing HSC migrations, it is possible that Cbfβ regulates expression of chemokines in a model analogous to cMyb-Cxcl12 identified by Zhang et al.

As with egression of HSCs from the DA to the CHT, chemokine signaling is also a vital component of the hematopoietic niche that enhances CHT colonization. By analyzing the CHT of 72 hpf zebrafish embryos, Blaser et al. observed that the chemokine Cxcl8 and its receptor Cxcr1 are positive regulators of HSC colonization (Blaser et al. 2017) (Fig. 4.3b). Mechanistically, Cxcr1 signals in a positive feedback loop to enhance CHT residency time and endothelial cell "cuddling" (an HSC-niche interaction described above), resulting in an increased mitotic rate and expanded pool of HSCs. Using a parabiotic zebrafish system they were able to show that Cxcr1 acts non-autonomously to promote HSC engraftment by directly altering the vascular niche.

Cytokine signaling from granulocyte-colony-stimulating factor (G-Csf) has also been reported to positively regulate the expansion of embryonic HSCs in the developing CHT (Stachura et al. 2013). The G-Csf receptor *gcsfr* is expressed starting as early as 6 hpf and up through 72 hpf, indicating it is active during both waves of hematopoiesis. The two ligands of Gcsfr, *gsfa* and *gsfb* are expressed at low levels at 6 hpf with increasing levels over time during development through 72 hpf. By modulating Gcsf levels through gain and loss of function experiments, Stachura et al. demonstrated that the number of HSCs in the CHT at 48 hpf directly correlated with Gcsf levels (Fig. 4.3b). Clinically, G-CSF is used to promote granulopoiesis and to promote HSC mobilization, thus it is possible that embryonic Gcsf could also play a role in developmental HSC migrations.

WNT (Wingless/INT) signaling is a well characterized pathway in regeneration and stem cell formation, but recent work from the Traver and Willert labs has demonstrated a new role for the WNT pathway in HSC migration to secondary sites such as the CHT (Grainger et al. 2016). In their study, Grainger and colleagues depicted how early Wnt signals, specifically Wnt9a, from the developing aorta, prior to 20 hpf, are required for HSCs to undergo an expansion event at 31 hpf. HSC loss when *wnt9a* levels are diminished persists to later stages of embryonic hematopoiesis, including CHT seeding, which the authors show is due to an accumulation of G1-arrested endothelial cells, preventing the initial HSC amplification needed to drive later expansion and seeding.

4.8   Conclusions and Perspectives

During development HSCs acquire the skills they will need throughout the life of an organism: self-renew, differentiate, regenerate, and migrate. These attributes are mediated via the integration of stem cell intrinsic programs and extrinsic microenvironmental signals. Using the imaging and genetic approaches afforded in the zebrafish system, researchers have defined niche components involved in hemogenic endothelial induction as well as HSC emergence, proliferation, mobilization, and engraftment. Although each stage of HSC development occurs in distinct anatomical locations, common themes are emerging among all of the developmental niches. Myeloid effector cells, endothelial cells, somite-derived and neural crest-derived cells serve as niche constituents beginning in the earliest stages of HSC ontogeny and into adulthood. The HSC-supportive cells are also developing and will therefore provide different signals depending on context. For example, granulocytes and macrophages provide inflammatory signaling which promotes HSC emergence, but macrophages also aid in HSC mobilization from the DA via secretion of MMPs (Espin-Palazon et al. 2014; He et al. 2015; Li et al. 2014; Travnickova et al. 2015).

The preserved juxtaposition of HSCs and niche constituents from emergence through adulthood also suggests that the signals employed during development have conserved roles in adult HSC biology. For example, interferon signaling is involved in HSC specification and then re-used in adulthood to promote HSC proliferation and differentiation (Baldridge et al. 2010; Essers et al. 2009; Li et al. 2014). The chemokine CXCL12 is a well-known regulator of HSC mobilization and proliferation in adults (Karpova and Bonig 2015), and it has similar conserved roles in HSC development (Nguyen et al. 2014; Zhang et al. 2011). Although the developmentally-defined biomechanical forces of blood flow are not known to play a role in adult HSCs, the enrichment of HSCs in the bone marrow positioned adjacent to arterioles, which have different blood flow properties than sinusoids, suggests that it is possible that biomechanical properties have a yet to be determined role in adult HSC homeostasis. Thus, despite the seemingly different characteristics of adult and developing HSCs, studies of both microenvironments provide complementary and surprisingly conserved information about the extrinsic regulation of HSCs.

The mounting data on the HSC niche clearly shows that it is a highly complex microenvironment with many revelations still to be learned. The discoveries on the contribution of the nervous system and neural crest cells to the HSC niche highlight the need for researchers to keep an open mind when conceptualizing the microenvironment. Zebrafish are a great system to make unbiased and novel discoveries regarding the niche. Conditional and inducible cell ablation strategies will allow researchers to determine which niche constituents are used and when. Tissue-specific mutagenesis approaches will now permit the assignment of specific signals to specific cell types. Lineage-tracing can provide information about the origins of niche cells from development into adulthood. By capitalizing on these advanced experimental strategies in zebrafish, researchers should be able to translate findings in the developmental HSC niche into approaches to benefit the treatment of blood diseases as well as in vitro HSC production.

References

Adamo L, Naveiras O, Wenzel PL, McKinney-Freeman S, Mack PJ, Gracia-Sancho J, Suchy-Dicey A, Yoshimoto M, Lensch MW, Yoder MC, Garcia-Cardena G, Daley GQ (2009) Biomechanical forces promote embryonic haematopoiesis. Nature 459:1131–1135

Aggarwal BB, Gupta SC, Kim JH (2012) Historical perspectives on tumor necrosis factor and its superfamily: 25 years later, a golden journey. Blood 119:651–665

Artavanis-Tsakonas S, Rand MD, Lake RJ (1999) Notch signaling: cell fate control and signal integration in development. Science 284:770–776

Baldridge MT, King KY, Boles NC, Weksberg DC, Goodell MA (2010) Quiescent haematopoietic stem cells are activated by IFN-gamma in response to chronic infection. Nature 465:793–797

Baldridge MT, King KY, Goodell MA (2011) Inflammatory signals regulate hematopoietic stem cells. Trends Immunol 32:57–65

Barnes NM, Sharp T (1999) A review of central 5-HT receptors and their function. Neuropharmacology 38:1083–1152

Bertrand JY, Chi NC, Santoso B, Teng S, Stainier DY, Traver D (2010) Haematopoietic stem cells derive directly from aortic endothelium during development. Nature 464:108–111

Blank U, Karlsson S (2015) TGF-beta signaling in the control of hematopoietic stem cells. Blood 125:3542–3550

Blaser BW, Moore JL, Hagedorn EJ, Li B, Riquelme R, Lichtig A, Yang S, Zhou Y, Tamplin OJ, Binder V, Zon LI (2017) CXCR1 remodels the vascular niche to promote hematopoietic stem and progenitor cell engraftment. J Exp Med 214:1011–1027

Boisset JC, van Cappellen W, Andrieu-Soler C, Galjart N, Dzierzak E, Robin C (2010) In vivo imaging of haematopoietic cells emerging from the mouse aortic endothelium. Nature 464:116–120

Bresciani E, Carrington B, Wincovitch S, Jones M, Gore AV, Weinstein BM, Sood R, Liu PP (2014) CBFbeta and RUNX1 are required at 2 different steps during the development of hematopoietic stem cells in zebrafish. Blood 124:70–78

Bronner ME, Simoes-Costa M (2016) The neural crest migrating into the twenty-first century. Curr Top Dev Biol 116:115–134

Burns CE, DeBlasio T, Zhou Y, Zhang J, Zon L, Nimer SD (2002) Isolation and characterization of runxa and runxb, zebrafish members of the runt family of transcriptional regulators. Exp Hematol 30:1381–1389

Butko E, Distel M, Pouget C, Weijts B, Kobayashi I, Ng K, Mosimann C, Poulain FE, McPherson A, Ni CW, Stachura DL, Del Cid N, Espin-Palazon R, Lawson ND, Dorsky R, Clements WK, Traver D (2015) Gata2b is a restricted early regulator of hemogenic endothelium in the zebrafish embryo. Development 142:1050–1061

Butko E, Pouget C, Traver D (2016) Complex regulation of HSC emergence by the Notch signaling pathway. Dev Biol 409:129–138

Carroll KJ, Esain V, Garnaas MK, Cortes M, Dovey MC, Nissim S, Frechette GM, Liu SY, Kwan W, Cutting CC, Harris JM, Gorelick DA, Halpern ME, Lawson ND, Goessling W, North TE (2014) Estrogen defines the dorsal-ventral limit of VEGF regulation to specify the location of the hemogenic endothelial niche. Dev Cell 29:437–453

Clements WK, Kim AD, Ong KG, Moore JC, Lawson ND, Traver D (2011) A somitic Wnt16/Notch pathway specifies haematopoietic stem cells. Nature 474:220–224

Connolly DT, Heuvelman DM, Nelson R, Olander JV, Eppley BL, Delfino JJ, Siegel NR, Leimgruber RM, Feder J (1989) Tumor vascular permeability factor stimulates endothelial cell growth and angiogenesis. J Clin Invest 84:1470–1478

Crane GM, Jeffery E, Morrison SJ (2017) Adult haematopoietic stem cell niches. Nat Rev Immunol 17(9):573–590

Damm EW, Clements WK (2017) Pdgf signalling guides neural crest contribution to the haematopoietic stem cell specification niche. Nat Cell Biol 19:457–467

Davidson AJ, Zon LI (2004) The 'definitive' (and 'primitive') guide to zebrafish hematopoiesis. Oncogene 23:7233–7246

Dekker RJ, van Soest S, Fontijn RD, Salamanca S, de Groot PG, VanBavel E, Pannekoek H, Horrevoets AJ (2002) Prolonged fluid shear stress induces a distinct set of endothelial cell genes, most specifically lung Kruppel-like factor (KLF2). Blood 100:1689–1698

Del Rey A, Chrousos GP, Besedovsky HO (2008) The hypothalamus-pituitary-adrenal axis. Elsevier, Amsterdam; Boston. xvii, 394 p

Dieterlen-Lievre F (1975) On the origin of haemopoietic stem cells in the avian embryo: an experimental approach. J Embryol Exp Morphol 33:607–619

Dimmeler S, Fleming I, Fisslthaler B, Hermann C, Busse R, Zeiher AM (1999) Activation of nitric oxide synthase in endothelial cells by Akt-dependent phosphorylation. Nature 399:601–605

Espin R, Roca FJ, Candel S, Sepulcre MP, Gonzalez-Rosa JM, Alcaraz-Perez F, Meseguer J, Cayuela ML, Mercader N, Mulero V (2013) TNF receptors regulate vascular homeostasis in zebrafish through a caspase-8, caspase-2 and P53 apoptotic program that bypasses caspase-3. Dis Model Mech 6:383–396

Espin-Palazon R, Stachura DL, Campbell CA, Garcia-Moreno D, Del Cid N, Kim AD, Candel S, Meseguer J, Mulero V, Traver D (2014) Proinflammatory signaling regulates hematopoietic stem cell emergence. Cell 159:1070–1085

Essers MA, Offner S, Blanco-Bose WE, Waibler Z, Kalinke U, Duchosal MA, Trumpp A (2009) IFNalpha activates dormant haematopoietic stem cells in vivo. Nature 458:904–908

Faustman D, Davis M (2010) TNF receptor 2 pathway: drug target for autoimmune diseases. Nat Rev Drug Discov 9:482–493

Feng CG, Weksberg DC, Taylor GA, Sher A, Goodell MA (2008) The p47 GTPase Lrg-47 (Irgm1) links host defense and hematopoietic stem cell proliferation. Cell Stem Cell 2:83–89

Grainger S, Richter J, Palazon RE, Pouget C, Lonquich B, Wirth S, Grassme KS, Herzog W, Swift MR, Weinstein BM, Traver D, Willert K (2016) Wnt9a Is required for the aortic amplification of nascent hematopoietic stem cells. Cell Rep 17:1595–1606

Groenendijk BC, Van der Heiden K, Hierck BP, Poelmann RE (2007) The role of shear stress on ET-1, KLF2, and NOS-3 expression in the developing cardiovascular system of chicken embryos in a venous ligation model. Physiology (Bethesda) 22:380–389

Habeck H, Odenthal J, Walderich B, Maischein H, Schulte-Merker S, screen c T (2002) Analysis of a zebrafish VEGF receptor mutant reveals specific disruption of angiogenesis. Curr Biol 12:1405–1412

Haffter P (1996) The identification of genes with unique and essential functions in the development of the zebrafish, Danio rerio. Development. 123:1–36

He Q, Zhang C, Wang L, Zhang P, Ma D, Lv J, Liu F (2015) Inflammatory signaling regulates hematopoietic stem and progenitor cell emergence in vertebrates. Blood 125:1098–1106

Ho RK, Kimmel CB (1993) Commitment of cell fate in the early zebrafish embryo. Science 261:109–111

Holderfield MT, Hughes CC (2008) Crosstalk between vascular endothelial growth factor, notch, and transforming growth factor-beta in vascular morphogenesis. Circ Res 102:637–652

Hotokezaka Y, van Leyen K, Lo EH, Beatrix B, Katayama I, Jin G, Nakamura T (2009) alphaNAC depletion as an initiator of ER stress-induced apoptosis in hypoxia. Cell Death Differ 16:1505–1514

Hwang WY, Fu Y, Reyon D, Maeder ML, Tsai SQ, Sander JD, Peterson RT, Yeh JR, Joung JK (2013) Efficient genome editing in zebrafish using a CRISPR-Cas system. Nat Biotechnol 31:227–229

Jaffredo T, Lempereur A, Richard C, Bollerot K, Gautier R, Canto PY, Drevon C, Souyri M, Durand C (2013) Dorso-ventral contributions in the formation of the embryonic aorta and the control of aortic hematopoiesis. Blood Cells Mol Dis 51:232–238

Jing L, Tamplin OJ, Chen MJ, Deng Q, Patterson S, Kim PG, Durand EM, McNeil A, Green JM, Matsuura S, Ablain J, Brandt MK, Schlaeger TM, Huttenlocher A, Daley GQ, Ravid K, Zon LI (2015) Adenosine signaling promotes hematopoietic stem and progenitor cell emergence. J Exp Med 212:649–663

Kalev-Zylinska ML, Horsfield JA, Flores MV, Postlethwait JH, Vitas MR, Baas AM, Crosier PS, Crosier KE (2002) Runx1 is required for zebrafish blood and vessel development and expression of a human RUNX1-CBF2T1 transgene advances a model for studies of leukemogenesis. Development 129:2015–2030

Karpova D, Bonig H (2015) Concise review: CXCR4/CXCL12 signaling in immature hematopoiesis—lessons from pharmacological and genetic models. Stem Cells 33:2391–2399

Kawakami K, Takeda H, Kawakami N, Kobayashi M, Matsuda N, Mishina M (2004) A transposon-mediated gene trap approach identifies developmentally regulated genes in zebrafish. Dev Cell 7:133–144

Khan JA, Mendelson A, Kunisaki Y, Birbrair A, Kou Y, Arnal-Estape A, Pinho S, Ciero P, Nakahara F, Ma'ayan A, Bergman A, Merad M, Frenette PS (2016) Fetal liver hematopoietic stem cell niches associate with portal vessels. Science 351:176–180

Kim AD, Melick CH, Clements WK, Stachura DL, Distel M, Panakova D, MacRae C, Mork LA, Crump JG, Traver D (2014) Discrete Notch signaling requirements in the specification of hematopoietic stem cells. EMBO J 33:2363–2373

Kirstein-Miles J, Scior A, Deuerling E, Morimoto RI (2013) The nascent polypeptide-associated complex is a key regulator of proteostasis. EMBO J 32:1451–1468

Kissa K, Herbomel P (2010) Blood stem cells emerge from aortic endothelium by a novel type of cell transition. Nature 464:112–115

Kissa K, Murayama E, Zapata A, Cortes A, Perret E, Machu C, Herbomel P (2008) Live imaging of emerging hematopoietic stem cells and early thymus colonization. Blood 111:1147–1156

Kobayashi Y (2008) The role of chemokines in neutrophil biology. Front Biosci 13:2400–2407

Kobayashi I, Kobayashi-Sun J, Kim AD, Pouget C, Fujita N, Suda T, Traver D (2014) Jam1a-Jam2a interactions regulate haematopoietic stem cell fate through Notch signalling. Nature 512:319–323

Kobayashi H, Kobayashi CI, Nakamura-Ishizu A, Karigane D, Haeno H, Yamamoto KN, Sato T, Ohteki T, Hayakawa Y, Barber GN, Kurokawa M, Suda T, Takubo K (2015) Bacterial c-di-GMP affects hematopoietic stem/progenitors and their niches through STING. Cell Rep 11:71–84

Kunisaki Y, Bruns I, Scheiermann C, Ahmed J, Pinho S, Zhang D, Mizoguchi T, Wei Q, Lucas D, Ito K, Mar JC, Bergman A, Frenette PS (2013) Arteriolar niches maintain haematopoietic stem cell quiescence. Nature 502:637–643

Kwan W, Cortes M, Frost I, Esain V, Theodore LN, Liu SY, Budrow N, Goessling W, North TE (2016) The central Nervous system regulates embryonic HSPC production via stress-Responsive glucocorticoid receptor signaling. Cell Stem Cell 19:370–382

Kyba M, Daley GQ (2003) Hematopoiesis from embryonic stem cells: lessons from and for ontogeny. Exp Hematol 31:994–1006

Lauring B, Sakai H, Kreibich G, Wiedmann M (1995) Nascent polypeptide-associated complex protein prevents mistargeting of nascent chains to the endoplasmic reticulum. Proc Natl Acad Sci U S A 92:5411–5415

Lee JS, Yu Q, Shin JT, Sebzda E, Bertozzi C, Chen M, Mericko P, Stadtfeld M, Zhou D, Cheng L, Graf T, MacRae CA, Lepore JJ, Lo CW, Kahn ML (2006) Klf2 is an essential regulator of vascular hemodynamic forces in vivo. Dev Cell 11:845–857

Lee Y, Manegold JE, Kim AD, Pouget C, Stachura DL, Clements WK, Traver D (2014) FGF signalling specifies haematopoietic stem cells through its regulation of somitic Notch signalling. Nat Commun 5:5583

Li Y, Esain V, Teng L, Xu J, Kwan W, Frost IM, Yzaguirre AD, Cai X, Cortes M, Maijenburg MW, Tober J, Dzierzak E, Orkin SH, Tan K, North TE, Speck NA (2014) Inflammatory signaling regulates embryonic hematopoietic stem and progenitor cell production. Genes Dev 28:2597–2612

Lim SE, Esain V, Kwan W, Theodore LN, Cortes M, Frost IM, Liu SY, North TE (2017) HIF1 alpha-induced PDGFRbeta signaling promotes developmental HSC production via IL-6 activation. Exp Hematol 46:83–95.e6

Lin Z, Kumar A, SenBanerjee S, Staniszewski K, Parmar K, Vaughan DE, Gimbrone MA Jr, Balasubramanian V, Garcia-Cardena G, Jain MK (2005) Kruppel-like factor 2 (KLF2) regulates endothelial thrombotic function. Circ Res 96:e48–e57

Massague J, Gomis RR (2006) The logic of TGFbeta signaling. FEBS Lett 580:2811–2820

Medvinsky A, Dzierzak E (1996) Definitive hematopoiesis is autonomously initiated by the AGM region. Cell 86:897–906

Mendez-Ferrer S, Lucas D, Battista M, Frenette PS (2008) Haematopoietic stem cell release is regulated by circadian oscillations. Nature 452:442–447

Monteiro R, Pinheiro P, Joseph N, Peterkin T, Koth J, Repapi E, Bonkhofer F, Kirmizitas A, Patient R (2016) Transforming growth factor beta drives hemogenic endothelium programming and the transition to hematopoietic stem cells. Dev Cell 38:358–370

Murayama E, Kissa K, Zapata A, Mordelet E, Briolat V, Lin HF, Handin RI, Herbomel P (2006) Tracing hematopoietic precursor migration to successive hematopoietic organs during zebrafish development. Immunity 25:963–975

Murayama E, Sarris M, Redd M, Le Guyader D, Vivier C, Horsley W, Trede N, Herbomel P (2015) NACA deficiency reveals the crucial role of somite-derived stromal cells in haematopoietic niche formation. Nat Commun 6:8375

Nguyen PD, Hollway GE, Sonntag C, Miles LB, Hall TE, Berger S, Fernandez KJ, Gurevich DB, Cole NJ, Alaei S, Ramialison M, Sutherland RL, Polo JM, Lieschke GJ, Currie PD (2014) Haematopoietic stem cell induction by somite-derived endothelial cells controlled by meox1. Nature 512:314–318

North TE, Goessling W, Walkley CR, Lengerke C, Kopani KR, Lord AM, Weber GJ, Bowman TV, Jang IH, Grosser T, Fitzgerald GA, Daley GQ, Orkin SH, Zon LI (2007) Prostaglandin E2 regulates vertebrate haematopoietic stem cell homeostasis. Nature 447:1007–1011

North TE, Goessling W, Peeters M, Li P, Ceol C, Lord AM, Weber GJ, Harris J, Cutting CC, Huang P, Dzierzak E, Zon LI (2009) Hematopoietic stem cell development is dependent on blood flow. Cell 137:736–748

Orkin SH, Zon LI (2008) Hematopoiesis: an evolving paradigm for stem cell biology. Cell 132:631–644

Parmar KM, Larman HB, Dai G, Zhang Y, Wang ET, Moorthy SN, Kratz JR, Lin Z, Jain MK, Gimbrone MA Jr, Garcia-Cardena G (2006) Integration of flow-dependent endothelial phenotypes by Kruppel-like factor 2. J Clin Invest 116:49–58

Pierce H, Zhang D, Magnon C, Lucas D, Christin JR, Huggins M, Schwartz GJ, Frenette PS (2017) Cholinergic signals from the CNS regulate G-CSF-mediated HSC mobilization from bone marrow via a glucocorticoid signaling relay. Cell Stem Cell 20:648–658.e4

Pillay LM, Mackowetzky KJ, Widen SA, Waskiewicz AJ (2016) Somite-derived retinoic acid regulates Zebrafish hematopoietic stem cell formation. PLoS One 11:e0166040

Pombo-de-Oliveira MS, Koifman S, Brazilian L, Collaborative Study Group of Infant Acute (2006) Infant acute leukemia and maternal exposures during pregnancy. Cancer Epidemiol Biomark Prev 15:2336–2341

Pouget C, Peterkin T, Simoes FC, Lee Y, Traver D, Patient R (2014) FGF signalling restricts haematopoietic stem cell specification via modulation of the BMP pathway. Nat Commun 5:5588

Robertson AL, Avagyan S, Gansner JM, Zon LI (2016) Understanding the regulation of vertebrate hematopoiesis and blood disorders—big lessons from a small fish. FEBS Lett 590:4016–4033

Sawamiphak S, Kontarakis Z, Stainier DY (2014) Interferon gamma signaling positively regulates hematopoietic stem cell emergence. Dev Cell 31:640–653

Simard M, Drolet R, Blomquist CH, Tremblay Y (2011) Human type 2 17beta-hydroxysteroid dehydrogenase in umbilical vein and artery endothelial cells: differential inactivation of sex steroids according to the vessel type. Endocrine 40:203–211

Sousa JB, Diniz C (2017) The adenosinergic system as a therapeutic target in the vasculature: new ligands and challenges. Molecules 22

Stachura DL, Svoboda O, Campbell CA, Espin-Palazon R, Lau RP, Zon LI, Bartunek P, Traver D (2013) The zebrafish granulocyte colony-stimulating factors (Gcsfs): 2 paralogous cytokines and their roles in hematopoietic development and maintenance. Blood 122:3918–3928

Takizawa H, Regoes RR, Boddupalli CS, Bonhoeffer S, Manz MG (2011) Dynamic variation in cycling of hematopoietic stem cells in steady state and inflammation. J Exp Med 208:273–284

Takizawa H, Boettcher S, Manz MG (2012) Demand-adapted regulation of early hematopoiesis in infection and inflammation. Blood 119:2991–3002

Tamplin OJ, Durand EM, Carr LA, Childs SJ, Hagedorn EJ, Li P, Yzaguirre AD, Speck NA, Zon LI (2015) Hematopoietic stem cell arrival triggers dynamic remodeling of the perivascular niche. Cell 160:241–252

Travnickova J, Tran Chau V, Julien E, Mateos-Langerak J, Gonzalez C, Lelievre E, Lutfalla G, Tavian M, Kissa K (2015) Primitive macrophages control HSPC mobilization and definitive haematopoiesis. Nat Commun 6:6227

Tulchinsky D, Hobel CJ, Yeager E, Marshall JR (1972) Plasma estrone, estradiol, estriol, progesterone, and 17-hydroxyprogesterone in human pregnancy I Normal pregnancy. Am J Obstet Gynecol 112:1095–1100

Walmsley M, Ciau-Uitz A, Patient R (2002) Adult and embryonic blood and endothelium derive from distinct precursor populations which are differentially programmed by BMP in Xenopus. Development 129:5683–5695

Wang L, Zhang P, Wei Y, Gao Y, Patient R, Liu F (2011) A blood flow-dependent klf2a-NO signaling cascade is required for stabilization of hematopoietic stem cell programming in zebrafish embryos. Blood 118:4102–4110

Yang Q, Liu X, Zhou T, Cook J, Nguyen K, Bai X (2016) RNA polymerase II pausing modulates hematopoietic stem cell emergence in zebrafish. Blood 128:1701–1710

Zhang P, Liu F (2011) In vivo imaging of hematopoietic stem cell development in the zebrafish. Front Med 5:239–247

Zhang Y, Jin H, Li L, Qin FX, Wen Z (2011) cMyb regulates hematopoietic stem/progenitor cell mobilization during zebrafish hematopoiesis. Blood 118:4093–4101

Zhou L, Nguyen AN, Sohal D, Ying Ma J, Pahanish P, Gundabolu K, Hayman J, Chubak A, Mo Y, Bhagat TD, Das B, Kapoun AM, Navas TA, Parmar S, Kambhampati S, Pellagatti A, Braunchweig I, Zhang Y, Wickrema A, Medicherla S, Boultwood J, Platanias LC, Higgins LS, List AF, Bitzer M, Verma A (2008) Inhibition of the TGF-beta receptor I kinase promotes hematopoiesis in MDS. Blood 112:3434–3443

Chapter 5
Spinal Cord Stem Cells In Their Microenvironment: The Ependyma as a Stem Cell Niche

Nicolás Marichal, Cecilia Reali, Omar Trujillo-Cenóz, and Raúl E. Russo

Abstract The ependyma of the spinal cord is currently proposed as a latent neural stem cell niche. This chapter discusses recent knowledge on the developmental origin and nature of the heterogeneous population of cells that compose this stem cell microenviroment, their diverse physiological properties and regulation. The chapter also reviews relevant data on the ependymal cells as a source of plasticity for spinal cord repair.

Keywords Central canal • Progenitor cells • Neural stem cells • Radial glia • Spinal cord • Plasticity • Regeneration

5.1 Introduction

In the adult mammalian brain, stem cells persist in the subventricular zone (SVZ) and the dentate gyrus (DG) of the hippocampus where together with other cell types, they form complex three dimensional microenvironments or neurogenic niches (Alvarez-Buylla and García-Verdugo 2002). The identification of similar microenvironmental entities supporting stem cells in the spinal cord has been less conclusive. However, the idea that the region surrounding the central canal (CC) -commonly believed as a layer of homogeneous epithelial-like cells- is a

N. Marichal
Neurofisiología Celular y Molecular, Instituto de Investigaciones Biológicas Clemente Estable, Montevideo, Uruguay

Institute of Physiological Chemistry, University Medical Center, Johannes Gutenberg University Mainz, Mainz, Germany

C. Reali • O. Trujillo-Cenóz • R.E. Russo (✉)
Neurofisiología Celular y Molecular, Instituto de Investigaciones Biológicas Clemente Estable, Montevideo, Uruguay
e-mail: rrusso@iibce.edu.uy; russoblanc@gmail.com

© Springer International Publishing AG 2017
A. Birbrair (ed.), *Stem Cell Microenvironments and Beyond,*
Advances in Experimental Medicine and Biology 1041,
DOI 10.1007/978-3-319-69194-7_5

potential stem cell niche sharing some features with those in the brain has gained ground. In this chapter, we focus on the properties of cells within this spinal cord stem cell niche in mammals and non-mammalian vertebrates. We first overview the current knowledge about the developmental origin of the cells that line the CC. Next, we address their morphological, molecular and functional properties in the postnatal spinal cord and discuss the potential of ependymal cells isolated from their natural microenvironment. We finally review recent data on the signals that modulate the biology of cells within this niche and discuss the reaction of ependymal cells in diverse models of spinal cord injuries or pathologies with implications for self-repair.

5.2　The Origin of Cells Within the CC Niche

In the caudal portion of the neural tube, progenitors are organized in dorso-ventral spatial domains (Jessell 2000; Briscoe et al. 2000) that initially produce distinct types of neurons to later generate oligodendrocytes and astrocytes. Six ventral domains can be identified in the ventral part of the neural tube based on the combinatorial expression of various transcription factors (ventral to dorsal): the floor plate domain (FP, Foxa2,Shh), p3 (Nkx2.2, Nkx6.1), pMN (Olig2, Pax6, Nkx6.1), p2 (Nkx6.1, Pax6, Irx3), p1 (Dbx2, Nkx6.2, Pax6, IrX3), and p0 (Dbx1, Dbx2, Pax6, Irx3) (p meaning progenitor and MN motoneuron; Briscoe et al. 2000; Lee and Pfaff 2001; Dessaud et al. 2008). The study of the dynamic expression of neural progenitor genes in distinct domains of the neuroepithelium indicate that adult ependymal cells arise from progenitors in the pMN and p2 domains, as they retain expression of Nkx6.1 (p3, pMN and p2 domains) but not Nkx2.2 (p3 domain) (Fu et al. 2003). Moreover, lineage-tracing experiments using a tamoxifen-inducible Cre-recombinase inserted into the Olig2 locus demonstrated that some ependymal cells are produced within the pMN domain (Masahira et al. 2006). More recent experiments with different transgenic Cre-recombinase expressing mice support the view that the mature ependymal region in the spinal cord comprises cells with different embryonic origins: dorsal ependymal cells originate from the embryonic p2 domain whereas ventral cells derive from pMN progenitors (Yu et al. 2013).

The exact timing of birth of ependymal cells has been explored with 5-bromo-2-deoxyuridine pulses applied at different stages of embryonic life and various survival periods. In the brain, the majority of ependymal cells appeared to be generated between E14 and E16 and derived from radial glial cells (Spassky et al. 2005). In the spinal cord, they seem to appear later. Most ependymal cells in the rat originate initially around E18 and continue until postnatal days 8 and 15 (Sevc et al. 2011). The increment in cell number accompanies the closure and transformation of the primitive lumen into the post-embryonic CC. The transcription factor FoxJ1 in the brain is involved in the differentiation of ependimocytes from radial glia (Jacquet et al. 2009). A recent study using a FoxJ1-CreER transgenic mouse dates the earliest birth of ependymal cells at E15.5 with the CC fully populated by these cells at P10

(Li et al. 2016). Ependymal cells in the mouse spinal cord continue dividing postnatally to become quiescent 9 weeks after birth (Sabourin et al. 2009).

Besides ependymal cells, a particular type of neuron contributes to the lining of the CC. The presence of cells with neuronal characteristics in this area was proposed by studies conducted at the beginning of the last century (reviewed by Vigh-Teichmann and Vigh 1983). They were called cerebrospinal fluid contacting neurons (CSFcNs) because their peculiar morphology with a prominent process that contacts the CC lumen (Vigh and Vigh-Teichmann 1998; Vigh et al. 1977, 1983) and their cell bodies located sub-ependymally. Pioneering studies performed in the rat spinal cord showed that neurogenesis is only present during the period between E11 and E16 (Nornes and Das 1972, 1974). The production of CSFcNs in the rat spinal cord starts in E12 and is maintained until E22, peaking between E14 and E15 (Kutna et al. 2014). CSFcNs have intriguing features (described in detail below) resembling immature neurons in adult neurogenic niches (Marichal et al. 2009). In the mouse spinal cord, CSFcNs are produced from two distinct dorsoventral regions during embryonic life. Most CSFcNs derive from progenitors circumscribed to the late-p2 and the oligodendrogenic (pOL, called pMN during the early neurogenic period, see above) domains, whereas a second subset of CSFcNs arises from cells bordering the FP (Petracca et al. 2016). These cells start to be generated at E10 and continue to arise until E14–E16, a developmental stage in which most spinal neurons have been already produced.

Taken together, both ependymal cells and neurons contacting the CC seem to be generated in late stages of spinal cord development, a fact that may be related with the postnatal maintenance of some peculiar anatomical, molecular and functional features of embryonic progenitors and neuroblasts.

5.3 The Nature of Neural Stem Cells

The first progenitors in the neural tube are neuroephitelial cells which at the onset of neurogenesis become radial glia (RG), the founders of most neurogenic lineages during development (Kriegstein and Alvarez-Buylla 2009). Both neuroepithelial cells and RG have a pronounced polarity with an apical pole bearing a single primary cilium protruding into the ventricular lumen and a distal process in contact with the pia (Kriegstein and Alvarez-Buylla 2009). This polarity is critical to determine and regulate the phenotype of neural stem cells (Alvarez-Buylla et al. 2001; Pinto and Götz 2007). For example, the apical pole of RG contains components like the centrosome and various key proteins (e.g., prominin, PAR3) whose inheritance during division determines the fate of daughter cells. In the adult mammalian brain, it is currently accepted that progenitors are a subtype of astrocyte (called B cells) that retain key features of both neuroepithelial cells and RG (Doetsch et al. 1997; Horner and Palmer 2003; Ming and Song 2005; Lledo et al. 2006; Lim et al. 2008). The cell bodies of B cells in the subventricular zone (SVZ) are intermingled with multiciliated ependymal cells. Each B cell exhibits an irregular process bearing a single primary cilium (9 + 0) projecting into the ventricle lumen. These cells form tube–like structures ensheathing the so called A-type cells when migrating towards the olfactory bulb (Lois et al. 1996). Recent fine structural studies performed in mice (Cebrián-Silla et al. 2017), indicated that some B cells show envelope-limited chromatin sheets, a rare specialization of the nuclear membrane initially described in blood neutrophils (Davies and Small 1968). These unusual nuclear compartments have been associated with the isolation of telomeres and epigenetic modifications. When dealing with other commonly accepted adult neurogenic region like the hippocampus, the unequivocal identification of a cell type as a "primary precursor" is less clear. According to Kempermann et al. (2008) several cell types "are involved in the course of adult hippocampal neurogenesis". The common feature is that precursor cells found in the dentate gyrus are of astroglial nature (Seri et al. 2001). However, not all neural progenitors share these anatomical features, such as intermediate progenitors in the SVZ of the developing cortex which have multipolar processes that do not contact the ventricle or pial surface (Kriegstein and Alvarez-Buylla 2009). In fact, progenitors are heterogeneous cells expressing different molecules that seem related with their lineage potential (Pinto and Götz 2007). For example, expression of brain lipid binding protein (BLBP) seems to determine RG as bi-potent or multipotent progenitors (Pinto and Götz 2007; Hartfuss et al. 2001; Anthony et al. 2007). The transcription factor Pax6 is a major regulator of the subpopulation of neurogenic RG (Götz et al. 1998; Heins et al. 2002; Bel-Vialar et al. 2007) and progenitors in the adult mammalian brain (Kohwi et al. 2005; Maekawa et al. 2005). In the developing spinal cord Pax6 interacts in a combinatorial manner with other transcription factors such as Olig2, Nkx2.2 and Sox9 to control neurogenesis and gliogenesis (Heins et al. 2002; Lee and Pfaff 2001; Rowitch 2004; Nacher et al. 2005; Guillemot 2007; Sugimori et al. 2007). The importance of the combination of key transcription factors in determining the biology of progenitors is highlighted by the possibility of reprogramming somatic cells to pluripotent stem cells with just a handful of factors (Yamanaka 2012).

5.4 Progenitor Cells in the Spinal Cord: The CC as a Stem Cell Niche

The identity of stem cells in the adult spinal cord has been difficult to establish and remains controversial. Although Horner et al. (2000) described proliferating cells in the grey and white matter of the rat spinal cord, other studies showed that the vast majority of stem cells resides within the ependyma (Mothe and Tator 2005; Sabourin et al. 2009). In most text-books the ependyma is depicted as a layer of ephitelial cells (Peters et al. 1991). Nevertheless, this oversimplified view has been challenged by both classical (Ramón y Cajal 1909) and recent studies. Both in mammals and non-mammalian vertebrates, the ependymal region is a complex structure consisting of diverse kinds of cells arranged in lateral and dorsal-ventral domains (Schnapp et al. 2005; Trujillo-Cenóz et al. 2007; Marichal et al. 2012). In murine, the cell

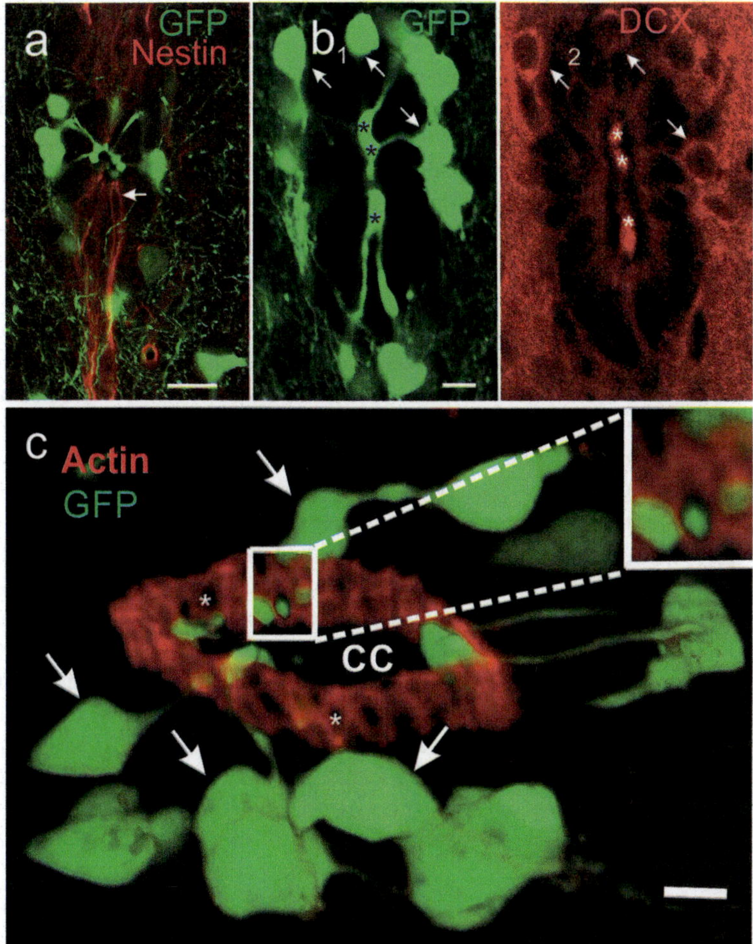

Fig. 5.1 (**a**) Nestin + processes (*in red*) from cell bodies far apart from the ependymal channel lumen invade the layer composed of ependymocytes and CSFcNs (GFP). (**b–c**) The CSFcNs of genetically modified mice expressing GFP under the control of GATA3 transcription factor (GATA3-GFP). GFP is expressed in both cell bodies (*arrows*) and intra-lumen enlarged apical processes. These peculiar neurons also express DCX, a marker of immature neurons. (**c**), 3D image of a stalk of 54 optical sections covering the CC a GATA3-GFP mouse (CSFcNs are pointed by *arrows*). The actin skeleton of the apical compartments of the ependymal cells appear as a red fluorescent network (phalloidin conjugated fluorophore). The enlarged rectangle shows a cross section of a neuron prolongation passing through an apparent "hole" of the actin network. Calibration bars: (**a**), 20 µm; (**b**) and (**c**), 5 µm

mantle lining the CC harbors no less than three morphologically different cell types: (a) ciliated ependymocytes with short basal processes entering the adjacent nervous tissue, (b) ciliated ependymocytes with long processes reaching the pial surface (RG-like cells) and (c) typical CSFcNs (Fig. 5.1a–c) The first kind of ependymal cell predominates in the lateral domains of the CC while RG-like cells are mainly concentrated in both polar regions. In contrast, CSFcNs do not have preferential locations (Fig. 5.1b). It is pertinent to discuss here the proposed presence of a fourth kind of cell termed "tanycytes" (Horstmann 1954). These cells, firstly described in the brain of cartilaginous fishes (Horstmann 1954), have the aspect of RG and are particularly abundant in the floor and ventrolateral walls of the third ventricle of mammals. Taking into account their location and other morphological details, they are usually subdivided into alpha and beta subtypes (Goodman and Hajihosseini 2015). Even though some authors have used the same term to identify a particular kind of ependymal cell in the spinal cord of mice (Meletis et al. 2008), cells with a morphological or molecular phenotype of tanycytes in the rat or turtle spinal cord has not been reported (Trujillo-Cenóz et al. 2007; Marichal et al. 2012). It is worth noting that the CC is far from being an empty channel. In addition to the cilia and shorter microvilli projections, the CC contains cellular material of diverse origin. In the particular case of mice, the bulbous terminals of the numerous CSFcNs appear as the most conspicuous well organized structures within the CC lumen. Together with these structurally complex cell compartments, there are also numerous cell projections arising from the ependymocytes apical membranes. In rodents, TEM studies show that intermingled with motile cilia there are vesicles of dissimilar sizes and shapes which seem to be floating freely in the CSF. In other cases, membrane extrusions attached by a thin process to their parent cells can be observed. This data strongly suggest that the release of cell membranes into the CSF is a common mechanism to introduce different kinds of molecules into the CSF. Marzesco et al. (2005) have proposed that membrane bounded particles are the carriers within the CC of the stem cell marker prominin 1. In other vertebrates like turtles, the CC contains together with vesicular material, a clearly visible Reissner's ribbon occupying most of the channel lumen. Nevertheless, in rodents the condensation of Reissner's ribbon glycoproteins only becomes apparent within the phylum terminalis. At this level, condensation of the proteins takes the form of the so-called massa caudalis (Molina et al. 2001).

As already mentioned, the basal pole of the ependimocytes also bear short and long processes that invade adjacent neuropile areas vanishing the apparent boundaries between the ependymal layer and the surrounding spinal tissue. The longest processes reach the pial surface whereas the shorter ones contact neighbor capillaries or end intermingled with glial and neuronal processes composing of adjacent neuropile zones. As found in other well recognized stem cell niches, progenitor cells are related to their neighbors at the level of their apical poles by zonnula occludens and gap junctions (Russo et al. 2008; Marichal et al. 2012).

Besides structural differences, several neural stem/progenitor cell markers such as Sox2, CD15, CD133, nestin, vimentin, BLBP and GFAP are expressed by subpopulations of cells lining the CC (Meletis et al. 2008; Hamilton et al. 2009; Hugnot and Franzen 2011). As in the brain (Spassky et al. 2005), most cells in the lateral domains of the CC express the ependymal cell marker S100β, with a sub-population that co-express the RG marker 3CB2 or vimentin and had a basal process projecting away from the CC, suggesting a progenitor cell nature (Pinto and Götz 2007).

Indeed, many S100β+/3CB2+/vimentin + cells express PCNA –indicating that they are cycling cells- with few undergoing division as indicated by pH3 expression (Eisch and Mandyam 2007). The expression of nestin -a marker of neuroepithelial cells and RG (Pinto and Götz 2007)- defines a second domain of heterogeneous cells contacting the poles of the CC (Fig. 5.1a). In adult mice, nestin is expressed preferentially on cells contacting the dorsal pole of the CC (Hamilton et al. 2009). Although the perikarya of these cells in neonatal rats lay at various distances from the CC, their centrosomes are always located in apical endfeet some of which bear a single cilium with a 9 + 0 organization, a structural signature of neural stem cells (Alvarez-Buylla et al. 2001; Kriegstein and Alvarez-Buylla 2009). The fact that pH3 nuclei belonging to nestin + cells are always found close to the CC lumen supports the idea that RG nuclei in the post-natal spinal cord move apically to divide as described for neurogenic RG in the embryo (Kriegstein and Alvarez-Buylla 2009). Progenitors in adult neurogenic niches express GFAP in addition to nestin (Ma et al. 2008). However, the ependyma in the rat lacked GFAP immunoreactivity, in contrast with GFAP-GFP transgenic mice which bear GFAP+ cells contacting the dorsal pole (Sabourin et al. 2009). The discrepancy between data obtained in rats and mice may be species specific or age related (neonatal versus adult). Another feature suggesting that midline domains may not be identical in their potential is the fact that cells contacting the ventral but not the dorsal pole express the astrocyte and ependymal cell marker S100β (Marichal et al. 2012). In line with this, GFAP RG-like cells located on the dorsal pole of the ependyma of adult mice have been proposed to be the only cells capable to generate several passages of neurospheres producing astrocytes, oligodendrocytes and neurons (Sabourin et al. 2009, discussed in more detail below). Another possibility is that nestin+/S100β + cells on the ventral pole may be a transitional stage between RG and ependymal cells as described during brain development (Spassky et al. 2005).

The occurrence of a well-developed actin skeleton completes the structural and molecular architecture of the ependymal region. 3D reconstructions using phalloidin-conjugated fluorophores have revealed that the apical portions of the ependymal cells are surrounded by a dense network of actin fibers usually associated with the zonula adherens (Fig. 5.1c). The same technical approach discloses a dense actin network close to the inner layer of the plasma membrane of all ependymal cells. The use of genetically modified mice whose CSFcNs express GFP combined with actin staining have demonstrated that the apical dendrites of these nerve cells are surrounded by the actin rich apical segments of ependymocytes (Fig. 5.1 c, unpublished). Experimental studies have revealed that intraventricular perfusion of cytochalasin B produce a pronounced disarrangement of actin microfilaments in ependymocytes. In turn, they became rounded up and loosened from their neighbors (Mestres and Garfia 1980). It is reasonable to speculate that migration of the ependymal cells induced by different kind of injuries should be preceded by disorganization of the normal actin network.

5.5 Membrane Properties of Cells in the CC Stem Cell Niche

The membrane properties of cells lining the CC has been characterized both in the spinal cord of turtles and rats (Russo et al. 2004, 2008; Marichal et al. 2009, 2012). Both in the turtle and neonatal rats, the cells in the lateral domains of the CC had electrophysiological properties similar to those of progenitors during cortical development: (a) low input resistances, (b) passive responses, (c) hyperpolarized resting membrane potentials and (d) extensive gap junction coupling via Cx43 (Bittman et al. 1997; Noctor et al. 2002). The cluster of electrically and metabolically coupled ependymal cells in the lateral aspects of the CC in turtles matches the expression of BLBP and the transcription factor Pax6 and Nkx6.1, suggesting these cells are multipotent progenitors (Pinto and Götz 2007). In line with this electrophysiological and molecular signature, these progenitors have a higher rate of proliferation to those located in dorsal or ventral domains (Russo et al. 2008) and can generate both glial cells and neurons (Fernández et al. 2002). Whereas progenitors in the lateral domains of rodents share some properties of those in non-mammalian vertebrates such as the predominance of passive membrane properties, extensive coupling via Cx43 and higher rates of proliferation than midline domains, they lack the expression of BLBP and Pax6, a fact that may be related to their inability to generate new neurons in the post-natal life (Marichal et al. 2009).

In contrast to cells on lateral domains, RG in dorsal and ventral domains are not electrically coupled via Cx43 and function as individual units, both in reptiles (Russo et al. 2008) and mammals (Marichal et al. 2012). Unlike neurogenic RG in the developing cortex (Noctor et al. 2002), cells in the midline domains of the CC have active membrane properties. In turtles, a subset of cells in the midline with the morphological phenotype of RG display a conductance that is active at resting membrane potentials and deactivates slowly when the membrane is hyperpolarized (Reali et al. 2011). The mechanisms and functional relevance of this voltage-gated conductance remains to be explored. Similarly, RG in the post-natal spinal cord of rodents had complex electrophysiological phenotypes displaying various combinations of a delayed rectifier (I_{KD}), A-type (I_A) and/or calcium currents. The presence of I_{KD} is a common feature among adult progenitors since it has been reported in hippocampal nestin+ type 2 cells (Filippov et al. 2003) and GFAP+ cells in the SVZ (Liu et al. 2006). Although I_A was not found in the adult SVZ (Liu et al. 2006), progenitors from the embryonic (Smith et al. 2008) and neonatal (Stewart et al. 1999) SVZ as well as human stem cells (Schaarschmidt et al. 2009) express I_A. The phenotype of midline RG with conspicuous I_{KD} and I_A is remarkably similar to that of oligodendrocyte progenitors (Chittajallu et al. 2004), raising the possibility they are bipolar precursors committed to the oligodendrocyte lineage (Levine et al. 2001). It remains to be explored whether these cells can differentiate in oligodendrocyte progenitor cells with a multipolar morphology and expression of NG2 or PDGFRα.

The complex repertoire of K$^+$ currents may regulate fundamental properties of ependymal progenitor-like cells. I_{KD} channels are major regulators of cell proliferation (Ghiani et al. 1999; MacFarlane and Sontheimer 2000a; Chittajallu et al. 2002) and I_A channels are essential for proliferation of multipotent human neural stem cells (Schaarschmidt et al. 2009). Thus, K^+ channels in midline RG may be part of epigenetic mechanisms that regulate proliferation. In addition, I_A has been implied in the differentiation of oligodendrocyte precursors (Sontheimer et al. 1989) and rat spinal cord astrocytes (MacFarlane and Sontheimer 2000b). Thus, another possibility is that K^+ currents participate in the transition from RG to post-mitotic spinal cells.

A minority of midline RG have I_{Ca} strong enough to sustain a low threshold spike, a phenotype described in some floor plate cells (Frischknecht and Randall 1998). Ca^{2+} electrogenesis plays a central role during development by regulating events such as neural induction (Webb et al. 2005) and neuronal differentiation (Spitzer et al. 2004). RG displaying I_{Ca} could be precursors showing the first signs of differentiation into CSFcNs (Marichal et al. 2009).

5.6 Spinal Stem Cells Outside Their Niche

Until now, we described the features of the different components of the ependymal stem cell region in their normal endogenous microenvironment. However, more than two decades ago, the properties showed by ependymal cells isolated and plated in different cultures media provided the first evidences of their stem-cell characteristics.

Weiss et al. (1996) developed an in vitro protocol for primary cultures of cells taken from the adult spinal cord and found the presence of growth factor-responsive stem cells. These cells generated spheres with self-renewal/expansion properties and multipotency giving rise to neurons, astrocytes, and oligodendrocytes. After this study, Frisen's group also tested whether ependymal cells specifically showed stem cells properties in vitro. For this, they labeled ependymal cells by injecting 1,1′-Dioctadecyl-3,3,3′,3′-Tetramethylindocarbocyanine Perchlorate (DiI)in the lateral ventricles of the brain (Johansson et al. 1999). Neurospheres derived from DiI-labeled primary cells expressed neuronal, astrocytic and oligodendrocyte molecular markers.

More recently, the same group generated two transgenic mouse lines expressing tamoxifen-dependent Cre recombinase (CreER) under the control of FoxJ1 (HFH4) or nestin regulatory sequences (Meletis et al. 2008). These mice allowed to study the fate map of the spinal cord ependymal cells (Meletis et al. 2008). FoxJ1 is a transcription factor involved in the formation of motile cilia active in the spinal cord only in cells that line the CC. Primary cultures showed that the recombined neurospheres from both nestin-CreER and FoxJ1-CreER mice could be serially passaged and were multipotent and differentiated into neurons, astrocytes, and oligodendrocytes. Despite being an excellent strategy to make the linage tracing of ependymal cells, the general expression of these genes in the cells surrounding the CC (excluding the CSFcNs) does not allow to discriminate between the heterogeneous populations of cells contacting the CC (described above). Using a GFAP-GFP transgenic mice line, Sabourin et al. (2009) showed that the majority of neurosphere forming cells are dorsally located GFAP+ cells lying ependymally and subependymally that extend radial processes toward the pial surface. A posterior study also claimed that GFAP-expressing cells lining the CC participate to the generation of multipotent neurospheres in vitro. However, they showed restricted self-renewal properties compared with GFAP-negative ependymal-derived neurospheres or GFAP-expressing neural stem cells from the SVZ (Fiorelli et al. 2013).

Taken together, the remarkable multipotent stem cell properties of spinal cord ependymal cells outside their niche make them an attractive source for the replacement of glia and neurons lost after injury or neurodegenerative diseases. The precise identity of the cell population with highest in vitro stemness within the spinal cord ependymal niche remains controversial and future work needs to be done to solve this problem.

5.7   Regulation of the CC Stem Cell Niche

Stem cells in neurogenic niches of the adult brain are regulated by a plethora of factors such as age and activity (Kempermann 2008), hormones (Lucassen et al. 2008) and neurotransmitters (Jang et al. 2008). The decrease of neurogenesis with age in mammals seems to be related with the decline of the activity of progenitor cells via distinct mechanisms in the SVZ and the hippocampus (Molofsky et al. 2006; Hattiangady and Shetty 2008). Similarly, the mitotic activity of spinal ependymal cells gradually declines as the animal ages to stop about 9 weeks after birth (Sabourin et al. 2009). The molecular mechanisms of the proliferation halt during early postnatal life in this spinal stem cell niche has not been yet unveiled.

During embryogenesis, the biology of progenitors and newborn cells is tightly regulated by activity via the action of diverse neurotransmitter systems (Ben-Ari and Spitzer 2010; Wang and Kriegstein 2009). Progenitors in adult stem cell niches in the brain have been shown to be regulated by γ-amino butyric acid (GABA), glutamate, acetylcholine, dopamine, serotonin and nitric oxide (Jang et al. 2008). In the SVZ for example, GABA released from newborn neurons inhibits the proliferation of neighboring progenitors making a feedback control system to adjust neurogenesis to functional demands (Lo Turco et al. 1995; Haydar et al. 2000; Liu et al. 2005). Although less is known about this kind of regulation in spinal stem cell niches, a similar set of neurotransmitters -either produced by cellular components of the niche or by surrounding axonal fibers- may influence the behavior of CC-contacting progenitors (Reali et al. 2011; Corns et al. 2013, 2015; Marichal et al. 2016). Both in low vertebrates and mammals, progenitor-like cells in the ependyma (Fig. 5.2a) are surrounded by CSFcNs which have the molecular signature of immature neurons (expression of DCX and PSA-NCAM) and synthetize GABA (Fig. 5.2b; Roberts et al. 1995; Stoeckel et al. 2003; Reali et al. 2011). In addition, GABAergic terminals are present around the CC (Fig. 5.2a; Trujillo-Cenóz et al. 2007;

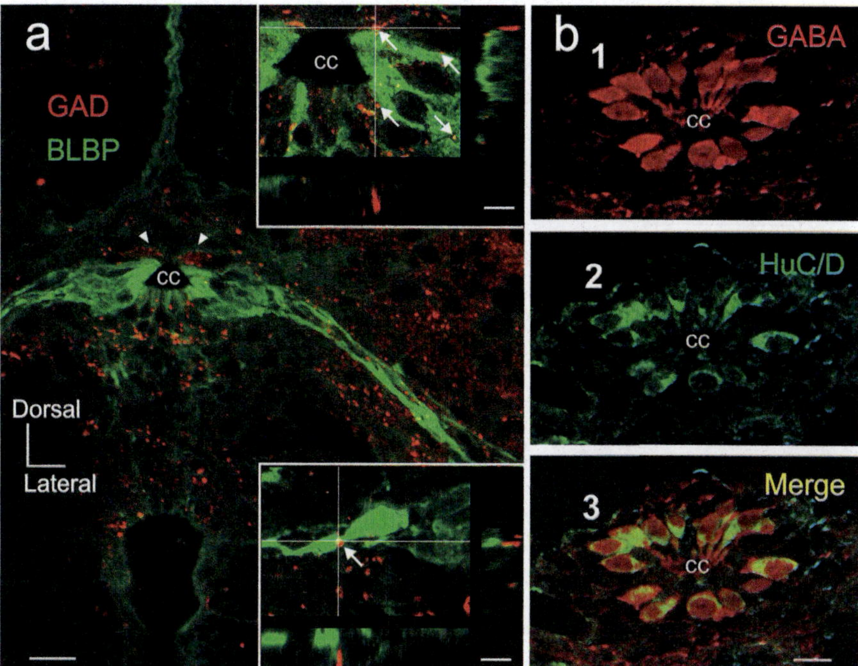

Fig. 5.2 GABA around the central canal (CC). (**a**) Immunohistochemistry for GAD-65/67 (*red*) and brain lipid binding protein (BLBP, *green*) reveals a plexus of GAD+ terminals surrounding the CC. A few CC-contacting cells on the dorso-lateral aspect of the CC are reactive for GAD (*arrowheads*). Many GAD+ terminals are in close apposition with either proximal (*upper inset, arrows*) or distal (*lower inset, arrow*) processes of BLBP+ cells. (**b**) In contrast to GAD expression, the CC is surrounded by a large number of cells containing GABA (*1*) which co-express HuC/D (*2 and 3*). Main panel in A and B are confocal optical sections. Upper and lower insets are stacks of ten optical sections. Scale bars: (**a**), main panel 20 μm, upper and lower insets, 10 μm; (**b**), 10 μm. Modified with permission from The Journal of Physiology (Reali et al. 2011)

Reali et al. 2011) and the fact that CSFcNs receive functional GABAergic contacts (Russo et al. 2004) suggest that there is an active GABAergic signaling in this stem cell niche. Indeed, functional studies in turtles (Reali et al. 2011) have shown that in clusters of gap junction coupled BLBP+ progenitors lining the CC (Fig. 5.3a), GABA generates currents with components mediated by GABA transporters (GAT, Fig. 5.3b1–3) and GABA$_A$ receptors (Fig. 5.3c1–3). Uncoupling BLBP+ progenitors with carbenoxolone suggests that individual progenitors react differently to GABA with various combinations of GABA transporter- and ionic-induced currents. GABA also depolarizes ependymal cells of the spinal cord of juvenile rats (Corns et al. 2013), suggesting that GABAergic signaling on CC-contacting progenitors is a phylogenetically preserved trait.

CSFcNs in close contact with ependymal cells also have functional GABA$_A$ receptors (Russo et al. 2004; Marichal et al. 2009; Reali et al. 2011). Gramicidin perforated patch recordings showed that GABA$_A$ receptor activation generate

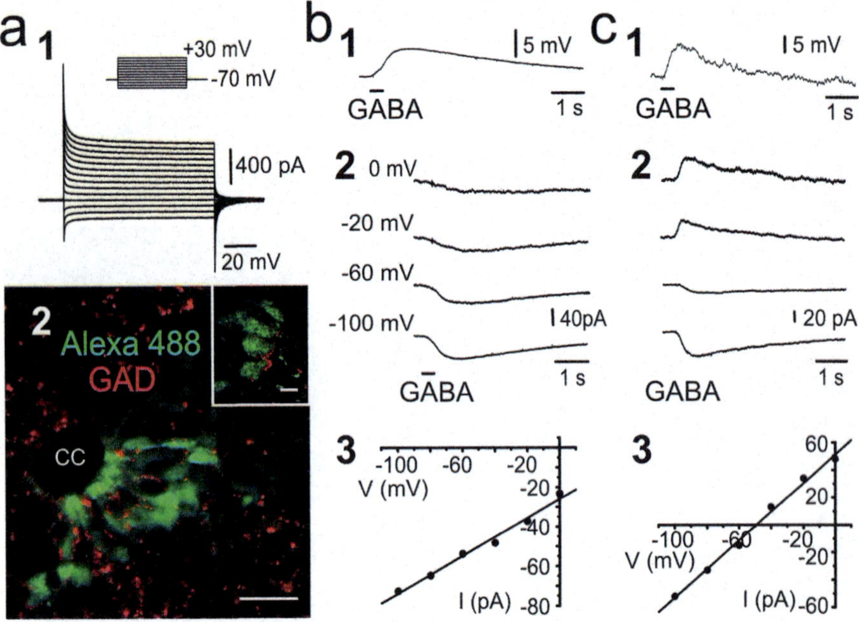

Fig. 5.3 GABA signals on CC-contacting progenitors. (**a**) Current responses to a series of voltage steps in a CC-contacting cell (*1*). The cell is dye coupled with neighbouring cells (*2*). Notice the close proximity of GAD terminals to the cluster of dye coupled cells (*2*, main panel and inset). (**b**) Progenitors are depolarized by transient application of GABA (400 ms, 1 mM; (*1*). In some progenitors GABA-induced inward currents (*2*) do not reverse at membrane potentials from −100 to +20 mV (*3*). (**c**) In other clusters of progenitors, the depolarization induced by GABA (*1*) is generated by currents (*2*) that have clear reversal potentials (*3*). (**a**)2, confocal optical section. Scale bar in (**a**)2: main panel, 20 μm; inset, 3 μm. Modified with permission from The Journal of Physiology (Reali et al. 2011)

responses ranging from depolarizations that elicited spike firing (Fig. 5.4a) to hyperpolarization from rest (Marichal et al. 2009; Reali et al. 2011). GABA-induced depolarization in neurons is a common feature in the developing brain (Ben-Ari 2002; Ben-Ari and Spitzer 2010) and migrating neuroblasts in the SVZ and rostral migratory stream (Wang et al. 2004). Both in the SVZ and the dentate gyrus, GABA provides an excitatory drive to newborn neurons because of high activity of NKCC1 (Bordey 2007). As neurons mature, GABA action switches from excitation to inhibition because down-regulation of NKCC1 and/or increase of KCC2 activity (Rivera et al. 1999; Ganguly et al. 2001). Similarly, the excitatory action of GABA in CSFcNs stems from a depolarized E_{Cl} because a predominance of NKCC1 over KCC2 (Fig. 5.4b, c; Reali et al. 2011). In line with the idea that CFScNs may be at different stages of maturation, cells hyperpolarized by GABA generally fired repetitively -an electrophysiological phenotype of more differentiated neurons (Spitzer et al. 2004; Russo and Hounsgaard 1999)- and have measurable KCC2 activity. $GABA_A$ receptor activation induces an increase in intracellular Ca^{2+} that requires

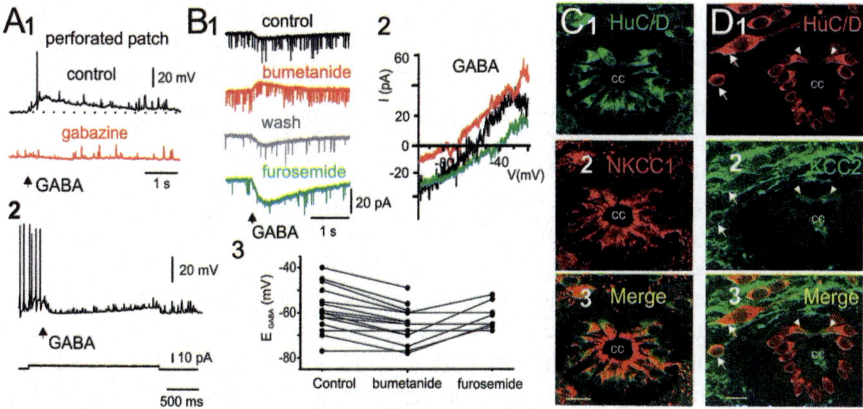

Fig. 5.4 GABA as a dual neurotransmitter: excitation and inhibition. (**a**) In some cases the GABA induces a induced depolarization strong enough to generate spikes (*1*, gramicidin perforated patch). In a different sub-population of CSFcNs GABA generated a hyperpolarization that stops firing (*2*, gramicidin perforated patch). (**b**) Currents induced by GABA at −70 mV (1) in control (top trace) and in the presence of bumetanide (20 μM, second trace from top). Notice that in the presence of bumetanide, GABA generates an outward current, an effect that reversed upon wash out (third trace from top). Blocakde of KCC2 with of furosemide (100 μM, fourth trace from top) increased the amplitude of the inward GABA-induced current. The I/V plots of GABA-induced currents shows the shifts in E$_{GABA}$ induced by bumetanide and furosemide (2). The scatter plot in 3 shows the changes in E$_{GABA}$ induced by bumetanide and furosemide for different cells. (**c**) Most HuC/D cells (*1*) express NKCC1 (*2* and *3*). (**d**) Double immunohistochemistry for HuC/D (*1*) and KCC2 (*2*). KCC2 is strongly expressed in neurons outside the CC region (*1–3 arrows*) whereas in the ependyma is weakly expressed in some HuC/D+ cells located dorsally (*1–3, arrowheads*). (**c**) and (**d**), confocal optical sections. Scale bars: (**c**) and (**d**), 20 μm. Modified with permission from The Journal of Physiology (Reali et al. 2011)

extracellular Ca^{2+}, in both progenitors and immature neurons of the turtle CC (Reali et al. 2011). These studies show that GABAergic signaling around the CC shares fundamental properties with those in the embryo and adult neurogenic niches, suggesting that GABA may be part of the mechanisms regulating the CC stem cell niche. Future research has to solve the source of ambient GABA around the CC and the functional role of GABAergic signaling on CC-contacting progenitors and CSFcNs.

Another transmitter that has shown to have a role in the ependymal stem cell niche is acetylcholine (Ach). The CC of rats is surrounded by cholin acetyl transferase (ChaT) fibers and Ach depolarizes ependymal cells and CSFcNs via activation of α7- and non-α7 nicotinic Ach receptors (Corns et al. 2015). Furthermore, activation of nicotinic receptors increases proliferation of ependymal cells both in vitro and in vivo (Corns et al. 2015). This study suggests that cholinergic signaling may be a key regulator of the CC stem cell niche. Interestingly, inflammation during experimental autoimmune encephalomyelitis produces a decrease of ependymal cell proliferation, an effect that is relieved by administration of nicotine (Gao et al. 2015).

ATP may be important to shape the properties of spinal progenitors, particularly in the context of spinal cord injury. Purinergic signaling has an important role during development by regulating processes such as progenitor cell proliferation, migration, differentiation and synapse formation (Zimmermann 2006). During the development of the cortex, ATP released through connexin hemichannels in RG activates P2Y1 receptors generating a Ca^{2+} wave by IP3 mediated Ca^{2+} release that propagates among neighboring RG (Weissman et al. 2004). These Ca^{2+} waves regulate the proliferation of RG and have been proposed as a mechanism for synchronizing the cell cycle of a cohort of progenitors (Weissman et al. 2004). Purinergic signaling leads to the expansion of the ventricular zone stem cell niche (Lin et al. 2007) and can initiate important events as transient release of ATP triggers the development of the eye by inducing the expression of Pax6 via P2Y1 receptor activation (Massé et al. 2007). In addition, nucleotide signaling seems also to participate in adult neurogenesis as the ATP hydrolyzing ectonucleotidase NTPDase2 is a hallmark of adult neurogenic niches (Abbracchio et al. 2006). However, the detailed function of ATP signaling in regulating adult stem cell niches is still poorly understood.

Different kinds of insults are linked to increased levels of extracellular ATP. For example, after spinal cord injury (SCI) ATP levels increase around the lesion epicenter (Wang et al. 2004). It has been proposed that ATP may act as a diffusible "danger signal" to alert about damage and start repair (Abbracchio et al. 2009). Purinergic signaling has been also implied in the secondary expansion of tissue damage after SCI (Wang et al. 2004; Peng et al. 2009). CSFcNs have P2X2 receptors (Stoeckel et al. 2003) that when activated by ATP generates a powerful excitation (Marichal et al. 2009).

The P2X7 receptor merits particular attention in relation to CNS injury. P2X7 receptors have the peculiarity of having a rather low sensitivity being activated at 100 μM to 3 mM of ATP whereas other P2X receptors have EC50 of 1–10 μM (Khakh and North 2006; Surprenant and North 2009). In addition, P2X7 are highly permeable to Ca^{2+} and when activated for several seconds become permeable to the large cation N-methyl-D-glucamine (Surprenant et al. 1996). Ependymal cells in both the medial (Fig. 5.5a, b) and lateral (Fig. 5.5c, d) domains of the CC have functional ionotropic P2X7 receptors (Masahira et al. 2006). The activation of P2X7 receptors by ATP or its selective agonist BzATP generates a slow inward current and a Ca^{2+} wave that propagates bidirectionally from the site of ATP application (Fig. 5.5). It is possible that P2X7 receptors may be a key component of the response of the ependymal stem cell niche to spinal cord injury. The activation of P2X7 receptors in the distal processes lying within injured tissue would generate a Ca^{2+} wave propagating towards the CC, generating a local Ca^{2+} increase in key cellular compartments such as the nucleus and the apical process of CC-contacting progenitors (Fig. 5.6). Ca^{2+} transients may modulate nuclear gene expression, activating or repressing function-specific transcription factors that may affect events such as proliferation, differentiation and migration (Glaser et al. 2013; Miras-Portugal et al. 2015) of ependymal cells. For example, interference of Ca^{2+} signaling by blockade or genetic knockdown of purinergic receptors impairs the migration of intermediate neuronal progenitors to the subventricular zone (Liu et al. 2008). Ca^{2+} waves propagated to or generated at the apical pole of

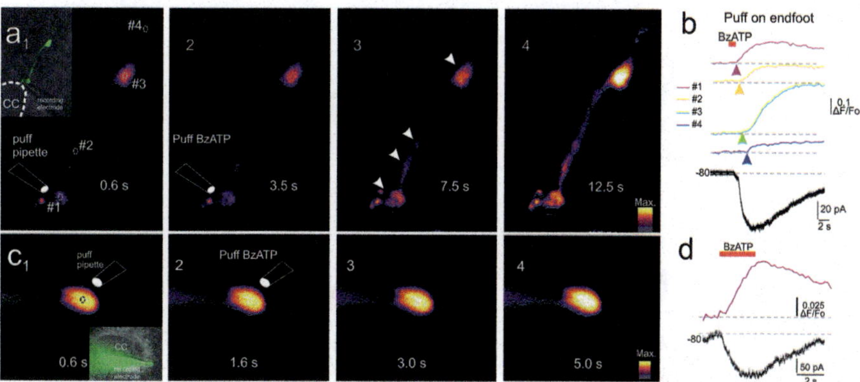

Fig. 5.5 ATP signaling in the CC. (**a**) RG in the dorsal pole of the CC filled with Fluo-4 before (*1*), during (*2*) and after (*3, 4*) application of BzATP (1 s, 1 mM) on the apical process that contacts the CC. The cell morphology is shown in the inset in *6*. The arrowheads indicate the sequence of Ca^{2+} increase along the cell. (**b**) ΔF/F0 analysis of ROIs in A1 (*open circles*) shows the time course of $(Ca^{2+})_i$ increase after BzATP application. (**c**) Time lapse imaging of an ependymocyte filled with Fluo-4 in the lateral domain of the CC (inset in *1*) before (*1*), during (*2*) and after (*3, 4*) application of BzATP (1 mM). (**d**) ΔF/F0 in the ROI (circle in A1) and the simultaneous inward current (black trace) generated by BzATP. (**a**) and (**b**), pseudocolor images in a living slice. Modified with permission from Purinergic Signalling (Marichal et al. 2009)

CC-contacting RG may generate changes in this cellular compartment with major functional consequences (Fig. 5.6).Whether the increase in Ca^{2+} induced by P2X7 receptor activation is beneficial by awakening the "dormant" progenitors in the ependyma or detrimental by activation of caspases inducing apoptosis (Gandelman et al. 2013) is a key question to solve in future studies.

5.8   A Perspective: The Ependyma as a Source of Plasticity for Spinal Cord Repair

Anamniotes like cyclostomes (Rovainen 1976; Wood and Cohen 1979; Armstrong et al. 2003; Shifman et al. 2007), certain fish (Hooker 1932; Coggeshall and Youngblood 1983; Dervan and Roberts 2003; Takeda et al. 2007; Reimer et al. 2008), anuran larvae (Michel and Reier 1979; Beattie et al. 1997; Gibbs and Szaro 2006) and tailed amphibians (Piatt and Piatt 1958; Sims 1962; Butler and Ward 1965; Davis et al. 1990; Chevallier et al. 2004; McHedlishvili et al. 2007) have remarkable endogenous mechanisms of regeneration that lead to functional recovery after spinal cord injury (Diaz Queiroz and Echeverri 2013; Lee-Liu et al. 2013). We have shown that this ability is partly shared by the fresh-water turtle *Trachemys dorbignyi*, an amniote vertebrate. Turtles spontaneously reconnect their transected spinal cord with the formation of a cellular bridge that serves as a permissive scaffold for regenerating axons (Rehermann et al. 2009). This is in sharp contrast to

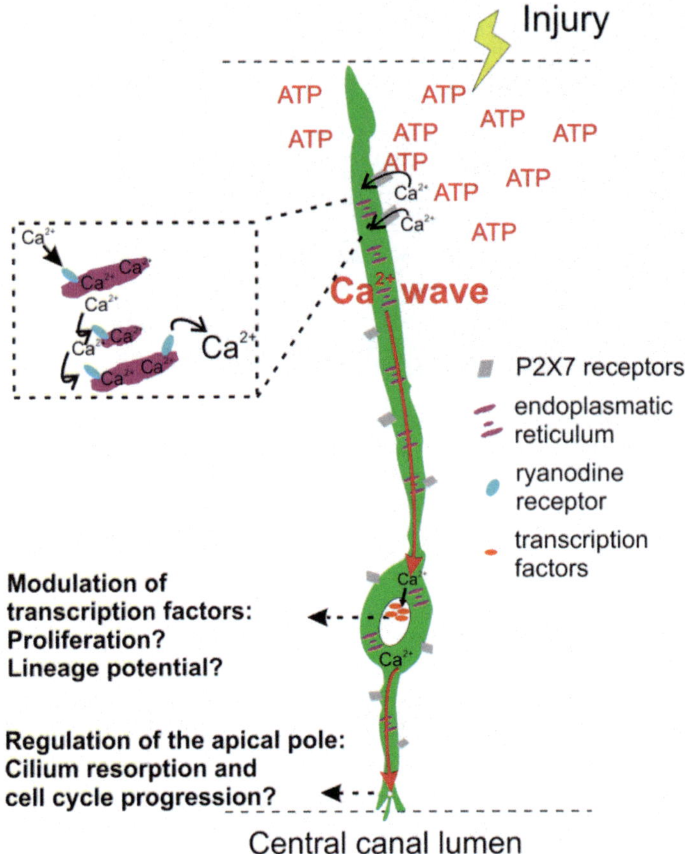

Fig. 5.6 Cartoon showing the possible involvement of P2X7 receptors and Ca²⁺ waves in the reaction of CC-contacting prgenitors after spinal cord injury. Putative mechanisms are proposed

mammals in which a "glial scar" blocks the passage of incoming axons (Ramón y Cajal 1913; Silver and Miller 2004; Thuret et al. 2006). Studies made on lizards (Egar et al. 1970), tailed amphibians (Singer et al. 1979; Zhang et al. 2000; McHedlishvili et al. 2007) and immature eels (Dervan and Roberts 2003) have shown that regeneration of the injured spinal cord is orchestrated by the plastic reaction of ependymal cells. One of the major events after injury in species with and without endogenous repair capabilities is an increase in the proliferation of ependymal cells (Mothe and Tator 2005; Meletis et al. 2008). In turtles, the proliferative reaction is spatially focused around the lesion epicenter engaging mostly the domains of BLBP/Pax6 progenitors (Rehermann et al. 2011). The capability of RG expressing BLBP to contribute to neural repair has been demonstrated in rodents in which transplanted embryonic RG "bridge spinal cord lesions and promote functional recovery" (Hasegawa et al. 2004). The presence of BLBP cells within the bridge suggest the possibility that some of the progenitors detach from the CC and invade the cellular bridge (Rehermann et al. 2009).

Cells belonging to the bridge matrix also behave as oligodendrocyte precursors and/or premyelinating oligodendrocytes (Levine et al. 2001) enveloping regenerating axons. Then, it is likely that analogous phenomena may occur in the embryonic-like microenvironment supporting neural repair in the turtle spinal cord.

The emerging view when comparing data from turtles with those obtained from anamniotes is that in the latter the ependymal layer as a whole, has retained some properties of the embryonic neural tube. For example, in Xenopus tadpoles there is a close axonal-ependymal association during early stages of the regeneration process (Michel and Reier 1979) and in axolotls the cells lining the CC regenerate the spinal cord inducing a multipotent blastema (Schnapp et al. 2005; Tanaka and Ferreti 2009). In turtles, however, the bridge region lacks a distinguishable CC. The BLBP+ cells of the ependymal layer appear as the most likely candidate to give rise to the abundant pre-myelinating oligodendrocytes that envelop the incoming axons. The origin and functional role of GFAP+ cells in the cellular bridge is still uncertain. However, unlike in mammals, GFAP+ cells do not interfere with regenerating axons because they appear aligned with axon bundles. Therefore, turtles appear as a unique amniote model system occupying a peculiar intermediate place between the anamniotes with complete regenerating capabilities and mammals, with very restricted capabilities to restore damaged spinal circuitry.

How tissue damage can activate ependymal cells to trigger endogenous repair remains a fundamental issue to solve. Experiments performed in zebrafish indicate that regenerative properties seem to be linked to a protein encoded by the connective tissue growth factor that helps forming a glial bridge allowing the transit of growing axons through the lesion site (Mokalled et al. 2016). In contrast, our recent wide scale genome study performed in the turtle *Trachemys scripta elegans* showed the "absence of a group of genes exclusive of regenerating taxa" suggesting that anatomical and functional recovery results from cellular and molecular mechanisms involving the "expression patterns of genes shared by all amniotes" (Valentin-Kahan et al. 2017).

Although mammals lost the ability for self-repair, some cells in the CC still react to injury by proliferating and migrating toward the lesion (Beattie et al. 1997; Johansson et al. 1999; Mothe and Tator 2005), where most become astrocytes within the core of the scar (Meletis et al. 2008). However, a number of ependyma-derived cells become oligodendrocytes that interact with axonal sprouts (Meletis et al. 2008). Ependyma-derived cells seem to play a central role in the formation of the scar thereby limiting the extension of damage (Sabelström et al. 2013). Astrocyte-like and oligodendrocyte-like cells derived from the ependyma concentrate in the core of the scar and release growth factors that improve the survival of neurons around the lesion (Sabelström et al. 2013). However, a recent report casts some doubts about the actual contribution of ependymal cells to scar formation (Ren et al. 2017) suggesting that more research is needed to understand the potential of the CC as a source for repair.

Acknowledgments This work was partly supported by grant FCE 103356 from ANII and grant #167 from Wings for Life Spinal Cord Research Foundation to RER. The authors would like to thank the kind donation of GATA3-GFP transgenic mice by Dr. Stavros Malas, The Cyprus Institute of Neurology and Genetics, Cyprus.

References

Abbracchio MP, Burnstock G, Boeynaems JM, Barnard EA, Boyer JL, Kennedy C, Knight GE, Fumagalli M, Gachet C, Jacobson KA, Weisman GA (2006) International Union of Pharmacology. Update and subclassification of the P2Y G protein-coupled nucleotide receptors: from molecular mechanisms and pathophysiology to therapy. Pharmacol Rev 58:281–341

Abbracchio MP, Burnstock G, Verkhratsky A, Zimmermann H (2009) Purinergic signalling in the nervous system: an overview. Trends Neurosci 32:19–29

Alvarez-Buylla A, García-Verdugo JM (2002) Neurogenesis in adult subventricular zone. J Neurosci 22:629–634

Alvarez-Buylla A, García-Verdugo JM, Tramontin AD (2001) A unified hypothesis on the lineage of neural stem cells. Nat Rev Neurosci 2:287–293

Anthony TE, Mason HA, Gridley T, Fishell G, Heintz N (2007) Brain lipid binding protein is a target of Notch signaling in radial glial cells. Genes Dev 19:1028–1033

Armstrong J, Zhang L, McClelland AD (2003) Axonal regeneration of descending and ascending spinal projection neurons in spinal cord-transected larval lamprey. Exp Neurol 180:156–166

Beattie MS, Bresnahan JC, Komon J, Tovar CA, Van Meter M, Anderson DK, Faden AI, Hsu CY, Noble LJ, Salzman S, Young W (1997) Endogenous repair after spinal cord contusion injuries in the rat. Exp Neurol 148:453–463

Bel-Vialar S, Medevielle F, Pituello F (2007) The on/off of Pax6 controls the tempo of neuronal differentiation in the developing spinal cord. Dev Biol 305:659–673

Ben-Ari Y (2002) Excitatory actions of GABA during development: the nature of the nurture. Nat Rev Neurosci 3:728–739

Ben-Ari Y, Spitzer NC (2010) Phenotypic checkpoints regulate neuronal development. Trends Neurosci 33:485–492

Bittman K, Owens DF, Kriegstein AR, LoTurco JJ (1997) Cell coupling and uncoupling in the ventricular zone of developing neocortex. J Neurosci 17:7037–7044

Bordey A (2007) Enigmatic GABAergic networks in adult neurogenic zones. Brain Res Brain Res Rev 53:124–134

Briscoe J, Pierani A, Jessell TM, Ericson J (2000) A homeodomain protein code specifies progenitor cell identity and neuronal fate in the ventral neural tube. Cell 101:435–445

Butler EG, Ward MB (1965) Reconstitution of the spinal cord following ablation in urodele larvae. J Exp Zool 160:47–65

Cebrián-Silla A, Alfaro-Cervelló C, Herranz-Pérez V, Kaneko N, Hwi Park D, Sawamoto K, Alvarez-Buylla A, Lim DA, García-Verdugo JM (2017) Unique organization of the nuclear envelope in the post-natal quiescent neural stem cells. Stem Cell Rep 9:203–216

Chevallier S, Landry M, Nagy F, Cabelguen JM (2004) Recovery of bimodal locomotion in the spinal-transected salamander, Pleurodeles waltlii. Eur J Neurosci 20:1995–2007

Chittajallu R, Chen Y, Wang H, Yuan X, Ghiani CA, Heckman T, McBain CJ, Gallo V (2002) Regulation of Kv1 subunit expression in oligodendrocyte progenitor cells and their role in G1/S phase progression of the cell cycle. Proc Natl Acad Sci U S A 99:2350–2355

Chittajallu R, Aguirre A, Gallo V (2004) NG2-positive cells in the mouse white and grey matter display distinct physiological properties. J Physiol 561:109–122

Coggeshall RE, Youngblood CS (1983) Recovery from spinal transection in fish: regrowth of axons past the transection. Neurosci Lett 38:227–231

Corns LF, Deuchars J, Deuchars SA (2013) GABAergic responses of mammalian ependymal cells in the central canal neurogenic niche of the postnatal spinal cord. Neurosci Lett 553:57–62

Corns LF, Atkinson L, Daniel J, Edwards IJ, New L, Deuchars J, Deuchars SA (2015) Cholinergic enhancement of cell proliferation in the postnatal neurogenic niche of the mammalian spinal cord. Stem Cells 33:2864–2876

Davies HG, Small JV (1968) Structural units in chromatin and their orientation on membranes. Nature 217:1122–1125

Davis BM, Ayers JL, Koran L, Carlson J, Anderson MC, Simpson SB Jr (1990) Time course of salamander spinal cord regeneration and recovery of swimming: HRP retrograde pathway tracing and kinematic analysis. Exp Neurol 108:198–213

Dervan AG, Roberts BL (2003) Reaction of spinal cord central canal cells to cord transaction and their contribution to cord regeneration. J Comp Neurol 458:293–306

Dessaud E, McMahon AP, Briscoe J (2008) Pattern formation in the vertebrate neural tube: a sonic hedgehog morphogen-regulated transcriptional network. Development 135:2489–2503

Diaz Queiroz JP, Echeverri K (2013) Spinal cord regeneration: where fish, frogs and salamanders lead the way, can we follow? Biochem J 451:353–364

Doetsch F, García-Verdugo JM, Alvarez-Buylla A (1997) Cellular composition and three-dimensional organization of the subgerminal zone in the adult mammalian brain. J Neurosci 17:5046–5061

Egar M, Simpson SB, Singer M (1970) The growth and differentiation of the regenerating spinal cord of the lizard Anolis carolinensis. J Morphol 131:131–152

Eisch AJ, Mandyam CD (2007) Adult neurogenesis: can analysis of cell cycle proteins move us "Beyond BrdU"? Curr Pharm Biotechnol 8:147–165

Fernández A, Radmilovich M, Trujillo-Cenóz O (2002) Neurogenesis and gliogenesis in the spinal cord of turtles. J Comp Neurol 453:131–144

Filippov V, Kronenberg G, Pivneva T et al (2003) Subpopulation of nestin-expressing progenitor cells in the adult murine hippocampus shows electrophysiological and morphological characteristics of astrocytes. Mol Cell Neurosci 23:373–382

Fiorelli R, Cebrian-Silla A, Garcia-Verdugo JM, Raineteau O (2013) The adult spinal cord harbors a population of GFAP-positive progenitors with limited self-renewal potential. Glia 61:2100–2113

Frischknecht F, Randall AD (1998) Voltage- and ligand-gated ion channels in floor plate neuroepithelia of the rat. Neuroscience 85:1135–1149

Fu H, Qi Y, Tan MIN, Cai J, Hu X, Liu Z, Qiu M (2003) Molecular mapping of the origin of postnatal spinal cord ependymal cells: evidence that adult ependymal cells are derived from Nkx6.1 + ventral neural progenitor cells. J Comp Neurol 456:237–244

Gandelman M, Levy M, Cassina P, Barbeito L, Beckman JS (2013) P2X7 receptor-induced death of motor neurons by a peroxynitrite/FAS-dependent pathway. J Neurochem 126:382–388

Ganguly K, Schinder AF, Wong ST, Poo M (2001) GABA itself promotes the developmental switch of neuronal GABAergic responses from excitation to inhibition. Cell 105:521–532

Gao Z, Nissen JC, Legakis L, Tsirka SE (2015) Nicotine modulates neurogenesis in the central canal during experimental autoimmune encephalomyelitis. Neuroscience 297:11–21

Ghiani CA, Yuan X, Eisen AM, Knutson PL, DePinho RA, McBain CJ, Gallo V (1999) Voltage-activated K+ channels and membrane depolarization regulate accumulation of the cyclin-dependent kinase inhibitors p27(Kip1) and p21(CIP1) in glial progenitor cells. J Neurosci 19:5380–5392

Gibbs KM, Szaro BG (2006) Regeneration of descending projections in Xenopus laevis tadpole spinal cord demonstrated by retrograde double labeling. Brain Res 1088:68–72

Glaser T, Resende RR, Ulrich H (2013) Implications of purinergic receptor-mediated intracellular calcium transients in neural differentiation. Cell Commun Signal 11:12. https://doi.org/10.1186/1478-811X-11-12

Goodman T, Hajihosseini MK (2015) Hypothalamic tanycytes—masters and servants of metabolic, neuroendocrine, and neurogenic functions. Front Neurosci 9:387. https://doi.org/10.3389/fnins.2015.00387

Götz M, Stoykova A, Gruss P (1998) Pax6 controls radial glia differentiation in the cerebral cortex. Neuron 21:1031–1044

Guillemot F (2007) Spatial and temporal specification of neural fates by transcription factor codes. Development 134:3771–3780

Hamilton LK, Truong MK, Bednarczyk MR, Aumont A, Fernandes KJ (2009) Cellular organization of the central canal ependymal zone, a niche of latent neural stem cells in the adult mammalian spinal cord. Neuroscience 164:1044–1056

Hartfuss E, Galli R, Heins N, Götz M (2001) Characterization of CNS precursor subtypes and radial glia. Dev Biol 229:15–30

Hasegawa K, Chang Y-W, Li H, Berlin Y, Ikeda O, Kane-Goldsmith N, Grumet M (2004) Embryonic radial glia bridge spinal cord lesions and promote functional recovery following spinal cord injury. Exp Neurol 193:394–410

Hattiangady B, Shetty AK (2008) Aging does not alter the number or phenotype of putative stem/progenitor cells in the neurogenic region of the hippocampus. Neurobiol Aging 29:129–147

Haydar TF, Wang F, Schwartz ML, Rakic P (2000) Differential modulation of proliferation in the neocortical ventricular and subventricular zones. J Neurosci 20:5764–5774

Heins N, Malatesta P, Cecconi F, Nakafuku M, Tucker KL, Hack MA, Chapouton P, Barde A, Götz M (2002) Glial cells generate neurons: the role of the transcription factor Pax6. Nat Neurosci 5:308–315

Hooker D (1932) Spinal cord regeneration in the young rainbow fish, Lebistes reticulatus. J Comp Neurol 56:277–297

Horner PJ, Palmer TD (2003) New roles for astrocytes: the nightlife of an 'astrocyte'. La vida loca! Trends Neurosci 26:597–603

Horner PH, Power AE, Kempermann G, Kuhn GH, Palmer TD, Winkler J, Thal LJ, Gage FH (2000) Proliferation and differentiation of progenitor cells throughout the intact adult rat spinal cord. J Neurosci 20:2218–2228

Horstmann E (1954) Die Faserglia des Selachiergehirns. Zellforsch 39:588–617

Hugnot JP, Franzen R (2011) The spinal cord ependymal region: a stem cell niche in the caudal central nervous system. Front Biosci 16:1044–1059

Jacquet BV, Salinas-Mondragon R, Liang H, Therit B, Buie JD, Dykstra M, Campbell K, Ostrowski LE, Brody SL, Ghashghaei HT (2009) FoxJ1-dependent gene expression is required for differentiation of radial glia into ependymal cells and a subset of astrocytes in the postnatal brain. Development 136:4021–4031

Jang MH, Song H, Ming GL (2008) Regulation of adult neurogenesis by neurotransmitters. In: Gage FH, Kempermann G, Song H (eds) Adult neurogenesis. Cold Spring Harbor Laboratory Press, New York, pp 397–421

Jessell TM (2000) Neuronal specification in the spinal cord: inductive signals and transcriptional codes. Nat Rev Genet 1:20–29

Johansson CB, Momma S, Clarke DL, Risling M, Lendahl U, Frisén J (1999) Identification of a neural stem cell in the adult mammalian central nervous system. Cell 96:25–34

Kempermann G (2008) Activity dependency and aging in the regulation of adult neurogenesis. In: Gage FH, Kempermann G, Song H (eds) Adult neurogenesis. Cold Spring Harbor Laboratory Press, New York, pp 341–362

Kempermann G, Song H, Gage F (2008) Neurogenesis in the adult hippocampus. In: Gage FH, Kempermann G, Song H (eds) Adult neurogenesis. Cold Spring Harbor Laboratory Press, New York, pp 159–174

Khakh BS, North RA (2006) P2X receptors as cell-surface ATP sensors in health and disease. Nature 442:527–532

Kohwi M, Osumi N, Rubenstein JL, Alvarez-Buylla A (2005) Pax6 is required for making specific subpopulations of granule and periglomerular neurons in the olfactory bulb. J Neurosci 25:6997–7003

Kriegstein A, Alvarez-Buylla A (2009) The glial nature of embryonic and adult neural stem cells. Annu Rev Neurosci 32:149–184

Kutna V, Sevc J, Gombalov Z, Matiasova A, Daxnerova Z (2014) Enigmatic cerebrospinal fluid-contacting neurons arise even after the termination of neurogenesis in the rat spinal cord during embryonic development and retain their immature-like characteristics until adulthood. Acta Histochem 116:278–285

Lee SK, Pfaff SL (2001) Transcriptional networks regulating neuronal identity in the developing spinal cord. Nat Neurosci 4:1183–1191

Lee-Liu D, Edwards-Faret G, Tapia VS, Larraín J (2013) Spinal cord regeneration: Lessons for mammals from non-mammalian vertebrates. Genesis 51:529–544

Levine JM, Reynolds R, Fawcett JW (2001) The oligodendrocyte precursor cell in health and disease. Trends Neurosci 24:39–47

Li X, Floriddia EM, Toskas K, Fernandes KJL, Guérout N, Barnabé-Heider F (2016) Regenerative potential of ependymal cells for spinal cord injuries over time. EBioMedicine 13:55–65

Lim DA, Huang Y-C, Alvarez-Buylla A (2008) Adult subventricular zone and olfactory bulb neurogenesis. In: Gage FH, Kempermann G, Song H (eds) Adult neurogenesis. Cold Spring Harbor, New York

Lin JH, Takano T, Arcuino G, Wang X, Hu F, Darzynkiewicz Z, Nunes M, Goldman SA, Nedergaard M (2007) Purinergic signaling regulates neural progenitor cell expansion and neurogenesis. Dev Biol 302:356–366

Liu X, Wang Q, Haydar TF, Bordey A (2005) Nonsynaptic GABA signaling in postnatal subventricular zone controls proliferation of GFAP-expressing progenitors. Nat Neurosci 8:1179–1187

Liu X, Bolteus AJ, Balkin DM, Henschel O, Bordey A (2006) GFAP-expressing cells in the postnatal subventricular zone display a unique glial phenotype intermediate between radial glia and astrocytes. Glia 54:394–410

Liu X, Hashimoto-Torii K, Torii M, Haydar TF, Rakic P (2008) The role of ATP signalling in the migration of intermediate neuronal progenitors to the neocortical subventricular zone. Proc Natl Acad Sci U S A 105:11802–11807

Lledo PM, Alonso M, Grubb MS (2006) Adult neurogenesis and functional plasticity in neuronal circuits. Nat Rev Neurosci 7:179–193

Lois C, Garcia-Verdugo JM, Alvarez-Buylla A (1996) Chain migration of neuronal precursors. Science 271:978–981

LoTurco JJ, Owens DF, Heath MJ, Davis MB, Kriegstein AR (1995) GABA and glutamate depolarize cortical progenitor cells and inhibit DNA synthesis. Neuron 15:1287–1298

Lucassen PJ, Oomen CA, van Dam AM, Czéh B (2008) Regulation of hippocampal neurogenesis by systemic factors including stress, glucocorticoids, sleep and inflammation. In: Gage FH, Kempermann G, Song H (eds) Adult neurogenesis. Cold Spring Harbor Laboratory Press, New York, pp 363–396

Ma DK, Ming G-l, Gage FH, Song H (2008) Neurogenic niches in the adult mammalian brain. In: Gage FH, Kempermann G, Song H (eds) Adult neurogenesis. Cold Spring Harbor, New York, pp 207–226

MacFarlane SN, Sontheimer H (2000a) Modulation of Kv1.5 currents by Src tyrosine phosphorylation: potential role in the differentiation of astrocytes. J Neurosci 20:5245–5253

MacFarlane SN, Sontheimer H (2000b) Changes in ion channel expression accompany cell cycle progression of spinal cord astrocytes. Glia 30:39–48

Maekawa M, Takashima N, Arai Y, Nomura T, Inokuchi K, Yuasa S, Osumi N (2005) Pax6 is required for production and maintenance of progenitor cells in postnatal hippocampal neurogenesis. Genes Cells 10:1001–1014

Marichal N, García G, Radmilovich M, Trujillo-Cenóz O, Russo RE (2009) Enigmatic central canal contacting cells: immature neurons in "standby mode"? J Neurosci 29:10010–10024

Marichal N, García G, Radmilovich M, Trujillo-Cenóz O, Russo RE (2012) Spatial domains of progenitor-like cells and functional complexity of a stem cell niche in the neonatal rat spinal cord. Stem Cells 30:2020–2031

Marichal N, Fabbiani G, Trujillo-Cenóz O, Russo RE (2016) Purinergic signalling in a latent stem cell niche of the rat spinal cord. Purinergic Signal 12:331–341

Marzesco AM, Janich P, Wilsch-Bräuninger M, Dubreuil V, Langenfeld K, Corbeil D, Huttner WB (2005) Release of extracellular membrane particles carrying the stem cell marker prominin-1 (CD133) from neural progenitors and other epithelial cells. J Cell Sci 118:2849–2858

Masahira N, Takebayashi H, Ono K, Watanabe K, Ding L, Furusho M, Ogawa Y, Nabeshima Y, Alvarez-Buylla A, Shimizu K, Ikenaka K (2006) Olig2-positive progenitors in the embryonic spinal cord give rise not only to motoneurons and oligodendrocytes, but also to a subset of astrocytes and ependymal cells. Dev Biol 293:358–369

Massé K, Bhamra S, Eason R, Dale N, Jones EA (2007) Purine-mediated signalling triggers eye development. Nature 449:1058–1062

McHedlishvili L, Epperlein HH, Telzerow A, Tanaka EM (2007) A clonal analysis of neural progenitors during axolotl spinal cord regeneration reveals evidence for both spatially restricted and multipotent progenitors. Development 134:2083–2093

Meletis K, Barnabé-Heider F, Carlén M, Evergren E, Tomilin N, Shupliakov O, Frisén J (2008) Spinal cord injury reveals multilineage differentiation of ependymal cells. PLoS Biol 6:1494–1507

Mestres P, Garfia A (1980) Effects of cytochalasin B on the ependyma. Scan Electron Microsc 3:465–474

Michel ME, Reier PJ (1979) Axonal-ependymal associations during early regeneration of the transected spinal cord in Xenopus laevis tadpoles. J Neurocytol 8:529–548

Ming GL, Song H (2005) Adult neurogenesis in the mammalian central nervous system. Annu Rev Neurosci 28:223–250

Miras-Portugal MT, Gomez-Villafuertes R, Gualix J, Diaz-Hernandez JI, Artalejo AR, Ortega F, Delicado EG, Perez-Sen R (2015) Nucleotides in neuroregeneration and neuroprotection. Neuropharmacology 104:243–254

Mokalled MH, Patra C, Dickson AL, Endo T, Stainier DYR, Poss KD (2016) Injury-induced ctgf directs glial bridging and spinal cord regeneration in Zebrafish. Science 354:630–634

Molina B, Rodríguez EM, Peruzzo B, Caprile T, Nualart F (2001) Spatial distribution of Reissner's fiber glycoproteins in the filum terminale of the rat and rabbit. Microsc Res Tech 52:552–563

Molofsky AV, Slutsky SG, Joseph NM, He S, Pardal R, Krishnamurthy J, Sharpless NE, Morrison SJ (2006) Increasing p16INK4a expression decreases forebrain progenitors and neurogenesis during ageing. Nature 443:448–452

Mothe AJ, Tator CH (2005) Proliferation, migration, and diferentiation of endogenous ependymal region stem/progenitor cells following minimal spinal cord inury in the adult rat. Neuroscience 131:177–187

Nacher J, Varea E, Blasco-Ibanez JM, Castillo-Gomez E, Crespo C, Martinez-Guijarro FJ, McEwen BS (2005) Expression of the transcription factor Pax 6 in the adult rat dentate gyrus. J Neurosci Res 81:753–761

Noctor SC, Flint AC, Weissman TA, Wong WS, Clinton BK, Kriegstein AR (2002) Dividing precursor cells of the embryonic cortical ventricular zone have morphological and molecular characteristics of radial glia. J Neurosci 22:3161–3173

Nornes HO, Das GD (1972) Temporal pattern of neurogenesis in spinal cord: cytoarchitecture and directed growth of axons. Proc Natl Acad Sci U S A 69:1962–1966

Nornes HO, Das GD (1974) Temporal pattern of neurogenesis in spinal cord of rat. I. An autoradiographic study—time and sites of origin and migration and settling patterns of neuroblasts. Brain Res 73:121–138

Peng W, Cotrina ML, Han X, Yu H, Bekar L, Blum L, Takano T, Tian GF, Goldman SA, Nedergaard M (2009) Systemic administration of an antagonist of the ATP-sensitive receptor P2X7 improves recovery after spinal cord injury. Proc Natl Acad Sci U S A 106:12489–12493

Peters A, Palay SL, Webster H dF (1991) The ependyma. In: The fine structure of the nervous system. Neurons and their supporting cells. Oxford University Press, Oxford, pp 312–327

Petracca YL, Sartoretti MM, Di Bella DJ, Marin-Burgin A, Carcagno AL, Schinder AF, Lanuza GM (2016) The late and dual origin of cerebrospinal fluid-contacting neurons in the mouse spinal cord. Development 143:880–891

Piatt J, Piatt M (1958) Transection of the spinal cord in the adult frog. Anat Rec 131:81–95

Pinto L, Götz M (2007) Radial glial cell heterogeneity-the source of diverse progeny in the CNS. Prog Neurobiol 83:2–23

Ramón y Cajal S (1909) Histologie du Systeme Nerveux de l'homme et des vertébres, vol I. (Edited by Consejo superior de Investigaciones Científicas, 1952). Maloine, Paris

Ramón y Cajal SR (1913) Estudios sobre la degeneración y regeneración del sistema nervioso. T I-II. Degeneración y regeneración de los centros nerviosos. Nicolás Moya, Madrid

Reali C, Fernández A, Radmilovich M, Trujillo-Cenóz O, Russo RE (2011) GABAergic signalling in a neurogenic niche of the turtle spinal cord. J Physiol 589:5633–5647

Rehermann MI, Marichal N, Russo R, Trujillo-Cenoz O (2009) Neural Reconnection in the transected spinal cord of the freshwater turtle Trachemys dorbignyi. J Comp Neurol 515:197–214

Rehermann MI, Santiñaqui FF, López-Carro B, Russo R, Trujillo-Cenoz O (2011) Cell proliferation and cytoarchitectural remodelling in the fresh-water turtle Trachemys dorbignyi. Cell Tissue Res 344:415–433

Reimer MM, Sörensen I, Kuscha V, Frank RE, Liu C, Becker CG, Becker T (2008) Motor neuron regeneration in adult zebrafish. J Neurosci 28:8510–8516

Ren Y, Ao Y, O'Shea TM, Burda JE, Bernstein AM, Brumm AJ, Muthusamy N, Ghashghaei HT, Carmichael ST, Cheng L, Sofroniew MV (2017) Ependymal cell contribution to scar formation after spinal cord injury is minimal, local and dependent on direct ependymal injury. Sci Rep 7:41122. https://doi.org/10.1038/srep41122

Rivera C, Voipio J, Payne JA, Ruusuvuori E, Lahtinen H, Lamsa K, Pirvola U, Saarma M, Kaila K (1999) The K+/Cl- co-transporter KCC2 renders GABA hyperpolarizing during neuronal maturation. Nature 397:251–255

Roberts BL, Maslam S, Scholten G, Smit W (1995) Dopaminergic and GABAergic cerebrospinal fluid-contacting neurons along the central canal of the spinal cord of the eel and trout. J Comp Neurol 354:423–437

Rovainen CM (1976) Regeneration of Müller and Mauthner axons after spinal cord transection in larval lampreys. J Comp Neurol 168:545–554

Rowitch DH (2004) Glial specification in the vertebrate neural tube. Nat Rev Neurosci 5:409–419

Russo RE, Hounsgaard J (1999) Dynamics of intrinsic electrophysiological properties in spinal cord neurones. Prog Biophys Mol Biol 72:329–365

Russo RE, Fernández A, Reali C, Radmilovich M, Trujillo-Cenóz O (2004) Functional and molecular clues reveal precursor-like cells and immature neurones in the turtle spinal cord. J Physiol 3:831–838

Russo RE, Reali C, Radmilovich M, Fernández A, Trujillo-Cenóz O (2008) Connexin 43 delimits functional domains of neurogenic precursors in the spinal cord. J Neurosci 28:3298–3309

Sabelström H, Stenudd M, Réu P, Dias DO, Elfineh M, Zdunek S, Damberg P, Göritz C, Frisén J (2013) Resident neural stem cells restrict tissue damage and neuronal loss after spinal cord injury in mice. Science 342:637–640

Sabourin JC, Ackema KB, Ohayon D, Guichet PO, Perrin FE, Garces A, Ripoll C, Charite J, Simonneu L, Ketenmann H, Zine A, Pivat A, Valmier J, Pattyn A, Hugnot JP (2009) A mesenchymal-like ZEB1+ niche harbors dorsal radial glial fibrillary acidic protein-positive stem cells in the spinal cord. Stem Cells 27:2722–2733

Schaarschmidt G, Wegner F, Schwarz SC, Schmidt H, Schwarz J (2009) Characterization of voltage-gated potassium channels in human neural progenitor cells. PLoS One 4:e6168. https://doi.org/10.1371/journal.pone.0006168

Schnapp E, Kragl M, Rubin L, Tanaka EM (2005) Hedgehog signaling controls dorsoventral patterning, blastema cell proliferation and cartilage induction during axolotl tail regeneration. Development 132:3243–3253

Seri B, García-Verdugo JM, McEwen BS, Alvarez-Buylla A (2001) Astrocytes give rise to new neurons in the adult mammalian hippocampus. J Neurosci 21:7153–7160

Sevc J, Daxnerova Z, Hanova V, Koval J (2011) Novel observations on the origin of ependymal cells in the ventricular zone of the rat spinal cord. Acta Histochem 113:156–162

Shifman MI, Jin LQ, Selzer M (2007) Regeneration in the lamprey spinal cord. In: Becker CG, Becker T (eds) Model organisms in spinal cord regeneration. Wiley-VCH Verlag, Weinheim, pp 229–262

Silver J, Miller JH (2004) Regeneration beyond the glial scar. Nat Rev Neurosci 5:146–156

Sims RT (1962) Transection of the spinal cord in developing Xenopus laevis. Embryol Exp Morphol 10:115–126

Singer M, Nordlander RTH, Egar M (1979) Axonal guidance during embryogenesis and regeneration in the spinal cord of the newt: the blue print hypothesis of neural pathway patterning. J Comp Neurol 185:1–22

Smith DO, Rosenheimer JL, Kalil RE (2008) Delayed rectifier and A-type potassium channels associated with Kv 2.1 and Kv 4.3 expression in embryonic rat neural progenitor cells. PLoS One 3:e1604

Sontheimer H, Trotter J, Schachner M, Kettenmann H (1989) Channel expression correlates with differentiation stage during the development of oligodendrocytes from their precursor cells in culture. Neuron 2:1135–1145

Spassky N, Merkle FT, Flames N, Tramontin AD, Garcia-Verdugo JM, Alvarez-Buylla A (2005) Adult ependymal cells are postmitotic and are derived from radial glial cells during embryogenesis. J Neurosci 25:10–18

Spitzer NC, Root CM, Borodinsky LN (2004) Orchestrating neuronal differentiation: patterns of Ca2+ spikes specify transmitter choice. Trends Neurosci 27:415–421

Stewart RR, Zigova T, Luskin MB (1999) Potassium currents in precursor cells isolated from the anterior subventricular zone of the neonatal rat forebrain. J Neurophysiol 81:95–102

Stoeckel ME, Uhl-Bronner S, Hugel S, Veinante P, Klein MJ, Mutterer J, Freund-Mercier MJ, Schlichter R (2003) Cerebrospinal fluid-contacting neurons in the rat spinal cord, a gamma-aminobutyric acidergic system expressing the P2X2 subunit of purinergic receptors, PSA-NCAM, and GAP-43 immunoreactivities: light and electron microscopic study. J Comp Neurol 457:159–174

Sugimori M, Nagao M, Bertrand N, Parras CM, Guillemot F, Nakafuku M (2007) Combinatorial actions of patterning and HLH transcription factors in the spatiotemporal control of neurogenesis and gliogenesis in the developing spinal cord. Development 134:1617–1629

Surprenant A, North RA (2009) Signaling at purinergic P2X receptors. Annu Rev Physiol 71:333–359

Surprenant A, Rassendren F, Kawashima E, North RA, Buell G (1996) The cytolytic P2Z receptor for extracellular ATP identified as a P2X receptor (P2X7). Science 272:735–738

Takeda A, Goris RC, Funakoshi K (2007) Regeneration of descending projections to the spinal cord neurons after spinal hemisection in the goldfish. Brain Res 1155:17–23

Tanaka EM, Ferretti P (2009) Considering the evolution of regeneration in the central nervous system. Nat Rev Neurosci 10:713–723

Thuret S, Moon LD, Gage FH (2006) Therapeutic interventions after spinal cord injury. Nat Rev Neurosci 7:628–643

Trujillo-Cenóz O, Fernández A, Radmilovich M, Realli C, Russo R (2007) Cytological organization of the central gelatinosa in the turtle spinal cord. J Comp Neurol 502:291–308

Valentin-Kahan A, García-Tejedor GB, Robello C, Trujillo-Cenóz O, Russo RE, Alvarez-Valin F (2017) Gene expression profiling in the injured spinal cord of Trachemys scripta elegans: an amniote with self-repair capabilities. Front Mol Neurosci 10:17. https://doi.org/10.3389/fnmol.2017.00017

Vigh B, Vigh-Teichmann I (1998) Actual problems of the cerebrospinal fluid-contacting neurons. Microsc Res Tech 41:57–83

Vigh B, Vigh-Teichmann I, Aros B (1977) Special dendritic and axonal endings formed by the cerebrospinal fluid contacting neurons of the spinal cord. Cell Tissue Res 183:541–552

Vigh B, Vigh-Teichmann I, Manzano e Silva MJ, van den Pol AN (1983) Cerebrospinal fluid-contacting neurons of the central canal and terminal ventricle in various vertebrates. Cell Tissue Res 231:615–621

Vigh-Teichmann I, Vigh B (1983) The system of cerebrospinal. Arch Histol Jap 46:427–468

Wang DD, Kriegstein AR (2009) Defining the role of GABA in cortical development. J Physiol 587:1873–1879

Wang X, Arcuino G, Takano T, Lin J, Peng WG, Wan P, Li P, Xu Q, Liu QS, Goldman SA, Nedergaard M (2004) P2X7 receptor inhibition improves recovery after spinal cord injury. Nat Med 10:821–827

Webb SE, Moreau M, Leclerc C, Miller AL (2005) Calcium transients and neural induction in vertebrates. Cell Calcium 37:375–385

Weiss S, Dunne C, Hewson J, Wohl C, Wheatley M, Peterson AC, Reynolds BA (1996) Multipotent CNS stem cells are present in the adult mammalian spinal cord and ventricular neuroaxis. J Neurosci 16:7599–7609

Weissman TA, Riquelme PA, Ivic L, Flint AC, Kriegstein AR (2004) Calcium waves propagate through radial glial cells and modulate proliferation in the developing neocortex. Neuron 43:647–661

Wood MR, Cohen MJ (1979) Synaptic regeneration in identified neurons of the lamprey spinal cord. Science 206:344–347

Yamanaka S (2012) Induced pluripotent stem cells: past, present, and future. Cell Stem Cell 10:678–684

Yu K, McGlynn S, Matise MP (2013) Floor plate-derived sonic hedgehog regulates glial and ependymal cell fates in the developing spinal cord. Development 140:1594–1604

Zhang F, Clarke JDW, Ferretti P (2000) FGF-2 up-regulation and proliferation of neural progenitors in the regenerating amphibian spinal cord in vivo. Dev Biol 225:381–391

Zimmermann H (2006) Nucleotide signaling in nervous system development. Pflügers Arch 452:573–588

Chapter 6
Being a Neural Stem Cell: A Matter of Character But Defined by the Microenvironment

Evangelia Andreopoulou, Asterios Arampatzis, Melina Patsoni, and Ilias Kazanis

Abstract The cells that build the nervous system, either this is a small network of ganglia or a complicated primate brain, are called neural stem and progenitor cells. Even though the very primitive and the very recent neural stem cells (NSCs) share common basic characteristics that are hard-wired within their character, such as the expression of transcription factors of the SoxB family, their capacity to give rise to extremely different neural tissues depends significantly on instructions from the microenvironment. In this chapter we explore the nature of the NSC microenvironment, looking through evolution, embryonic development, maturity and even disease. Experimental work undertaken over the last 20 years has revealed exciting insight into the NSC microcosmos. NSCs are very capable in producing their own extracellular matrix and in regulating their behaviour in an autocrine and paracrine manner. Nevertheless, accumulating evidence indicates an important role for the vasculature, especially within the NSC niches of the postnatal brain; while novel results reveal direct links between the metabolic state of the organism and the function of NSCs.

E. Andreopoulou • M. Patsoni
Lab of Developmental Biology, Department of Biology, University of Patras, Patras, Greece
e-mail: evoievan94@gmail.com; melina.p@windowslive.com

A. Arampatzis
Wellcome Trust- MRC Cambridge Stem Cell Biology Institute, University of Cambridge, Cambridge, UK

School of Medicine, Aristotle University of Thessaloniki, Thessaloniki, Greece
e-mail: asteriosa@auth.gr

I. Kazanis (✉)
Lab of Developmental Biology, Department of Biology, University of Patras, Patras, Greece

Wellcome Trust- MRC Cambridge Stem Cell Biology Institute, University of Cambridge, Cambridge, UK
e-mail: ikazanis@upatras.gr

© Springer International Publishing AG 2017
A. Birbrair (ed.), *Stem Cell Microenvironments and Beyond*,
Advances in Experimental Medicine and Biology 1041,
DOI 10.1007/978-3-319-69194-7_6

Keywords Neural stem cell • Development • Evolution • Extracellular matrix • Vasculature • Cancer

6.1 Intrinsic vs. Extrinsic Regulation

The emergence of neural tissue has been an early event during evolution, having occurred in pre-bilaterian animals. Irrespective of being a small network of neurons, or an extremely elaborate neuroglial system, as in the avian and the mammalian CNS, the nervous tissue is built by neural progenitors. These can be either loosely specified cells within an epithelium, or more distinct neural stem cells that undergo a well-controlled scenario of symmetric and asymmetric divisions before generating differentiated cells (reviewed in Hartenstein and Stollewerk (2015)). By focusing on the basic similarities or the vast differences of neurogenic processes across evolution, it can be hypothesized that the identity and the properties of neural stem cells (NSCs; in this review the term will be applied for any cell directly or indirectly giving rise to mature neurons and glia) are largely inherent, or mostly dependent on micro-environmental cues; both statements being equally strong.

The conserved expression of transcription factors of the SoxB gene family, as well as the ability of both very primitive and very recent NSCs to generate neurons with shared characteristics (e.g. expression of synaptotagmin), indicate that being a NSC is most probably a matter of character, installed within the hard wiring of early neuroectodermal/neuroepithelial specification. This introduces a first question to be debated: Does the epithelial to neuroepithelial transformation depend on external cues? The classical view is that small parts of the mesoderm, outside the neuroepithelial anlage (for example the organizer or the node of amniotes) induce the neural fate through the Bone Morphogenetic Protein (BMP) and Int/Wingless (Wnt) signalling pathways (Hartenstein and Stollewerk 2015). Notably though, Van der Kooy and his team (Smukler et al. 2006) experimenting at the edges of physiological conditions, reported that mouse Embryonic Stem Cells (ESCs), when cultured in conditions lacking instructive cues, start a neural differentiation program, suggesting that this is the default pathway of early embryonic pluripotent cells. If that is the case, signals from the organizer are important not in installing neural identity on stem cells, but in liberating them form the activity of suppressors. Adding to the concept of a mighty inherent programming in NSCs, both embryonic and adult-derived NSCs can be cultured *in vitro* in very low densities (what is called clonal expansion) where they exhibit the fundamental properties that are observed in the organism. For example, in clones generated by embryonic cortical NSCs, neurons are generated before glial cells and with the appropriate temporal specification pattern (Okano and Temple 2009). In addition, the decreased expression of a single gene (encoding for the DNA-associated protein Trnp1) can produce gyrencephalia in the normally lissencephalic (i.e. lacking cortical folds) mice (Stahl et al. 2013); whilst over-expression only of Polo-like kinase 4 results in centrosome amplification in NSCs and subsequent microcephaly (Marthiens et al. 2013). Furthermore, a Rho guanosine triphosphatase–activating protein (called ARHGAP11B) has been singled out as a gene expressed only in human RGCs that has possibly contributed to their distinct behaviour as compared to mouse RGCs (Florio et al. 2015).

On the other hand, it is obvious that even the very primitive NSCs (meaning either those found in early animal species, or those operating during early embryonic stages of neurogenesis) do not behave in similar ways, exhibiting fundamentally different repertoires of neurogenesis, as dictated by the available genetic information and cues from the micro-environment. NSCs isolated from the adult mouse brain are still in a stage of genetic and epigenetic flexibility that allows them to revert to an ESC behaviour once implanted in the blastocyst microenvironment (Clarke et al. 2000), or to generate directly blood cells when exposed to the appropriate cues (Bjornson et al. 1999). In contrast, human glial progenitors grafted in the mouse hippocampus generate astrocytes that exhibit the human morphology, irrespective of the overwhelming mouse tissue environment (Han et al. 2013). The interplay between intrinsic and extrinsic cues in controlling the behaviour of NSCs can be clearly seen in ex vivo experimental assays. When mouse NSCs are cultured in growth factor-rich conditions as free-floating cells, they tend to generate 3D aggregates called neurospheres (Fig. 6.1). These are essentially amorphous complexes consisted of multipotent progenitors and more differentiated cells of neuronal and glial fate. When the same cells are cultured on 3D scaffolds, again in growth factor-rich conditions and under appropriate rotor spin, they generate microscopic brains (Camp et al. 2015; Qian et al. 2016).

In this chapter, we will focus mostly on the microenvironment of NSCs, describing the main cellular and extracellular elements with which NSCs are in direct or indirect contact throughout the different stages of their life: from embryonic development to the mature CNS as well as in cases of disease, especially during tumorigenesis. Whenever possible, we will also refer to the functional significance of these micro-environmental cues. The larger volume of information is derived from studies in rodents (mostly mice and rats); however, whenever available, information will be given for the human NSCs and for other model organisms. Before continuing, a reference should be made to the concept of the NSCs as "builders" of their microenvironment. NSCs have the extraordinary ability to perform significant aspects of their cell-generation program *in vitro*, and when operating *in vivo* they built the nervous system with a minimal contribution from other tissues (such as adipose, bone and connective tissue, or muscles). These observations suggest that NSCs combine the ability to perform a sophisticated program of symmetric and asymmetric divisions, with strictly controlled progression towards fate commitment, and to produce, at the same time, basic elements of the tissue, such as the extracellular matrix (ECM). Indeed, embryonic and adult NSCs gradually create and maintain a tissue populated almost exclusively by their daughter cells that -besides neurons- include cells exhibiting a range of structural, nutritional and even tissue-clearing activity (Lovelace et al. 2015). In parallel, NSCs synthesize basic ECM components, such as laminins and Tenascin-C (Kazanis et al. 2007, 2010; Lathia et al. 2007) and single-cell transcriptome analysis in mouse and human NSCs has revealed their ability to control self-renewal in an autocrine way (Pollen et al. 2015).

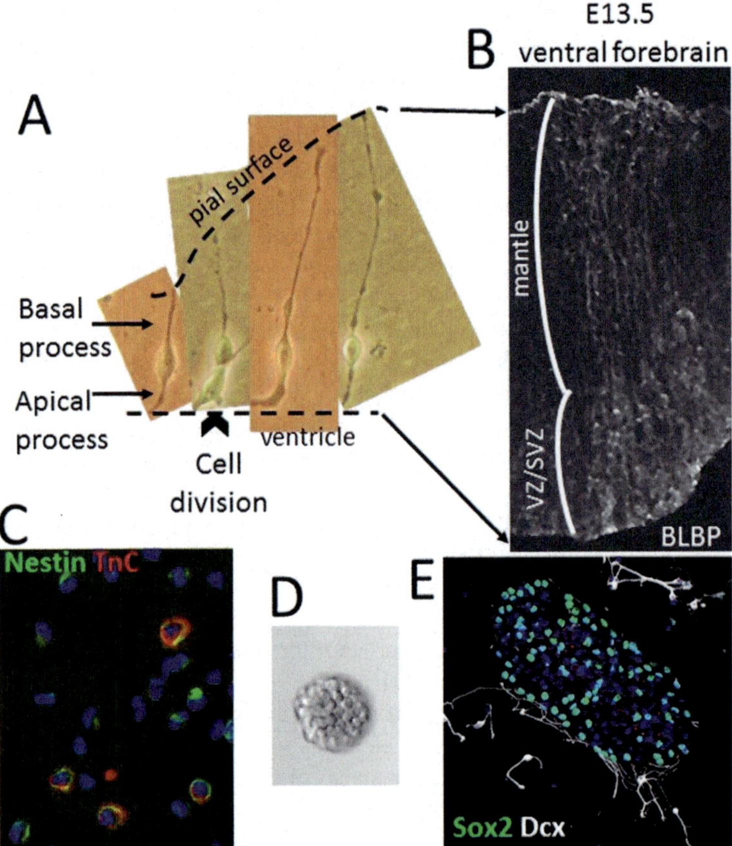

Fig. 6.1 (**a**) NSCs isolated from the embryonic cortex at E12.5 and kept in culture exhibit the characteristic structure of NEP cells and RGCs. They possess a short and a long process (apical and basal, respectively, in the tissue). Here, cells are shown at different times in culture, with the long process becoming more elongated over time, as would happen in the tissue in order to stay attached to the pial surface. Note a NSC cell that has divided, with the daughter cell remaining in contact with the basal process. (**b**) In a microphotograph of a section taken from an E13.5 mouse embryo and immunostained for the RGC marker BLBP, the three microenvironments to which a RGC is exposed can be seen: (1) cell bodies are stacked around the ventricle, in the ventricular/subventricular zones. (2) Basal processes cross the mantle, where immature neurons assume their final positions and differentiate. (3) The RGC endfeet reach the pial surface. (**c**) NSCs are able to produce much of the ECM components of the microenvironment. Here, isolated NSCs from the E13.5 mouse cortex have been immunostained for nestin (marker of NSCs, *in green*) and for the glycoprotein Tenascin-C (*in red*). (**d**) NSCs isolated from the embryonic or the adult mouse and rat brain can be kept in cultures, where they form 3D, free-floating aggregates called neurospheres. (**e**) When neurospheres are allowed to attach to a substrate and growth factors are removed from the medium, many progenitors start to differentiate in neurons (Sox2, marking nueral progenitors is in green and Dcx, marking immature neurons is in *white*)

6.2   The NSC Microenvironment During Embryogenesis

As already mentioned NSCs operate in microenvironments with limited presence of extra-neural elements; thus, they are surrounded and remain in direct contact mostly with other NSCs and their progeny. An interesting example of how these cells interact with each other at the very early stages of development in order to shape the nervous system is the phenomenon of lateral inhibition that has been described in *Drosophila melanogaster* (Hartenstein and Wodarz 2003). Within the homogeneous pool of neuroectodermal cells, that all express proneural genes, individual cells show activation of Notch ligands (such as Delta) that act upon their immediate neighbours and instruct them to downregulate proneural gene expression. Only these, stochastically selected, cells progress towards becoming NSCs that delaminate from the neuroectoderm and initiate a program of asymmetric neurogenic divisions (Isshiki et al. 2001).

6.2.1   One NSC, Three Microenvironments

During mammalian neurogenesis, NSC bodies are densely packed as a pseudolayered epithelium that forms the neural tube. In the developing mouse cerebral cortex and up to embryonic day (E) 11 these cells are called neuroepithelial (NEP). They express Sox2 and nestin and because they undergo mitosis only when positioned at the surface of the ventricle (initially the lumen of the neural tube that progressively expands to form the brain ventricles) their nuclei remain in constant movement, migrating away and towards the ventricle (a phenomenon called interkinetic nuclear migration) in order to allow enough space for cell division. NEP cells are bipolar, with a short apical process always remaining in contact with the ventricle and a longer basal process remaining in contact with the pial surface (Fig. 6.1a); however, recently an alternative architecture was described to occur in the ventral forebrain with the basal process ending on a blood vessel (Tan et al. 2016) (see also section 6.2.3). The length of this basal process constantly increases as new layers of neural stem and progenitor cells as well as of neurons and glia are formed. NEP cells divide almost exclusively in a symmetric self-renewing mode in order to expand their population, creating the so called ventricular zone (VZ). After E11, possibly in order to adjust to the ever increasing width of the cortex, NEP cells acquire more glial characteristics becoming radial glial cells (RGCs). These retain the bipolar shape, with the basal process becoming much longer (Fig. 6.1a), express additional markers such as the transcription factor Pax6 and the glial proteins BLBP (Fig. 6.1b) and GLAST and exhibit a wide range of cell-division types with the asymmetric, self-renewing being the dominant (Johansson et al. 2010). The tightly packed NEP cells and RGCs of the VZ can support their self-renewal in an autocrine manner, both in humans and rodents (Fietz et al. 2012). In addition, their cell bodies are surrounded by an ECM rich in laminins, Tenascin-C and glycosaminoglycans but poor in collagen IV, nidogen, perlecan and fibronectin (Akita et al. 2008; Fietz et al. 2012; Garcion et al. 2004; Lathia et al. 2007; Milošević et al. 2014). The exact role of different ECM molecules has yet to be fully elucidated, but Tenascin-C seems to control the NEP-to-RGC maturation process via orchestrating the activity of FGFs and BMPs (Garcion et al. 2004; Theocharidis et al. 2014) and glycosaminoglycans regulate proliferation (Sirko et al. 2010). Moreover, NSCs of the VZ bind to ECM via integrins and syndecan and these interactions are crucial for their periventricular positioning, for interkinetic nuclear migration and for regulating the angle of division (Loulier et al. 2009; Marthiens and ffrench-Constant 2009). The latter is of key importance in controlling the type of division, with symmetric divisions being dependent on the split of a small fragment of the apical pole of the cell membrane (Kosodo et al. 2004). The asymmetry of division of NSCs and the ensuing cell fate are also affected by N-cadherin and β-catenin, components of the molecular machinery of the adherens junctions that are formed between cells of the VZ (Draganova et al. 2015; Jiang and Nardelli 2015; Marthiens and Ffrench-Constant 2009). The existence of adherens junctions is important for the correct positioning of RGCs and when impaired, for example by the genetic perturbation of afadin expression -another of their components- it can lead to cortical malformations similar to human pathologies such as lissencephaly and double cortex (Yamamoto et al. 2015). Recently, a better image of how local extrinsic cues signal to NSCs via the ECM has been elucidated in the developing chick cortex. Wnt7alpha signalling among neighbouring cells was reported to be mediated through integrin/decorin (another ECM component) interactions in order to control proliferation and differentiation (Long et al. 2016). Finally, the location of NEP cells and RGCs around the neural tube/ventricle, especially the observation that mitosis occurs only at the surface of the ventricle (Fig. 6.1a), also suggests that factors from the CSF most probably act upon NSCs. Indeed, factors such as Wnts, BMPs (Lehtinen et al. 2011), IGFs, FGFs and Shh (that is sensed by the apically positioned single cilium of each NSC) regulate proliferation and specification of progenitors (briefly reviewed in Jiang and Nardelli (2015)). Lehtinen and colleagues (Lehtinen et al. 2011) reported expression of the receptor of IGF1 and 2 at the apical membrane of NSCs and showed that NSCs lack expression of these growth factors. They also confirmed that IGF1 and 2, produced by the embryonic choroid plexus, are crucial in supporting proliferation of embryonic NSCs. In the spinal cord, where the succession of neurogenesis and oligodendrogenesis has been investigated extensively, Shh secreted at the ventral neural tube instructs the generation of oligodendrocyte progenitors whilst BMPs from the dorsal domains instruct the generation of motor neurons (Mekki-Dauriac et al. 2002).

The bipolar morphology of NEP cells and especially of RGCs, allows them to act as scaffolds for the guided migration of their daughter cells that move towards the pial surface in order to occupy their correct position within the 6-layered cortex (Rakic 2003). Therefore, RGCs remain in constant and direct contact with migrating progeny (Cameron and Rakic 1994), a process controlled by integrins (Anton et al. 1999), although shifts between neighbouring RGC basal processes (tangential migration) occur under the control of ephrin signalling (Torii et al. 2009). Interestingly, at the end of neurogenesis, when the appropriate number of neurons has been produced, RGCs receive feedback instructing them to switch towards gliogenesis; cardiotrophin-1 produced by neurons has been reported to be one such signal (Barnabé-Heider et al. 2005).

The third compartment of each RGC, which shows functionally distinct interactions with the microenvironment, is the endfoot of the basal process at the pial surface. The pial surface is an area of the developing nervous system that is especially rich in ECM molecules, characterized by the expression of components such as nidogen, collagen IV and perlecan that are excluded from the main embryonic cortical tissue (Fietz et al. 2012; Lathia et al. 2007). The multiple interactions between integrins expressed by RGCs and different laminins of the basement membrane of the pial/meningeal surface are of paramount importance for the correct migration and positioning of newborn neurons. Defects in these interactions lead to pathologies reminiscent of cobblestone lissencephaly and double cortex in humans (Belvindrah et al. 2007a; Graus-Porta et al. 2001; Yamamoto et al. 2015). Notably, the laminin- integrin interactions seem to be dispensable for the correct proliferative behaviour of RGCs (Haubst et al. 2006), but very instrumental for their survival (Radakovits et al. 2009).

6.2.2 Evolution

During evolution, this prototype architecture of the embryonic cortex changed with the appearance of additional types of neural progenitors within the VZ, but most importantly with the emergence of novel stem and progenitor populations located in novel germinal zones (reviewed in Borrell and Calegari (2014)). In mammals, RGCs started dividing asymmetrically in order to generate intermediate (or basal) progenitors that migrate deeper into the tissue forming the Sub-Ventricular Zone (SVZ). These progenitors undergo a number of mitoses before terminally differentiating and their emergence allowed the formation of the 6-layered neocortex (Wilsch-Bräuninger et al. 2016). In a next evolutionary step, a third group of progenitors located even deeper in the tissue (outer SVZ/oSVZ) appeared in lissenecephalic species (such as rodents) but their population became much more prominent in primates. oSVZ cells generate larger clones than intermediate progenitors (Pollen et al. 2015) and their presence is correlated with the explosive expansion of neocortex accommodated by the formation of gyrencephalia (Wilsch-Bräuninger et al. 2016). In rodents, the transcriptome of the SVZ (of intermediate progenitors) is similar to that found at the cortical plate where newborn neurons mature (Fietz et al. 2012) and significantly distinct to that of the VZ. This, applies also to ECM components. In contrast, in the human developing cortex the ECM signature of all three progenitor pools seems to be similar, but significantly different from that of the cortical plate (Fietz et al. 2012). This might reflect size requirements as the human brain has scaled in a way that allowed for larger extracellular space (Herculano-Houzel 2012; Syková and Nicholson 2008). For example, the ECM molecule Tenascin-C, that in the mouse participates in the creation of the VZ environment but is not expressed in the SVZ parenchyma (Garcion et al. 2004), re-appears in the human SVZ, both in the basal progenitor and the oSVZ compartments (Pollen et al. 2015). The necessity for NSCs to build and/or operate in microenvironments that are specialized according to the future brain structure to be formed has been revealed in another recent transcriptome analysis in the ferret (De Juan Romero et al. 2015). When comparing NSC compartments between areas that give rise to folds or fissures, the expression of genes correlated with cell-to-cell interactions, such as cadherin 8, and with the response to growth factors, such as FGF receptors 2 and 3, was found to be significantly different (De Juan Romero et al. 2015).

6.2.3   Blood Vessels, Systemic Cues and Tissue Mechanics

As mentioned earlier, the CNS is a tissue constructed mostly by neural elements (neurons and glia), with minimal contribution of other tissues, with the exception of blood vessels (BVs). Although at the initial stages of neurogenesis (both in terms of evolution and of development) NSCs become specified and proliferate in the absence of vasculature, BVs become an important component of the NSC microenvironment (this has been reviewed in more detail in Koutsakis and Kazanis (2016)). Notably, although the levels of oxygenation of the embryonic CNS are not known in detail, the culture of embryonic stem cell-derived NSCs in low oxygen (3%; rather than in the usual 20% culture conditions), which is thought to be more relevant to normal situation, is reported to enhance survival and differentiation efficiency (Stacpoole et al. 2011). The processes of neurogenesis and angiogenesis are considered to be "coupled" due to their co-ordination during the seasonal changes in the size of certain nuclei in the brain of songbirds (Louissaint et al. 2002). Furthermore, endothelial cells can enhance neurogenesis in co-culture assays (Androutsellis-Theotokis et al. 2010; Shen 2004). However, the existence of more direct interactions between NSCs and BV mural cells (endothelial cells and pericytes), with functional implications in the developing nervous systems, is only now starting to be elucidated. The vascularisation of the forebrain begins after NEP cells have formed the VZ and it progresses first at the pial surface (around E9) and subsequently at the periventricular domains (around E11) (Tan et al. 2016; Vasudevan et al. 2008). Descriptive analyses have now convincingly shown that RGC basal processes remain in contact with BVs throughout the forebrain (Vasudevan et al. 2008) and that intermediate progenitors, which form the SVZ, are positioned and undergo mitoses in close proximity to BVs (Javaherian and Kriegstein 2009). The latter could be of significance in the rodents, in which intermediate progenitors (a) lack the capacity to produce ECM (in contrast to NSCs of the VZ) (Fietz et al. 2012), (b) lack processes that are in contact with the pial basement membrane. Therefore, the basement membrane of BVs (Tan et al. 2016) might be offering necessary extracellular cues. Recent experimental work revealed that specifically in the ventral telencephalon and in a gradually increasing proportion of RGCs, the basal process does not anchor at the pial surface, but rather to the basement mabrane of

BVs (Tan et al. 2016). Notably, this cell-to-cell communication is mediated by integrin/laminin interactions, as has been described for NSCs of the adult brain (Kazanis et al. 2010; Shen et al. 2008). Disruption of these interactions leads to decreased proliferation, aberrant formation of interneurons and reduced cortical synaptic inhibition (Tan et al. 2016), because interneurons are generated at the ventral telencephalon before migrating and populating the cortex (Alifragis et al. 2004).

The other, obvious, contribution of BVs to the microenvironment of NSCs is the transfer of components of the circulation; primarily oxygen, but also a myriad of other factors. It should be noted that in *Drosophila*, in which NSCs are bathed in the blood precursor called haemolymph (Limmer et al. 2014), the stereotypic successive changes in the expression of different transcription factors, which defines the cellular output of NSCs and which was thought to be entirely cell-autonomous, is partly regulated by ecdysone, a systemic steroid hormone (Syed et al. 2017). The haemolymph also provides insulin-like peptides; some of which are produced locally in the brain and some are of systemic origin (reviewed in Liu et al. (2014)). The same seems to apply for the coupling of the insulin/IGF metabolic pathway to NSC proliferation in mammals, with low quantities of both insulin and IGF-1 and 2 being able to cross the blood brain barrier and the occurrence of local synthesis (Liu et al. 2014). Notably, changes in local IGF expression lead to changes in brain growth and size (D'Ercole et al. 2002). Another family of hormones that was recently revealed to regulate proliferation of NSCs, with an evolutionary twist in their role, are thyroid hormones. Their activity on intermediate progenitors of the SVZ, via integrin $\alpha_v\beta_3$ expressed on their membrane, regulates proliferation; hence, the size of this progenitor pool and the size of neocortex (Stenzel et al. 2014).

During the last decade another area in which increasing amount of work has been performed is the assessment on how the biomechanics of the tissue can influence the behaviour of stem cells (reviewed in detail in Lin et al. (2016)). Results from *in vitro* assays have shown that specification of both embryonic stem cells and human induced pluripotent stem cells towards the neuronal fate is affected by the stiffness of the microenvironment (Keung et al. 2012; Kothapalli and Kamm 2013) and that oligodendrogenesis depends on the mechanical properties of the microenvironment (Jagielska et al. 2012). Importantly, *in vivo* analysis of the developing rodent brain using atomic force microscopy showed significant changes in the stiffness of the tissue over time (gradual increase in the VZ/SVZ and decrease in the cortical plate) and differences between areas (e.g. higher stiffness in the SVZ compared to the VZ) (Iwashita et al. 2014). These data suggest that the triple microenvironment to which NEP cells and RGCs are exposed during cortical development (as described in section 6.2.1) also vary in mechanical properties, although it is very difficult to differentiate if NSC behaviour is affected primarily by these differences, or just by the ECM composition variations that also control mechanical properties (Kothapalli and Kamm 2013). In practical terms, in order to achieve maximal results in directing cell cultures towards the optimal direction, the mechanical properties of the microenvironment (that include stiffness, elasticity, surface topography and composition) have to be combined with micro-patterning with specific growth factors, morphogens and ECM molecules (Kothapalli and Kamm 2013), altogether mimicking the complex *in vivo* situation.

6.3 The NSC Microenvironment of the Postnatal CNS

Contrary to the assumption that neurogenesis in mammals is completed by the end of the embryonic period, experiments continuously demonstrate that small scale NSC activity occurs in the adult brain as well. NSC-driven cell generation in the adult CNS is detected primarily in two specialized microenvironments: the Sub-Ependymal Zone of the lateral walls of the lateral ventricles (SEZ, also known as Ventricular-SVZ) and the Sub-Granular Zone (SGZ) of the dentate gyrus of the hippocampus (general review in (Kazanis 2012, 2013). According to the unified hypothesis of the lineage of NSCs, a fraction of RGCs during early postnatal stages gives rises to adult NSCs (Alvarez-Buylla et al. 2001). The SEZ is populated mostly by NSCs of ventral telencephalon origin, although some cortical origin has also been demonstrated (Willaime-Morawek et al. 2006), while the SGZ is populated by NSCs having migrated from the ventral hippocampus (Li et al. 2013).

6.3.1 In the Subependymal Zone

6.3.1.1 The SEZ System

The SEZ is located at the lateral walls of the lateral ventricles of the brain and NSCs cluster between BVs and the monolayer of ependymal cells that separate it from the ventricle (Fig. 6.1a, c). It contains quiescent NSCs, various progenitor types, differentiated cells and vessels and, due to its specific cytoarchitecture and its anatomical restriction, it is called a neurogenic niche. NSCs of astroglial morphology (also called B1 cells) divide rarely and asymmetrically to self-renew and to generate transit amplifying cells (type C cells) (Doetsch et al. 1999). Type B1 cells are the most multipotent cells of the SEZ and can exist in a quiescent or activated state (Mich et al. 2014; Pastrana et al. 2009). Type C cells then divide symmetrically approximately three times before becoming type A cells (Ponti et al. 2013). The more committed progeny of type C cells are: (a) type A cells (or neuroblasts) that are of neuronal commitment, express doublecortin and PSA-NCAM and migrate as chains towards the olfactory bulbs via the rostral migrtaory stream (RMS) where they differentiate into interneurons and contribute to the olfaction (Lazarini et al. 2014). (b) Oligodendrogenic precursor cells that migrate to the corpus callosum (Etxeberria et al. 2010; Kazanis et al. 2017).

6.3.1.2 Ependyma, Choroid Plexus and Cerebrospinal Fluid

The highly specialized architecture of the SEZ enables specific cell-cell interactions and the local activity of important modulators of NSC behaviour. Type B1 cells have small processes which penetrate the ependymal layer and communicate with the ventricle and longer processes that maintain contact with BVs (Mirzadeh et al.

2008; Tavazoie et al. 2008). We have suggested that ependymal cells are an important component of the niche, since their number is correlated with the number of NSCs and the ependymal expression on noggin (a BMP inhibitor) facilitates neurogenesis (Kazanis and ffrench-Constant 2012; Lim et al. 2000). CSF is constantly produced by the choroid plexus (CP), a monolayer of epithelial cells that lie on a highly vascularized stroma, floating within the brain ventricles (Marques et al. 2016). The adult human has about 150 mL of CSF which is renewed 2–3 times per day and NSCs remain in contact with the CSF which contains multiple soluble factors which modulate NSC properties (Delgado et al. 2014; Lehtinen et al. 2011; Lun et al. 2015; Silva-Vargas et al. 2016). The synthesis of CSF is controlled by mechanisms such as the blood-CSF barrier formed by the CP epithelial cells and the blood flow which in the CP is five times higher than that in the brain parenchyma. Comprehensive analysis of the adult CP transcriptome and secretome, both under physiological conditions and in disease (Marques et al. 2016; Silva-Vargas et al. 2016; Thouvenot et al. 2006), has unraveled the expression of several genes encoding key molecules known to modulate NSCs. Among these are insulin-like growth factor 2, several members of the fibroblast growth factor family, epidermal growth factor, transforming growth factor alpha, platelet-derived growth factors, bone morphogenetic proteins, sonic hedgehog, Wnts and axon guidance molecules such as Slits. Notably, the transcriptome of the CP changes significantly during adulthood; as a consequence, NSC medium conditioned with early postnatal CP is pro-proliferative for NSCs, but becomes anti-proliferative when conditioned with adult CP (Silva-Vargas et al. 2016).

6.3.1.3 Vasculature

As described above, type B1 cells retain a process connecting them with BVs of the neurogenic niche. NSC quiescence is controlled through direct cell contact with endothelial cells, via ephrinB2 and Jagged1 signalling (Ottone et al. 2014) and via diffusible factors such as neurotrophin-3 (Delgado et al. 2014). Moreover, the chemokine stromal-derived factor 1 (SDF-1) is expressed in endothelial cells and recruits type B1 and C cells into the vascular plexus via chemotaxis (Kokovay et al. 2010). Endothelial cells also secrete an EGF-like growth factor, called Betacellulin, which activates EGFRs on type B1 and C cells and ErbB-4 receptors on neuroblasts. It is therefore suggested that betacellulin acts *in vivo* on early as well as later stages of the SEZ neurogenic lineage and its role is the promotion of SEZ cell proliferation and OB neurogenesis (Gómez-Gaviro et al. 2012). Notably, the architecture of the vasculature within the SEZ presents specialized features, when compared with other periventricular areas, such as its higher density, its penetration very proximal to the ependymal layer and its higher levels of leakage (Culver et al. 2013; Kazanis et al. 2010; Tavazoie et al. 2008). Finally, the transit amplifying progenitors in the SEZ, similar to intermediate progenitors in the SVZ of the developing cortex, are positioned and undergo mitosis very close to BVs and especially at domains that lack pericytes and astrocytic endfeet (Shen et al. 2008; Tavazoie et al. 2008).

6.3.1.4 Cell-ECM Interactions

NSCs express low levels of ECM receptors such as integrin-α6β1, syndecan-1 and Lutheran, but their expression rises significantly when they become mitotically active and remains high in proliferating progenitors (Kazanis et al. 2010). When integrin-α6β1 function is disturbed *in vivo* by intracerebroventricular infusion of blocking antibodies, progenitors move away of BVs and their proliferation increases (Kazanis et al. 2010; Shen et al. 2008). We and others have shown that NSCs and local astrocytes produce many of the laminins, the tenascin-C and the chondroitin/dermatan sulfate chains that surround them within the SEZ (Akita et al. 2008; Kazanis et al. 2007, 2010). In addition, highly organized extravascular 3D structures, formed by laminins, collagen IV, heparan sulfates and perlecan, that are called fractones, have been shown to connect the meninges with the ependymal cell layer (Mercier et al. 2002). Fractones are in direct contact with NSCs and their progeny and a significant aspect of their function is to sequester and capture growth factors and morphogens such as FGF2 and BMP-4 (Douet et al. 2013; Kerever et al. 2007; Mercier and Douet 2014). Recently, another ECM component, called anosmin-1, was described to be of importance in regulating NSC proliferation through its binding to FGFR1 (García-González et al. 2016).

When type A cells migrate towards the olfactory bulbs through the rostral migratory stream, they do so within corridors rich in ECM components, such as laminins and tenascin-C, formed by specialized astrocytes and in close contact with BVs (Bovetti et al. 2007b; Todd et al. 2017). Their organized migration is controlled by cell-ECM interactions and is facilitated by the expression of matrix metalloproteinases by neuroblasts (Bovetti et al. 2007a). Defective β1 or β8-integrin function impairs migration (Belvindrah et al. 2007b; Mobley and McCarty 2011), overexpression of anosmin-1 enhances it (García-González et al. 2016) and the interaction between laminin-γ1 chains and soluble netrin-4 (produced by astrocytes) is necessary for the activation of integrin signalling (Staquicini et al. 2009). As soon as neuroblasts reach their target area they migrate tangentially and differentiate into interneurons, a process controlled by another ECM component, Tenascin-R (David et al. 2013). The mess frequent oligodendroglial progenitors produced by NSCs, migrate towards the corpus callosum again in close contact with BVs and under the control of netrin (Cayre et al. 2013).

6.3.1.5 Diffusible Factors

The extremely complicated processes that take place within the SEZ stem cell niche are controlled and co-ordinated by a vast range of factors, operating in small volume and at the same time. In most of the cases we don't know in detail the source of these factors, nor the exact way their activity is tuned with each other. An important morphogen during embryonic development of the nervous system is **Sonic Hedgehog (Shh)**. It has been shown to control proliferation of activated NSCs (in culture it shortens both G_1 and S-G_2/M phases) and by knocking out its receptor

Patched it led to a transient increase in proliferation followed by exhaustion of type B1 cells (Daynac et al. 2016). Shh signalling has been shown to be more significant in two domains of the SEZ: at the dorsal, subcallosal, ventricular wall, where it transiently controls oligodendrogenesis during early postnatal stages (Tong et al. 2015) and at the ventral SEZ where its activity depends on the function of primary cilia (Tong et al. 2014b).

Another family of molecules known to control proliferation and cell fate of stem and progenitor cells during development are **bone morphogenetic proteins (BMP)**. The low density lipoprotein receptor-related protein 2 (LRP2), an endocytic receptor for BMP4, is specifically expressed in ependymal cells of the SEZ and lack of expression in adult mice leads to impaired proliferation of neural progenitors (Gajera et al. 2010). Other experiments revealed that exogenous BMPs limit the EGF-induced proliferation of type C cells, while inhibition of BMP-SMAD signalling promoted activation of quiescent NSCs. Therefore, correct tuning of BMP and EGF activity is necessary for the regulation of NSC quiescence and transit amplifying progenitor mitotic activity (Joppé et al. 2015). Moreover, induced overexpression of Noggin, a secreted BMP inhibitor, in ependymal cells or in NSCs led to enhanced proliferation of transit amplifying progenitors and favoured oligodendrogenesis (Lim et al. 2000; Morell et al. 2015). The binding of BMPs to fractones (exemplified by BMP-4) has been suggested as one mechanism for regulating their local activity (Mercier and Douet 2014). Finally, in the ependymal of the post-injury spinal cord, anchoring of the BMP type 1b receptor subunit into lipid rafts is controlled by β1-integrin; thus, providing a mechanism of regulating BMP-dependent astrogliogenesis by NSCs (North et al. 2015).

In vitro studies have shown that different members of the **int/Wingless (Wnt)** family of morphogens can promote proliferation and either neuronal (Wnt3a, Wnt5a) or oligodendroglial (Wnt3a) differentiation of SEZ-derived NSC cultures (Ortega et al. 2013; Yu et al. 2006) and transcripts for these factors have been identified in the SEZ and adult OB (Shimogori et al. 2004). Notably, the transcription of Axin2, a target of the canonical Wnt signalling pathway, is active in type B, type C, but not type A cells; hence, suggesting distinct functions of Wnt signalling in different progenitors (Adachi et al. 2007). Non-canonical Wnt signalling has also been shown to function in the SEZ, since diversin is expressed in type A cells and its overexpression increases their proliferation (Ikeda et al. 2010).

Several **Eph** tyrosine kinase receptors and their **ephrin** ligands, as well as multiple components of the Notch signaling are present in the SEZ. Holmberg and colleagues (Holmberg et al. 2005) identified an ephrin-A2/Eph A7 feedback system, operating between progenitors of different maturity in the SEZ in order to control proliferation and cell generation capacity. Ephrin-A and B signaling has been shown to positively regulate NSC proliferation within the SEZ as well as the cellular architecture and the migration of neuroblasts in the RMS (Conover et al. 2000; Todd et al. 2017). In 2008, the intriguing ability of SEZ NSCs to regenerate low levels of ependymal cell damage was revealed (Luo et al. 2008). Few years later, EphB2 signaling was found to mediate this type of plasticity, with astrocytes becoming ependymal cells and vice versa (Nomura et al. 2010). Recently, the Eph/ephrinB2 and **Notch** signaling was also shown to regulate quiescence of NSCs at the cell contact between NSCs and endothelial cells, mediated by Jagged1 expression on endothelial cells (Ottone et al. 2014). Canonical Notch signaling has been shown to be crucial for regulating quiescence and activation of NSCs by additional experimental work. Deletion of Rbpj forced NSCs to progress in the lineage producing type C and A cells, eventually depleting their pool (Imayoshi et al. 2010). Moreover, conditional deletion of Notch1 affected activated NSCs, but spared quiescent NSCs (Basak et al. 2012). An interesting link between notch signalling and the activity of PEDF, a factor well-described to affect NSC self-renewal in the SEZ (Ramírez-Castillejo et al. 2006), was reported with PEDF inducing symmetric cell divisions downstream of notch (Andreu-Agulló et al. 2009).

An important aspect of the neurogenic activity of the SEZ is the migration of neuroblasts to the olfactory bulbs. The direction of movement of neuroblasts is guided by the direction of movement of the CSF, controlled by ependymal cilia (Sawamoto et al. 2006). Netrins have been identified as potent modulators of neuroblast migration, since netrin-1 is expressed by olfactory bulb cells and netrin receptors, such as neogenin and DCC, are expressed on type A cells (Astic et al. 2002; Murase and Horwitz 2002). Although the regulation of neuroblast migration and differentiation is still largely unknown, neogenin has also been shown to synchronize their migration and their terminal differentiation by affecting cell cycle kinetics, similar to cannabinoids' activity through the PKC-dependent phosphorylation of fascin (O'Leary et al. 2015; Sonego et al. 2013).

6.3.1.6 Neurotransmitters and Neuromodulators

Because cytogenesis in the SEZ is known only to supply new neurons and oligodendrocyte progenitors in the olfactory bulbs and the corpus callosum, respectively, neurotransmission wasn't expected to play a significant regulatory role. Nevertheless, augmenting evidence suggests the contrary. An interesting hypothesis is that **dopaminergic** regulation of proliferation in the SEZ allows the niche to communicate with the periphery. A downside of this system is that in pathologic conditions of loss of dopaminergic innervation, as observed in Parkinson's disease, NSC self-renewal and progenitor proliferation is disturbed (O'Keeffe et al. 2009b). Dopaminergic regulation appears to be EGF-dependent and FGF2-independent (O'Keeffe et al. 2009b), and most of the experimental data converge on a key role of D1, D2 and D3 receptors (Kim et al. 2010; Lao et al. 2013; O'Keeffe et al. 2009a). Specifically for D3 receptors, they are specifically expressed on type C cells (Kim et al. 2010) and their activation induces NSC self-renewal and type-C cell generation through Akt and ERK1/2 signaling (Lao et al. 2013). A role for **cholinergic** neurotransmission was also suggested when researchers identified a population of choline acetyltransferase (ChAT) positive neurons in the rodent SEZ with a morphology distinct to those of the striatum (Paez-Gonzalez et al. 2014). Neuroblast generation could be modulated using optogenetic tools in order to induce or block cholinergic activity, and it was shown that response to Ach was mediated through FGFRs (Paez-Gonzalez et al. 2014). Recently, cholinergic stimulation was shown to control the response of SEZ cells to experimental stroke (Wang et al. 2017). **Serotonergic** axons had been identified near the walls of the ventricles since the late 90's, described also to be in close association with ependymal cells (Mathew 1999). These observations were confirmed recently and serotonin receptors 2C and 5A were found to be expressed by type B cells. The intracerebroventricular infusion of 2C agonists and antagonists altered the proliferation of NSCs in the SEZ (Tong et al. 2014a) and the chronic activation of 1A and 2C receptors induced proliferation in the olfactory bulbs (Soumier et al. 2010).

Finally, the best described neurotransmitter to have a role in regulating neurogenesis in the SEZ is **gamma-amino butyric acid (GABA)** . In the postnatal SEZ, young neuroblasts spontaneously release GABA, which activates $GABA_\alpha$ receptors and depolarizes precursor cells resulting in the inhibition of cell proliferation and neuronal differentiation through the recruitment of the PI3K-related kinase signaling pathway and histone H2AX phosphorylation. Surprisingly, these changes can lead to long-lasting changes in stem cell numbers, the niche size, and neuronal output (Fernando et al. 2011). A year later a negative-feedback mechanism with which neuroblasts restrict their own production was described to involve GABAergic inhibition (Alfonso et al. 2012), since type B and C cells were shown to secrete the diazepam-binding inhibitor protein (DBI) which acts as a positive modulator of SEZ postnatal proliferation and neurogenesis by competitively inhibiting GABA binding to its receptors. Subsequently, the existence of significant heterogeneity in regard to GABA receptor subunit composition throughout the human SEZ was found (Dieriks et al. 2013). For example, expression of GABA(A)R α_2 and γ_2 units was specifically detected on cells proximal to large SEZ BVs, where the SEZ was much thicker.

6.3.2   In the Subgranular Zone (SGZ)

The neurogenic region of the hippocampus is restricted to the subranular zone (SGZ) close to the dentate gyrus (Kazanis 2012, 2013). Similarly to the SEZ, the SGZ is characterized by the existence of relatively quiescent NSCs of astroglial morphology that generate precursors of neuronal commitment (neuroblasts) which migrate to their target area in the granule neuron layers to differentiate in mature neurons, which integrate in an already existent neuronal network. The neurogenic system of the SGZ does not form an anatomically separate structure. However, NSCs are confined in a restricted area, surrounded by intermediate progenitors, astrocytes, mature granule cells and blood vessels; thus, the SGZ is also referred to as a stem cell niche (Palmer et al. 2000) (Fig. 6.2b). In the SGZ the number of cells being produced is smaller compared to that of the SEZ (Kazanis 2013), but their function is significantly different as they contribute to processes such as memory and learning. This is why impaired neurogenesis has been linked to the development of mental health and cognitive disorders (Aimone et al. 2011; Noonan et al. 2010; Sahay et al. 2011).

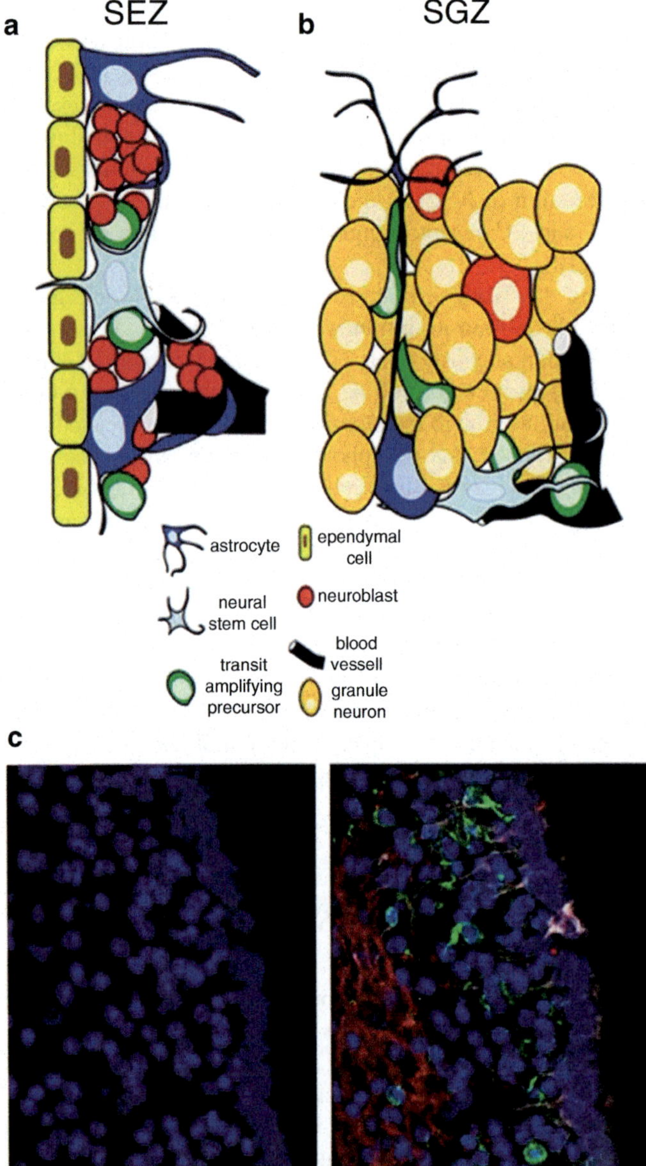

Fig. 6.2 Schematic illustrations of the cyto-architecture of the SEZ (in **a**) and of the SGZ (in **b**). Note the NSCs of astroglial morphology (*light blue cells*) and the presence of blood vessels in these neurovascular niches. (**c**) High magnification image of the human SEZ taken from a newborn baby. The ependymal cell layer is at the right of the image (*in white*), Dcx + neuroblasts are at the centre (*in green*) and an astrocyte-rich ribbon area at the left (GFAP+ cells *in red*); nuclei are counterstained with Dapi (*in blue*). [Schemas in A, B reproduced from: The neural stem cell microenvironment (August 31, 2008), StemBook, ed. The Stem Cell Research Community, StemBook, doi/https://doi.org/10.3824/stembook.1.15.1,   http://www.stembook.org,   under   a   Creative Commons Attribution 3.0 Unported License]

6.3.2.1   Cell-Cell and Cell-ECM Contact

The ECM composition of the SGZ has not been thoroughly investigated so far. *In vitro* assays of NSCs isolated from the rat hippocampus have revealed that they are able to produce their own ECM (laminin, fibronectin) and their adhesion to different substrates is controlled by integrins (Harper et al. 2010). The role of β1-integrin in the SGZ seems to mimic aspects described during development and in the SEZ. It is necessary for the structural integrity of the niche, for controlling proliferation of NSCs and for inhibiting astroglial differentiation (Brooker et al. 2016). Recent experimental work also indicated that the large ECM glycoprotein Reelin, which is mostly known to regulate migration of progenitors (Courtès et al. 2011) plays a role in controlling quiescence of NSCs, but is dispensable for progression into the lineage (Sibbe et al. 2015). Finally, it should be noted that the SGZ is a neurovascular niche (Palmer et al. 2000), with abundant BVs that offer access to an ECM rich basement membrane.

6.3.2.2   Diffusible Factors

NSCs isolated from the adult hippocampus can be kept in culture in the form of neurospheres (Fig. 6.1) , similar to NSCs of the embryonic brain and of the SEZ. These cells are kept in medium rich in FGF2 and EGF. Recently, it was found that hippocampal NSCs produce their own **ciliary neurotrophic factor (CNTF)**, which pushes them towards differentiation. In contrast, the presence of FGF2 down-regulates the expression of CNTF and inhibits spontaneous differentiation in vitro (He et al. 2012). The role of **Wnt signalling** has been investigated *in vitro* and *in vivo*, and the intriguing finding was that distinct functional outputs were generated depending on whether the Wnt/β-catenin, or the Wnt/Planar cell polarity pathway was activated. The first is crucial for cell fate determination and the latter for morphological maturation of neuroblasts (Schafer et al. 2015). NSCs and their daughter progenitor cells also express the Wnt receptor Frizzled-1 (FZD1). Conditional knockdown of FZD1 resulted in a range of defects, including increased astroglial differentiation in expense of neurogenesis and increased migration of newborn neurons (Mardones et al. 2016). Finally, interesting results have been generated by investigating **BMP** signalling in the hippocampal cytogenic niche. Expression of BMP target genes, such as ID3 revealed low levels of BMP activity in proliferating cells. In addition, overexpression of noggin into granule cells of the dentate gyrus resulted in increased proliferation of NSCs (Bonaguidi et al. 2008). Furthermore, ablation of BMP receptor–II enhanced proliferation and maturation of NSCs and of neuroblasts, whilst overexpression of BMP-4 caused cell cycle exit (Bond et al. 2014).

6.3.2.3   Metabolic Regulation

The contribution of hippocampal neurogenesis to learning, memory and mental health in general has invited high interest in the mechanisms by which the peripheral body conditions can be coupled with NSC activity (reviewed in Triviño-Paredes et al. (2016)). Nevertheless, it is still early days and no solid hypotheses can be formed. Many factors such as BDNF and orexin-A have been identified (Chieffi et al. 2017), but much more details have been generated regarding the role of oxygen. Low O_2 tensions induce expression of hypoxia-inducible factors (HIFs) which are heterodimers consisting of a constitutively expressed β-subunit and an oxygen-regulated α-subunit and regulate oxygen homeostasis and direct molecular responses to hypoxia (Semenza 2012). HIF1α promotes self-renewal (via Notch signalling) and blocks apoptosis and differentiation (Panchision 2009). Indeed, deep within the nervous tissue, and especially in the SGZ, normally hypoxic conditions seem to enhance Wnt/β-catenin signalling that—as described above- is essential for NSC self-renewal and lineage progress (Mazumdar et al. 2010). As soon as neuroblasts migrate away from this hypoxic microenvironment they become vulnerable to oxidative stress and apoptosis occurs. If HIF1α becomes stabilized, then apoptotic cell death is significantly reduced (Chatzi et al. 2015). Finally, a mechanism coupling levels of exogenous glucose to NSC activity was described. High glucose levels lead to reduced Hes-1 expression, through reduced CREB activity, and subsequent inactivation of NSCs; the same effect being achieved in CREB-deficient NSCs (Fusco et al. 2016).

6.3.2.4 Neurotransmission, Neuropeptides-Neurohormones

Various neurotransmitters and neurohormones have been identified as significant modulators of neurogenesis in the hippocampus. **Glutamate**, the major excitatory neurotransmitter, can act on NSCs mostly via the metabotropic glutamate receptor 5 (mGluR5) that is highly expressed on them. When norbin, which is a positive regulator of mGluR5, was knocked-out it led to impaired proliferation and maturation of newborn neurons without affecting cell-fate specification (Wang et al. 2015). Because norbin is expressed in granule cells, its effect seems to be non-cell autonomous for NSCs and is probably mediated by cell-cell contact. Importantly, norbin-deficient mice exhibit depressive-like behaviour. Moreover, a newly identified target of neural activity and especially of glutamatergic neurotransmission is the family of BMP/RA-inducible neural-specific proteins (BRINP) (Motomiya et al. 2007). BRINP-1 expression is increased in response to kainic acid stimulation of the hippocampus and in BRINP-1 deficient mice neurogenesis is significantly deregulated with the generation of more immature neurons and the emergence of symptoms relevant to human mental disorders such as schizophrenia and attention-deficit/hyperactivity disorder (Kobayashi et al. 2014). Another neurotransmitter that has been directly linked to NSC activity is **norepinephrine**, acting through β3 adrenergic receptors that are specifically expressed in Hes5-expressing progenitors in the SGZ. Notably, serotonin was found not to exert any effects of NSCs of the SGZ (Jhaveri et al. 2010).

Additional indications for the role of peripheral cues, such as of **neurohormones**, have been recently provided in cell culture assays. Ghrelin, which acts primarily in the hypothalamus but it has been also observed in other regions such as the hippocampus, exerts a mitogenic effect in cultured hippocampal NSCs. This is due to the enhanced progression from G1 to S phase mediated by the enhanced nuclear expression of E2F1 and the balanced activity on positive and negative cell cycle regulators (Chung and Park 2016). Finally, melatonin, which is secreted by the pineal gland and has a remarkably pleiotropic nature is able to enhance the proliferation of adult hippocampal NSC *in vitro*. Melatonin exerted its action via the melatonin receptor, a G-protein coupled-receptor, and resulted in phosphorylation of Raf, MEK, and ERK1/2 (Tocharus et al. 2014).

6.3.3 *Hypothalamus*

Although the main and better described neurogenic niches in the adult mammalian brain are located in the SEZ of the lateral ventricle and the SGZ of the hippocampal dentate gyrus, there is increasing evidence for adult neurogenesis also in the hypothalamus (Batailler et al. 2016; Kokoeva et al. 2007), a brain region that is known for its key role in the neuroendocrine regulations including reproduction, metabolism or food intake. Recent lineage-tracing studies have convincingly demonstrated the existence of neurogenic cells, capable of generating mainly neurons, within the pool of tanycytes, the non-ciliated ependymoglial cells lining the ventral part of the third ventricle (3V) (Haan et al. 2013; Lee et al. 2012; Robins et al. 2013). In light of the function of the hypothalamus, cytogenesis within this niche is expected to be regulated by extrinsic cues, linking it with metabolic and other behavioural conditions (Lee et al. 2012; Migaud et al. 2015; Perera et al. 2011). Until now, though, only limited progress has been made towards elucidating this regulation. Tanycytes are expressing FGF10; hence, suggesting a role of this growth factor in the hypothalamic niche (Haan et al. 2013). Other growth factors that when administered intracerebroventricularly have been shown to induce neurogenesis in the hypothalamus are BDNF (Pencea et al. 2001) and CNTF (Belsham et al. 2009). Finally, in sheep that are a seasonal species, neurogenesis in the hypothalamus shows response to seasonal stimuli (Batailler et al. 2016).

6.4 The NSC Microenvironment in the Aged and the Pathological Brain

6.4.1 *Ageing, Degeneration and the Cancer Stem Cell Hypothesis*

The NSC microenvironment changes significantly during ageing; however, very little is known about the details of these changes and on how they might affect NSCs. Within the ageing SEZ niche, the ECM structures called fractones seem to change (Kerever et al. 2015): their numbers are decreased but their size increases and their composition in heparan sulfate chain is altered; however, FGF2 capture, one major functional aspect, remains unchanged. The CNS microenvironment is known to change, especially in terms of its ECM components, in different pathological conditions such as Alzheimer's and vascular dementia (Lepelletier et al. 2017; Rosenberg 2017), autism (Mercier et al. 2011), as well as after stroke (Haddock et al. 2007). At the same time, accumulating evidence suggests that in such cases of pathology the behaviour of NSCs changes (Curtis et al. 2007; Darsalia et al. 2005, 2007; Ekonomou et al. 2010; Yamashita et al. 2006; Zhang et al. 2014) and much more intriguingly latent progenitors seem to be activated (Florio et al. 2015; Magnusson et al. 2014; Sirko et al. 2013). These data, in combination with the exciting finding that adult brain NSCs can be rejuvenated by systemic factors (Katsimpardi et al. 2014; Ruckh et al. 2012), support the expectation that, by deciphering the cross-talk between aged or disease-affected NSC with their microenvironment, this will lead to major breakthroughs in regenerative medicine.

Cancer stem cells (CSCs) exhibit tumor initiation potential (Deleyrolle et al. 2011; Piccirillo et al. 2012). CSCs express markers such as CD133+ (or prominin) and Nestin and are found in various brain cancers (glioblastoma, medulloblastomas, ependymomas and oligodendrogliomas) in close proximity with the aberrant tumor vasculature. Their interaction with endothelial cells seems to be crucial in maintaining the CSC pool (Bao et al. 2006; Calabrese et al. 2007), while various microenvironmental factors provide instructive cues that modulate CSC behavior (Lathia et al. 2011) using mechanisms such as cell-to-ECM interactions, growth factor signaling and cell-to-cell communication (Gilbertson and Rich 2007).

6.4.2 Glioma/Glioblastoma (GBM)

6.4.2.1 CSCs-ECM Interactions

Significant signaling crosstalk exists between ECM components and membrane proteins expressed on CSCs and vascular cells, for example between the laminin family of proteins and their receptors integrins. Integrins $\alpha v\beta 3$ and $\alpha 6\beta 1$, are enriched in CSCs located in the perivascular niche and normally in direct contact with endothelial cells and are essential for their proliferation and self-renewal while targeting of integrin $\alpha 6$ reduces tumor formation potential (Burgett et al. 2016; Lathia et al. 2010). Activation of integrin-mediated phosphoinositide 3-kinase (PI3-K) survival signaling has also been reported to facilitate glioma cell migration (Joy et al. 2003), while repression of integrin $\alpha 6$ promoter inhibits stemness of glioblastoma cells (Ying et al. 2014) via reduced expression by Kruppel-like transcription factor 9 (KLF9). Moreover, the pro-oncogenic activity of integrin $\alpha 6$ stems from its association with N-cadherin in order to achieve translocation of extracellular signal-regulated kinase (ERK) to the nucleus (Velpula et al. 2012). The ERK pathway has also been implicated in the promotion of glioma invasion properties through the expression of integrin α3 (overexpressed in invading CSCs and perivascular tumor cells) (Nakada et al. 2013). Another mechanism by which integrins (namely α6β1) inhibit pro-apoptotic cascades initiated by tumor necrosis factor (TNF) receptor 1 (TNF-R1), when the cells are attached to laminin, is the induction of the expression of the cellular FLICE inhibitory protein (cFLIP) (Huang et al. 2012). The production of laminins and other ECM components within the perivascular glioblastoma niche is dependent mainly on non-CSCs and endothelial cells (laminin α2, fibronectin and collagen IV) while CSCs express low levels of α2, α3 and α5 laminin chains (Lathia et al. 2012).

Tenascin-C extracellular matrix glycoprotein is another main component of the ECM of malignant brain neoplasms and is overexpressed in GBM tumors where it exhibits a strong association with cells in the perivascular compartment (Behrem et al. 2005; Brösicke et al. 2013; Lathia et al. 2012). Tenascin-C is secreted by tumor cells (Hirata et al. 2009) and has been characterized to be inducing neural progenitor cell migration (Ziu et al. 2006), as well as to have anti-anchorage effects on endothelial cells, visualized by distinct blood vessel characteristics (Xia et al. 2016). Its link with increased motility/migration of GSCs *in vitro* and *in vivo* is the anti-adhesive role it exerts through the modulation of the focal adhesion kinase pathway (Xia et al. 2016). Glioma invasiveness has also been attributed to increased expression and proteolytic activity of ADAM9 (a disintegrin and metalloproteinase 9) which involves the mitogen-activated protein kinase 8 pathway (Sarkar et al. 2015). Metalloproteinases are logical candidates for regulating invasiveness of tumor cells since ECM remodeling is a necessary condition; however, two other members of the ADAM family, ADAM10 and ADAM17 have also been shown to promote self-renewal of the GSC population (Bulstrode et al. 2012).

In addition to the biochemical cues provided by the ECM, mechanotransduction has been recognized as a pivotal element in cell-ECM interactions, modulating the tumorigenic phenotype because the rigidity of the microenvironment impacts tumor cell migration/proliferation properties. As ECM rigidity increases (stiff ECMs), glioblastoma cells migrate extensively and proliferate rapidly, while the opposite behavior is observed in soft ECMs. This efficient rigidity-sensing and propulsive force-generating system of tumor cells involves components of the cytoskeleton like myosin II (Umesh et al. 2014), α-actinins 1 and 4 (Sen et al. 2009), linkers to the ECM such as talin-1 (Sen et al. 2011) as well as sophisticated regulation of epidermal growth factor receptor signaling (Umesh et al. 2014). Surprisingly, CSCs are generally insensitive to mechanical inputs and don't present the typical non-motile phenotype in compliant ECMs as they generate low levels of myosin-dependent contractile force. This phenomenon is reversed and activation of myosin-dependent contractility through RhoA GTPase, Rho-associated kinase (ROCK), or myosin light chain kinase (MLCK) restores stiffness-dependent migration, leading to a less invasive phenotype and prolonged survival in mice orthotopically implanted with such CSCs (Wong et al. 2015).

6.4.2.2 CSCs-Vasculature Interactions

Extensive interplay between CSCs and endothelial cells (ECs) has been identified, one example already being examined regarding the generation of laminin α2 from endothelial cells which can bind to integrin receptors on the CSCs. EC mediated pathophysiological processes such as angiogenesis and hallmark functions of CSCs such as self-renewal are interconnected and elimination of endothelial cells can lead to detrimental effects for GSCs, driven in part from Notch signaling (Hovinga et al. 2010). The endothelial secretome can preserve CSCs properties, promoting mTOR-dependent survival (Maria Galan-Moya et al. 2011) or activating Notch signaling via nitric oxide (NO) produced by nitric oxide synthase 3 (NOS3) (Charles et al. 2010). High levels of NO can also be produced endogenously in GSCs, which have elevated nitric oxide synthase 2 (NOS2) expression, compared with non-GSCs and normal neural progenitor cells, thus promoting their tumorigenic profile (Eyler et al. 2011). Vice versa, CSCs produce proangiogenic factors such as vascular endothelial growth factor (VEGF) and stromal-derived factor 1 (SDF1), boosting EC activation and migration and supporting angiogenesis/vasculogenesis (Bao et al. 2006; Folkins et al. 2009; Oka et al. 2007; Ping et al. 2011). VEGF secretion can be promoted by the chemokine CXCL12 (SDF1), whose receptor's (CXCR4) expression is elevated in GSCs and activates a PI3K/Akt dependent pathway (Ping et al. 2011). Interestingly, VEGF is part of an autocrine VEGF–VEGFR2–Neuropilin-1 signaling pathway, indicating that CSCs can regulate, sustain and promote their own growth rate through the expression of vascular endothelial growth factor receptor 2 (VEGFR2) (Hamerlik et al. 2012).

In addition to the effect of secreted factors exchanged between ECs and CSCs, direct interactions between these two distinct cell types have been studied. Notch signaling plays a pivotal role in EC-CSC crosstalk and can be activated through Notch ligand independent way (e.g. nitric oxide) or through ligand dependent manner. Nestin is co-expressed with Notch receptors, Notch1 and Notch2, on CSCs which have elevated levels of Notch activity in conjunction with constitutive activation of the STAT3/NF-κB signaling pathway and up-regulation of STAT3- and NF-κB-dependent genes (Garner et al. 2013; Zhu et al. 2011). On the other hand, ECs express Notch ligands Delta-like 4 (DLL4) and Jagged-1 (JAG1) and enhance the self-renewal of adjacent Notch receptor-expressing stem cells properties through juxtacrine signaling (Zhu et al. 2011). Recently, the binding of integrin αvβ3, expressed on ECs to the RGD-peptide in the extracellular domain of L1CAM on CSCs, was shown to mediate an FGF2-induced cascade involving the activation of BMX, FAK, p130CAS and the downstream effectors ERK and JNK. It resulted in enhanced formation of ECs networks and their chemotactic attraction by FGF2; thereby, promoting angiogenesis (Burgett et al. 2016). Notch signaling has also been linked to glioblastoma radioresistance and the utilization of gamma-secretase inhibitors (GSIs) attenuates this protective effect, sensitizing CSCs to radiation damage (Wang et al. 2010). Finally, junctional adhesion molecule A (JAM-A) is expressed on CSCs and becomes suppressed during differentiation; being essential as an adhesion factor for cell maintenance (Lathia et al. 2014). JAM-A is a target of miR-145 (negative regulator), which is downregulated in CSCs (Alvarado et al. 2016). High JAM-A/low miR-145 expression signature has significant predictive value and presents clinical relevance as it indicates poor patient survival (Alvarado et al. 2016).

6.4.2.3 Diffusible Factors

A key factor of the glioma micro-environment in initiating tumorigenesis by GSCs is **Transforming Growth Factor-beta (TGFβ)**. TGFβ induces the self-renewal capacity of GSCs through the Smad-dependent induction of leukemia inhibitory factor (LIF) and the subsequent activation of the JAK-STAT pathway (Peñuelas et al. 2009). Moreover, TGFβ activity is mediated by the ID family of helix-loop-helix proteins (Anido et al. 2010). Expression levels of IDs are higher in tumor cells and ECs of higher grade glial neoplasms correlate with proliferation and angiogenesis patterns (Vandeputte et al. 2002). ID4 transforms astrocytes into a stem-cell like state through Notch signaling (Jeon et al. 2008) and ID3 promotes GSCs features in astrocytes (Jin et al. 2011). In addition, pAKT-pSmad5 signaling-driven ID3 induction by activation of **EGF signaling** along with the subsequent production of ID3-regulated cytokines (GRO1, **interleukin-6 (IL-6) and interleukin-8 (IL-8)**, enhances tumor formation, expansion and heterogeneity (Jin et al. 2011). Finally, CSCs are able to protect themselves from endogenous **BMPs** via the expression of the BMP antagonist Gremlin1. Gremlin1 blocks the pro-differentiation effects of BMPs via the p21WAF1/CIP1 complex and overexpression of Gremlin1 in non-CSCs decreases their endogenous BMP signaling to promote stem-like features (Yan et al. 2014).

6.4.2.4 Hypoxia

As with NSCs during development, in the adult brain hypoxic micro-environmental conditions maintain and expand the GSC phenotype (Bar et al. 2010; Heddleston et al. 2009; Li et al. 2009; Seidel et al. 2010; Soeda et al. 2009). HIFs have been shown to facilitate GSC survival and self-renewal (Li et al. 2009; Soeda et al. 2009). HIF2α induces VEGF expression exclusively in the stem cell subpopulation (Li et al. 2009; Seidel et al. 2010), whereas HIF1α transcriptionally regulates the VEGF promoter in GSCs and non-GSCs cells (Li et al. 2009) and represses Neuropilin-2 (NRP2) in non-GSC cells, increasing paracrine VEGF-induced EC function and inhibiting the potent anti-tumorigenic activity of semaphorin 3F (SEMA3F) (Coma et al. 2011). By stabilizing the intracellular domain of Notch, HIF1α synergizes with Notch and drives CSC maintenance (Qiang et al. 2012). Targeting HIF isoforms results in anti-proliferative and anti-tumorigenic effects while reducing cell migration/invasion in vitro and in vivo (Li et al. 2009; Méndez et al. 2010; Seidel et al. 2010).

6.4.3   Other Brain Cancer Types

Ependymoma and medulloblastoma CSCs interact with ECs in matrigel cultures and promote EC network formation (Calabrese et al. 2007). Upregulation of laminin α2 has been detected in a distinct group of invasive and metastatic posterior fossa ependymomas, linked to poor patient survival (Witt et al. 2011). Tumor perivascular cells in medulloblastoma were resistant to radiation-induced damage through the activation of the PI3K/Akt pathway (Hambardzumyan et al. 2008).

6.5   New Relevant Research Trends

There are three fields that are currently at the forefront of research but haven't produced solid results, yet. Firstly, the regulation of NSCs by exosomes released in the microenvironment by proximal cells (reviewed in Bátiz et al. (2016). Secondly, the cross-talk between the immune system and NSCs, that was initially thought to be one-way (immune cells to NSCs) (Pluchino et al. 2005), but seems to be bi-directional (Cossetti et al. 2014; Pluchino and Cossetti 2013). Thirdly, the effects of the gut microbiota to brain development and adult neurogenesis (Fung et al. 2017; Heberden 2016; Heijtz et al. 2011).

Acknowledgements IK was supported by funding from Action Medical Research, UK (GN2291). The authors would like to thank Ms. Kassandra- Danai Meri for providing the original image in Fig. 6.1e.

References

Adachi K, Mirzadeh Z, Sakaguchi M, Yamashita T, Nikolcheva T, Gotoh Y et al (2007) Beta-catenin signaling promotes proliferation of progenitor cells in the adult mouse subventricular zone. Stem Cells 25:2827–2836. https://doi.org/10.1634/stemcells.2007-0177

Aimone JB, Deng W, Gage FH (2011) Resolving new memories: a critical look at the dentate gyrus, adult neurogenesis, and pattern separation. Neuron 70:589–596. https://doi.org/10.1016/j.neuron.2011.05.010

Akita K, von Holst A, Furukawa Y, Mikami T, Sugahara K, Faissner A (2008) Expression of multiple chondroitin/dermatan sulfotransferases in the neurogenic regions of the embryonic and adult central nervous system implies that complex chondroitin sulfates have a role in neural stem cell maintenance. Stem Cells 26:798–809. https://doi.org/10.1634/stemcells.2007-0448

Alfonso J, Le Magueresse C, Zuccotti A, Khodosevich K, Monyer H (2012) Diazepam binding inhibitor promotes progenitor proliferation in the postnatal SVZ by reducing GABA signaling. Cell Stem Cell 10:76–87. https://doi.org/10.1016/j.stem.2011.11.011

Alifragis P, Liapi A, Parnavelas JG (2004) Lhx6 regulates the migration of cortical interneurons from the ventral telencephalon but does not specify their GABA phenotype. J Neurosci 24:5643–5648. https://doi.org/10.1523/JNEUROSCI.1245-04.2004

Alvarado AG, Turaga SM, Sathyan P, Mulkearns-Hubert EE, Otvos B, Silver DJ et al (2016) Coordination of self-renewal in glioblastoma by integration of adhesion and microRNA signaling. Neuro Oncol. https://doi.org/10.1093/neuonc/nov196

Alvarez-Buylla A, García-Verdugo JM, Tramontin AD (2001) A unified hypothesis on the lineage of neural stem cells. Nat Rev Neurosci 2:287–293. https://doi.org/10.1038/35067582.

Andreu-Agulló C, Morante-Redolat JM, Delgado AC, Fariñas I (2009) Vascular niche factor PEDF modulates Notch-dependent stemness in the adult subependymal zone. Nat Neurosci 12:1514–1523. https://doi.org/10.1038/nn.2437

Androutsellis-Theotokis A, Rueger MA, Park DM, Boyd JD, Padmanabhan R, Campanati L et al (2010) Angiogenic factors stimulate growth of adult neural stem cells. PLoS One 5:e9414. https://doi.org/10.1371/journal.pone.0009414

Anido J, Sáez-Borderías A, Gonzàlez-Juncà A, Rodón L, Folch G, Carmona MA et al (2010) TGF-β receptor inhibitors target the CD44high/Id1 high glioma-initiating cell population in human glioblastoma. Cancer Cell. https://doi.org/10.1016/j.ccr.2010.10.023

Anton ES, Kreidberg JA, Rakic P (1999) Distinct functions of alpha3 and alpha(v) integrin receptors in neuronal migration and laminar organization of the cerebral cortex. Neuron 22:277–289

Astic L, Pellier-Monnin V, Saucier D, Charrier C, Mehlen P (2002) Expression of netrin-1 and netrin-1 receptor, DCC, in the rat olfactory nerve pathway during development and axonal regeneration. Neuroscience 109:643–656

Bao S, Wu Q, Sathornsumetee S, Hao Y, Li Z, Hjelmeland AB et al (2006) Stem cell-like glioma cells promote tumor angiogenesis through vascular endothelial growth factor. Cancer Res. https://doi.org/10.1158/0008-5472.CAN-06-1010

Bar EE, Lin A, Mahairaki V, Matsui W, Eberhart CG (2010) Hypoxia increases the expression of stem-cell markers and promotes clonogenicity in glioblastoma neurospheres. Am J Pathol 177:1491–1502. https://doi.org/10.2353/ajpath.2010.091021

Barnabé-Heider F, Wasylnka JA, Fernandes KJL, Porsche C, Sendtner M, Kaplan DR et al (2005) Evidence that embryonic neurons regulate the onset of cortical gliogenesis via cardiotrophin-1. Neuron 48:253–265. https://doi.org/10.1016/j.neuron.2005.08.037

Basak O, Giachino C, Fiorini E, MacDonald HR, Taylor V (2012) Neurogenic subventricular zone stem/progenitor cells are notch1-dependent in their active but Not quiescent state. J Neurosci 32:5654–5666. https://doi.org/10.1523/JNEUROSCI.0455-12.2012

Batailler M, Derouet L, Butruille L, Migaud M (2016) Sensitivity to the photoperiod and potential migratory features of neuroblasts in the adult sheep hypothalamus. Brain Struct Funct 221:3301–3314. https://doi.org/10.1007/s00429-015-1101-0

Bátiz LF, Castro MA, Burgos PV, Velásquez ZD, Muñoz RI, Lafourcade CA, et al (2016) Exosomes as novel regulators of adult neurogenic niches. 9, 1–28. doi:https://doi.org/10.3389/fncel.2015.00501.

Behrem S, Zarković K, Eskinja N, Jonjić N (2005) Distribution pattern of tenascin-C in glioblastoma: correlation with angiogenesis and tumor cell proliferation. Pathol Oncol Res. doi:PAOR.2005.11.4.0229

Belsham DD, Fick LJ, Dalvi PS, Centeno M-L, Chalmers JA, Lee PKP et al (2009) Ciliary neurotrophic factor recruitment of glucagon-like peptide-1 mediates neurogenesis, allowing immortalization of adult murine hypothalamic neurons. FASEB J 23:4256–4265. https://doi.org/10.1096/fj.09-133454

Belvindrah R, Graus-Porta D, Goebbels S, Nave K-A, Muller U (2007a) Beta 1 integrins in radial glia but not in migrating neurons are essential for the formation of cell layers in the cerebral cortex. J Neurosci 27:13854–13865. https://doi.org/10.1523/JNEUROSCI.4494-07.2007

Belvindrah R, Hankel S, Walker J, Patton BL, Müller U (2007b) β1 Integrins control the formation of cell chains in the adult rostral migratory stream. J Neurosci 27:2704–2717. https://doi.org/10.1523/jneurosci.2991-06.2007

Bjornson CR, Rietze RL, Reynolds BA, Magli MC, Vescovi AL (1999) Turning brain into blood: a hematopoietic fate adopted by adult neural stem cells in vivo. Science 283:534–537

Bonaguidi MA, Peng C-Y, McGuire T, Falciglia G, Gobeske KT, Czeisler C et al (2008) Noggin expands neural stem cells in the adult hippocampus. J Neurosci 28:9194–9204. https://doi.org/10.1523/JNEUROSCI.3314-07.2008

Bond AM, Peng C-Y, Meyers EA, McGuire T, Ewaleifoh O, Kessler JA (2014) BMP signaling regulates the tempo of adult hippocampal progenitor maturation at multiple stages of the lineage. Stem Cells 32:2201–2214. https://doi.org/10.1002/stem.1688

Borrell V, Calegari F (2014) Mechanisms of brain evolution: regulation of neural progenitor cell diversity and cell cycle length. Neurosci Res. https://doi.org/10.1016/j.neures.2014.04.004

Bovetti S, Bovolin P, Perroteau I, Puche AC (2007a) Subventricular zone-derived neuroblast migration to the olfactory bulb is modulated by matrix remodelling. Eur J Neurosci 25:2021–2033. https://doi.org/10.1111/j.1460-9568.2007.05441.x

Bovetti S, Hsieh Y-C, Bovolin P, Perroteau I, Kazunori T, Puche AC (2007b) Blood vessels form a scaffold for neuroblast migration in the adult olfactory bulb. J Neurosci 27:5976–5980. https://doi.org/10.1523/JNEUROSCI.0678-07.2007

Brooker SM, Bond AM, Peng C-Y, Kessler JA (2016) β1-integrin restricts astrocytic differentiation of adult hippocampal neural stem cells. Glia 64:1235–1251. https://doi.org/10.1002/glia.22996

Brösicke N, Van Landeghem FKH, Scheffler B, Faissner A (2013) Tenascin-C is expressed by human glioma in vivo and shows a strong association with tumor blood vessels. Cell Tissue Res 354(2):409–430

Bulstrode H, Jones LM, Siney EJ, Sampson JM, Ludwig A, Gray WP et al (2012) A-Disintegrin and Metalloprotease (ADAM) 10 and 17 promote self-renewal of brain tumor sphere forming cells. Cancer Lett. https://doi.org/10.1016/j.canlet.2012.07.022

Burgett ME, Lathia JD, Roth P, Nowacki AS, Galileo DS, Pugacheva E et al (2016) Direct contact with perivascular tumor cells enhances integrin avβ3 signaling and migration of endothelial cells. Oncotarget 7:43852–43867

Calabrese C, Poppleton H, Kocak M, Hogg TL, Fuller C, Hamner B et al (2007) A perivascular niche for brain tumor stem cells. Cancer Cell. https://doi.org/10.1016/j.ccr.2006.11.020

Cameron RS, Rakic P (1994) Identification of membrane proteins that comprise the plasmalemmal junction between migrating neurons and radial glial cells. J Neurosci 14:3139–3155

Camp JG, Badsha F, Florio M, Kanton S, Gerber T, Wilsch-Bräuninger M et al (2015) Human cerebral organoids recapitulate gene expression programs of fetal neocortex development. Proc Natl Acad Sci U S A 112:15672–15677. https://doi.org/10.1073/pnas.1520760112.

Cayre M, Courtès S, Martineau F, Giordano M, Arnaud K, Zamaron A et al (2013) Netrin 1 contributes to vascular remodeling in the subventricular zone and promotes progenitor emigration after demyelination. Development 140:3107–3117. https://doi.org/10.1242/dev.092999

Charles N, Ozawa T, Squatrito M, Bleau AM, Brennan CW, Hambardzumyan D et al (2010) Perivascular nitric oxide activates notch signaling and promotes stem-like character in PDGF-induced glioma cells. Cell Stem Cell. https://doi.org/10.1016/j.stem.2010.01.001

Chatzi C, Schnell E, Westbrook GL (2015) Localized hypoxia within the subgranular zone determines the early survival of newborn hippocampal granule cells. Elife 4:e08722. https://doi.org/10.7554/eLife.08722.

Chieffi S, Messina G, Villano I, Messina A, Esposito M, Monda V et al (2017) Exercise influence on hippocampal function: possible involvement of orexin-A. Front Physiol 8:85. https://doi.org/10.3389/fphys.2017.00085

Chung H, Park S (2016) Ghrelin regulates cell cycle-related gene expression in cultured hippocampal neural stem cells. J Endocrinol 230:239–250. https://doi.org/10.1530/JOE-16-0126

Clarke DL, Johansson CB, Wilbertz J, Veress B, Nilsson E, Karlström H et al (2000) Generalized potential of adult neural stem cells. Science 288:1660–1663

Coma S, Shimizu A, Klagsbrun M (2011) Hypoxia induces tumor and endothelial cell migration in a semaphorin 3F- and VEGF-dependent manner via transcriptional repression of their common receptor neuropilin 2. Cell Adhes Migr 5:266–275

Conover JC, Doetsch F, Garcia-Verdugo JM, Gale NW, Yancopoulos GD, Alvarez-Buylla A (2000) Disruption of Eph/ephrin signaling affects migration and proliferation in the adult subventricular zone. Nat Neurosci 3:1091–1097. https://doi.org/10.1038/80606.

Cossetti C, Iraci N, Mercer TR, Leonardi T, Alpi E, Drago D et al (2014) Extracellular vesicles from neural stem cells transfer IFN-γ via Ifngr1 to activate Stat1 signaling in target cells. Mol Cell 56:193–204. https://doi.org/10.1016/j.molcel.2014.08.020

Courtès S, Vernerey J, Pujadas L, Magalon K, Cremer H, Soriano E et al (2011) Reelin controls progenitor cell migration in the healthy and pathological adult mouse brain. PLoS One 6:e20430. https://doi.org/10.1371/journal.pone.0020430

Culver JC, Vadakkan TJ, Dickinson ME (2013) A specialized microvascular domain in the mouse neural stem cell niche. PLoS One 8:e53546. https://doi.org/10.1371/journal.pone.0053546

Curtis MA, Eriksson PS, Faull RL (2007) Progenitor cells and adult neurogenesis in neurodegenerative diseases and injuries of the basal ganglia. Clin Exp Pharmacol Physiol 34:528–532. https://doi.org/10.1111/j.1440-1681.2007.04609.x

D'Ercole J a, Ye P, O'Kusky JR (2002) Mutant mouse models of insulin-like growth factor actions in the central nervous system. Neuropeptides 36:209–220. https://doi.org/10.1054/npep.2002.0893

Darsalia V, Heldmann U, Lindvall O, Kokaia Z (2005) Stroke-induced neurogenesis in aged brain. Stroke 36:1790–1795

Darsalia V, Kallur T, Kokaia Z (2007) Survival, migration and neuronal differentiation of human fetal striatal and cortical neural stem cells grafted in stroke-damaged rat striatum. Eur J Neurosci 26:605–614

David LS, Schachner M, Saghatelyan A (2013) The extracellular matrix glycoprotein tenascin-R affects adult but not developmental neurogenesis in the olfactory bulb. J Neurosci 33:10324–10339. https://doi.org/10.1523/JNEUROSCI.5728-12.2013

Daynac M, Tirou L, Faure H, Mouthon M-A, Gauthier LR, Hahn H et al (2016) Hedgehog controls quiescence and activation of neural stem cells in the adult ventricular-subventricular zone. Stem cell reports 7:735–748. https://doi.org/10.1016/j.stemcr.2016.08.016

De Juan Romero C, Bruder C, Tomasello U, Sanz-Anquela JM, Borrell V (2015) Discrete domains of gene expression in germinal layers distinguish the development of gyrencephaly. EMBO J 34:1859–1874. 10.15252/embj.

Deleyrolle LP, Harding A, Cato K, Siebzehnrubl FA, Rahman M, Azari H et al (2011) Evidence for label-retaining tumour-initiating cells in human glioblastoma. Brain. https://doi.org/10.1093/brain/awr081

Delgado AC, Ferrón SR, Vicente D, Porlan E, Perez-Villalba A, Trujillo CM et al (2014) Endothelial NT-3 delivered by vasculature and CSF promotes quiescence of subependymal neural stem cells through nitric oxide induction. Neuron 83:572–585. https://doi.org/10.1016/j.neuron.2014.06.015

Dieriks BV, Waldvogel HJ, Monzo HJ, Faull RLM, Curtis MA (2013) GABA(A) receptor characterization and subunit localization in the human sub-ventricular zone. J Chem Neuroanat 52:58–68. https://doi.org/10.1016/j.jchemneu.2013.06.001

Doetsch F, García-Verdugo JM, Alvarez-Buylla A (1999) Regeneration of a germinal layer in the adult mammalian brain. Proc Natl Acad Sci U S A 96:11619–11624

Douet V, Kerever A, Arikawa-Hirasawa E, Mercier F (2013) Fractone-heparan sulphates mediate FGF-2 stimulation of cell proliferation in the adult subventricular zone. Cell Prolif 46:137–145. https://doi.org/10.1111/cpr.12023

Draganova K, Zemke M, Zurkirchen L, Valenta T, Cantù C, Okoniewski M et al (2015) Wnt/β-Catenin signaling regulates sequential fate decisions of murine cortical precursor cells. Stem Cells 33:170–182. https://doi.org/10.1002/stem.1820

Ekonomou A, Ballard CG, Pathmanaban ON, Perry RH, Perry EK, Kalaria RN et al (2010) Increased neural progenitors in vascular dementia. Neurobiol Aging 32(12):2152–2161

Etxeberria A, Mangin J-M, Aguirre A, Gallo V (2010) Adult-born SVZ progenitors receive transient synapses during remyelination in corpus callosum. Nat Neurosci 13:287–289. https://doi.org/10.1038/nn.2500

Eyler CE, Wu Q, Yan K, MacSwords JM, Chandler-Militello D, Misuraca KL et al (2011) Glioma stem cell proliferation and tumor growth are promoted by nitric oxide synthase-2. Cell. https://doi.org/10.1016/j.cell.2011.06.006

Fernando RN, Eleuteri B, Abdelhady S, Nussenzweig A, Andang M, Ernfors P (2011) Cell cycle restriction by histone H2AX limits proliferation of adult neural stem cells. Proc Natl Acad Sci 108:5837–5842. https://doi.org/10.1073/pnas.1014993108

Fietz SA, Lachmann R, Brandl H, Kircher M, Samusik N, Schröder R et al (2012) Transcriptomes of germinal zones of human and mouse fetal neocortex suggest a role of extracellular matrix in progenitor self-renewal. Proc Natl Acad Sci U S A 109:11836–11841. https://doi.org/10.1073/pnas.1209647109

Florio M, Albert M, Taverna E, Namba T, Brandl H, Lewitus E et al (2015) Human-specific gene ARHGAP11B promotes basal progenitor amplification and neocortex expansion. Science 347:1465–1470. https://doi.org/10.1126/science.aaa1975

Folkins C, Shaked Y, Man S, Tang T, Lee CR, Zhu Z et al (2009) Glioma tumor stem-like cells promote tumor angiogenesis and vasculogenesis via vascular endothelial growth factor and stromal-derived factor 1. Cancer Res. https://doi.org/10.1158/0008-5472.CAN-09-0167

Fung TC, Olson CA, Hsiao EY (2017) Interactions between the microbiota, immune and nervous systems in health and disease. Nat Neurosci 20:145–155. https://doi.org/10.1038/nn.4476

Fusco S, Leone L, Barbati SA, Samengo D, Piacentini R, Maulucci G et al (2016) A CREB-Sirt1-Hes1 circuitry mediates neural stem cell response to glucose availability. Cell Rep 14:1195–1205. https://doi.org/10.1016/j.celrep.2015.12.092

Gajera CR1, Emich H, Lioubinski O, Christ A, Beckervordersandforth-Bonk R, Yoshikawa K, Bachmann S, Christensen EI, Götz M, Kempermann G, Peterson AS, Willnow TE, Hammes A. (2010) LRP2 in ependymal cells regulates BMP signaling in the adult neurogenic niche. J Cell Sci 123:192219–192230. doi:10.1242/jcs.065912

García-González D, Murcia-Belmonte V, Esteban PF, Ortega F, Díaz D, Sánchez-Vera I et al (2016) Anosmin-1 over-expression increases adult neurogenesis in the subventricular zone and neuroblast migration to the olfactory bulb. Brain Struct Funct 221:239–260. https://doi.org/10.1007/s00429-014-0904-8

Garcion E, Halilagic A, Faissner A, Ffrench-Constant C (2004) Generation of an environmental niche for neural stem cell development by the extracellular matrix molecule tenascin C. Development 131:3423–3432. https://doi.org/10.1242/dev.01202

Garner JM, Fan M, Yang CH, Du Z, Sims M, Davidoff AM et al (2013) Constitutive activation of signal transducer and activator of transcription 3 (STAT3) and nuclear factor κB signaling in glioblastoma cancer stem cells regulates the notch pathway. J Biol Chem. https://doi.org/10.1074/jbc.M113.477950

Gilbertson RJ, Rich JN (2007) Making a tumour's bed: glioblastoma stem cells and the vascular niche. Nat Rev Cancer 7:733–736. https://doi.org/10.1038/nrc2246

Gómez-Gaviro MV, Scott CE, Sesay AK, Matheu A, Booth S, Galichet C et al (2012) Betacellulin promotes cell proliferation in the neural stem cell niche and stimulates neurogenesis. Proc Natl Acad Sci U S A 109:1317–1322. https://doi.org/10.1073/pnas.1016199109

Graus-Porta D, Blaess S, Senften M, Littlewood-Evans A, Damsky C, Huang Z et al (2001) Beta1-class integrins regulate the development of laminae and folia in the cerebral and cerebellar cortex. Neuron 31:367–379

Haan N, Goodman T, Najdi-Samiei A, Stratford CM, Rice R, El Agha E et al (2013) Fgf10-Expressing tanycytes add new neurons to the appetite/energy-balance regulating centers of the postnatal and adult hypothalamus. J Neurosci 33:6170–6180. https://doi.org/10.1523/JNEUROSCI.2437-12.2013

Haddock G, Cross AK, Allan S, Sharrack B, Callaghan J, Bunning RAD et al (2007) Brevican and phosphacan expression and localization following transient middle cerebral artery occlusion in the rat. Biochem Soc Trans 35:692–694. https://doi.org/10.1042/BST0350692.

Hambardzumyan D, Becher OJ, Rosenblum MK, Pandolfi PP, Manova-Todorova K, Holland EC (2008) PI3K pathway regulates survival of cancer stem cells residing in the perivascular niche following radiation in medulloblastoma in vivo. Genes Dev. https://doi.org/10.1101/gad.1627008

Hamerlik P, Lathia JD, Rasmussen R, Wu Q, Bartkova J, Lee M et al (2012) Autocrine VEGF-VEGFR2-neuropilin-1 signaling promotes glioma stem-like cell viability and tumor growth. J Exp Med 209:507–520. https://doi.org/10.1084/jem.20111424

Han X, Chen M, Wang F, Windrem M, Wang S, Shanz S et al (2013) Forebrain engraftment by human glial progenitor cells enhances synaptic plasticity and learning in adult mice. Cell Stem Cell 12:342–353. https://doi.org/10.1016/j.stem.2012.12.015

Hartenstein V, Wodarz A. (2013) Initial neurogenesis in Drosophila. Wiley Interdiscip Rev Dev Biol 2:701–721. doi: 10.1002/wdev.111

Harper MM, Ye E-A, Blong CC, Jacobson ML, Sakaguchi DS (2010) Integrins contribute to initial morphological development and process outgrowth in rat adult hippocampal progenitor cells. J Mol Neurosci 40:269–283. https://doi.org/10.1007/s12031-009-9211-x

Hartenstein V, Stollewerk A (2015) Review the evolution of early neurogenesis. Dev Cell 32:390–407. https://doi.org/10.1016/j.devcel.2015.02.004

Haubst N, Georges-Labouesse E, De Arcangelis A, Mayer U, Götz M (2006) Basement membrane attachment is dispensable for radial glial cell fate and for proliferation, but affects positioning of neuronal subtypes. Development 133:3245–3254. https://doi.org/10.1242/dev.02486

He Z, Ding J, Zhang J, Liu Y, Gong C, Sun S et al (2012) Fibroblast growth factor-2 counteracts the effect of ciliary neurotrophic factor on spontaneous differentiation in adult hippocampal progenitor cells. J Huazhong Univ Sci Technolog Med Sci 32:867–871. https://doi.org/10.1007/s11596-012-1049-8

Heberden C (2016) Modulating adult neurogenesis through dietary interventions. Nutr Res Rev 29:163–171. https://doi.org/10.1017/S0954422416000081

Heddleston JM, Li Z, McLendon RE, Hjelmeland AB, Rich JN (2009) The hypoxic microenvironment maintains glioblastoma stem cells and promotes reprogramming towards a cancer stem cell phenotype. Cell Cycle. https://doi.org/10.4161/cc.8.20.9701

Heijtz RD, Wang S, Anuar F, Qian Y, Bjorkholm B, Samuelsson A et al (2011) Normal gut microbiota modulates brain development and behavior. Proc Natl Acad Sci 108:3047–3052. https://doi.org/10.1073/pnas.1010529108

Herculano-Houzel S (2012) The remarkable, yet not extraordinary, human brain as a scaled-up primate brain and its associated cost. Proc Natl Acad Sci U S A 109(Suppl):10661–10668. https://doi.org/10.1073/pnas.1201895109

Hirata E, Arakawa Y, Shirahata M, Yamaguchi M, Kishi Y, Okada T et al (2009) Endogenous tenascin-C enhances glioblastoma invasion with reactive change of surrounding brain tissue. Cancer Sci. https://doi.org/10.1111/j.1349-7006.2009.01189.x

Holmberg J, Armulik A, Senti K-A, Edoff K, Spalding K, Momma S et al (2005) Ephrin-A2 reverse signaling negatively regulates neural progenitor proliferation and neurogenesis. Genes Dev 19:462–471. https://doi.org/10.1101/gad.326905

Hovinga KE, Shimizu F, Wang R, Panagiotakos G, Van Der Heijden M, Moayedpardazi H et al (2010) Inhibition of notch signaling in glioblastoma targets cancer stem cells via an endothelial cell intermediate. Stem Cells. https://doi.org/10.1002/stem.429

Huang P, Rani MRS, Ahluwalia MS, Bae E, Prayson RA, Weil RJ et al (2012) Endothelial expression of TNF receptor-1 generates a proapoptotic signal inhibited by integrinα6β1 in glioblastoma. Cancer Res 72:1428–1437. https://doi.org/10.1158/0008-5472.CAN-11-2621

Ikeda M, Hirota Y, Sakaguchi M, Yamada O, Kida YS, Ogura T et al (2010) Expression and proliferation-promoting role of Diversin in the neuronally committed precursor cells migrating in the adult mouse brain. Stem Cells 28:2017–2026. https://doi.org/10.1002/stem.516

Imayoshi I, Sakamoto M, Yamaguchi M, Mori K, Kageyama R (2010) Essential roles of Notch signaling in maintenance of neural stem cells in developing and adult brains. J Neurosci 30:3489–3498. https://doi.org/10.1523/JNEUROSCI.4987-09.2010

Isshiki T, Pearson B, Holbrook S, Doe CQ (2001) Drosophila neuroblasts sequentially express transcription factors which specify the temporal identity of their neuronal progeny. Cell 106:511–521

Iwashita M, Kataoka N, Toida K, Kosodo Y (2014) Systematic profiling of spatiotemporal tissue and cellular stiffness in the developing brain. Development 141:3793–3798. https://doi.org/10.1242/dev.109637

Jagielska A, Norman AL, Whyte G, Vliet KJV, Guck J, Franklin RJM (2012) Mechanical environment modulates biological properties of oligodendrocyte progenitor cells. Stem Cells Dev 21:2905–2914. https://doi.org/10.1089/scd.2012.0189

Javaherian A, Kriegstein A (2009) A stem cell niche for intermediate progenitor cells of the embryonic cortex. Cereb Cortex 19(Suppl 1):i70–i77. https://doi.org/10.1093/cercor/bhp029

Jeon HM, Jin X, Lee JS, Oh SY, Sohn YW, Park HJ et al (2008) Inhibitor of differentiation 4 drives brain tumor-initiating cell genesis through cyclin E and notch signaling. Genes Dev. https://doi.org/10.1101/gad.1668708

Jhaveri DJ, Mackay EW, Hamlin AS, Marathe SV, Nandam LS, Vaidya VA et al (2010) Norepinephrine directly activates adult hippocampal precursors via 3-adrenergic receptors. J Neurosci 30:2795–2806. https://doi.org/10.1523/JNEUROSCI.3780-09.2010

Jiang X, Nardelli J (2015) Cellular and molecular introduction to brain development. Neurobiol Dis. https://doi.org/10.1016/j.nbd.2015.07.007

Jin X, Yin J, Kim SH, Sohn YW, Beck S, Lim YC et al (2011) EGFR-AKT-Smad signaling promotes formation of glioma stem-like cells and tumor angiogenesis by ID3-driven cytokine induction. Cancer Res. https://doi.org/10.1158/0008-5472.CAN-11-1330

Johansson PA, Cappello S, Götz M (2010) Stem cells niches during development-lessons from the cerebral cortex. Curr Opin Neurobiol. https://doi.org/10.1016/j.conb.2010.04.003

Joppé SE, Hamilton LK, Cochard LM, Levros L-C, Aumont A, Barnabé-Heider F et al (2015) Bone morphogenetic protein dominantly suppresses epidermal growth factor-induced proliferative expansion of adult forebrain neural precursors. Front Neurosci 9:407. https://doi.org/10.3389/fnins.2015.00407

Joy AM, Beaudry CE, Tran NL, Ponce F a, Holz DR, Demuth T et al (2003) Migrating glioma cells activate the PI3-K pathway and display decreased susceptibility to apoptosis. J Cell Sci 116:4409–4417. https://doi.org/10.1242/jcs.00712

Katsimpardi L, Litterman NK, Schein PA, Miller CM, Loffredo FS, Wojtkiewicz GR et al (2014) Vascular and neurogenic rejuvenation of the aging mouse brain by young systemic factors. Science 344:630–634. https://doi.org/10.1126/science.1251141

Kazanis I (2012) Can adult neural stem cells create new brains? Plasticity in the adult mammalian neurogenic niches: realities and expectations in the era of regenerative biology. Neurosci 18:15–27

Kazanis I (2013) Neurogenesis in the adult mammalian brain: how much do we need, how much do we have? Curr Top Behav Neurosci 15:3–29. https://doi.org/10.1007/7854_2012_227

Kazanis I, ffrench-Constant C (2012) The number of stem cells in the subependymal zone of the adult rodent brain is correlated with the number of ependymal cells and not with the volume of the niche. Stem Cells Dev 21:1090–1096. https://doi.org/10.1089/scd.2011.0130

Kazanis I, Belhadi A, Faissner A, ffrench-Constant C (2007) The adult mouse subependymal zone regenerates efficiently in the absence of tenascin-C. J Neurosci 27:13991–13996

Kazanis I, Lathia JD, Vadakkan TJ, Raborn E, Wan R, Mughal MR et al (2010) Quiescence and activation of stem and precursor cell populations in the subependymal zone of the mammalian brain are associated with distinct cellular and extracellular matrix signals. J Neurosci 30:9771–9781

Kazanis I, Evans KA, Andreopoulou E, Dimitriou C, Koutsakis C, Karadottir RT et al (2017) Subependymal zone-derived oligodendroblasts respond to focal demyelination but fail to generate myelin in young and aged mice. Stem Cell Reports 8:685–700. https://doi.org/10.1016/j.stemcr.2017.01.007

Kerever A, Schnack J, Vellinga D, Ichikawa N, Moon C, Arikawa-Hirasawa E et al (2007) Novel extracellular matrix structures in the neural stem cell niche capture the neurogenic factor fibroblast growth factor 2 from the extracellular milieu. Stem Cells 25:2146–2157. https://doi.org/10.1634/stemcells.2007-0082

Kerever A, Yamada T, Suzuki Y, Mercier F, Arikawa-Hirasawa E (2015) Fractone aging in the subventricular zone of the lateral ventricle. J Chem Neuroanat 66:52–60. https://doi.org/10.1016/j.jchemneu.2015.06.001.

Keung AJ, Asuri P, Kumar S, Schaffer DV, Hall S (2012) Soft microenvironments promote the early neurogenic differentiation but not self-renewal of human pluripotent stem cells. Integr Biol (Camb) 21:1049–1058. https://doi.org/10.1039/c2ib20083j

Kim Y, Wang W-Z, Comte I, Pastrana E, Tran PB, Brown J et al (2010) Dopamine stimulation of postnatal murine subventricular zone neurogenesis via the D3 receptor. J Neurochem 114:750–760. https://doi.org/10.1111/j.1471-4159.2010.06799.x

Kobayashi M, Nakatani T, Koda T, Matsumoto K-I, Ozaki R, Mochida N et al (2014) Absence of BRINP1 in mice causes increase of hippocampal neurogenesis and behavioral alterations relevant to human psychiatric disorders. Mol Brain 7:12. https://doi.org/10.1186/1756-6606-7-12

Kokoeva MV, Yin H, Flier JS (2007) Evidence for constitutive neural cell proliferation in the adult murine hypothalamus. J Comp Neurol 505:209–220. https://doi.org/10.1002/cne.21492

Kokovay E, Goderie S, Wang Y, Lotz S, Lin G, Sun Y et al (2010) Adult SVZ lineage cells home to and leave the vascular niche via differential responses to SDF1/CXCR4 signaling. Cell Stem Cell 7:163–173. https://doi.org/10.1016/j.stem.2010.05.019

Kosodo Y, Röper K, Haubensak W, Marzesco A-M, Corbeil D, Huttner WB (2004) Asymmetric distribution of the apical plasma membrane during neurogenic divisions of mammalian neuro-epithelial cells. EMBO J 23:2314–2324. https://doi.org/10.1038/sj.emboj.7600223

Kothapalli CR, Kamm RD (2013) 3D matrix microenvironment for targeted differentiation of embryonic stem cells into neural and glial lineages. Biomaterials 34:5995–6007. https://doi.org/10.1016/j.biomaterials.2013.04.042

Koutsakis C, Kazanis I (2016) How necessary is the vasculature in the life of neural stem and progenitor cells? Evidence from evolution, development and the adult nervous system. doi:https://doi.org/10.3389/fncel.2016.00035

Lao CL, Lu C-S, Chen J-C (2013) Dopamine D3 receptor activation promotes neural stem/progenitor cell proliferation through AKT and ERK1/2 pathways and expands type-B and -C cells in adult subventricular zone. Glia 61:475–489. https://doi.org/10.1002/glia.22449

Lathia JD, Patton B, Eckley DM, Magnus T, Mughal MR, Sasaki T et al (2007) Patterns of laminins and integrins in the embryonic ventricular zone of the CNS. J Comp Neurol. https://doi.org/10.1002/cne.21520

Lathia JD, Gallagher J, Heddleston JM, Wang J, Eyler CE, Macswords J et al (2010) Integrin alpha 6 regulates glioblastoma stem cells. Cell Stem Cell 6:421–432. https://doi.org/10.1016/j.stem.2010.02.018

Lathia JD, Heddleston JM, Venere M, Rich JN (2011) Cell stem cell minireview deadly teamwork: neural cancer stem Cells and the tumor microenvironment. Stem Cell 8:482–485. https://doi.org/10.1016/j.stem.2011.04.013.

Lathia JD, Li M, Hall PE, Gallagher J, Hale JS, Wu Q et al (2012) Laminin alpha 2 enables glioblastoma stem cell growth. Ann Neurol. https://doi.org/10.1002/ana.23674

Lathia JD, Li M, Sinyuk M, Alvarado AG, Flavahan WA, Stoltz K et al (2014) High-throughput flow cytometry screening reveals a role for junctional adhesion molecule a as a cancer stem cell maintenance factor. Cell Rep. https://doi.org/10.1016/j.celrep.2013.11.043

Lazarini F, Gabellec MM, Moigneu C, de Chaumont F, Olivo-Marin JC, Lledo PM (2014) Adult neurogenesis restores dopaminergic neuronal loss in the olfactory bulb. J Neurosci 34:14430–14442. https://doi.org/10.1523/JNEUROSCI.5366-13.2014

Lee DA, Bedont JL, Pak T, Wang H, Song J, Miranda-Angulo A et al (2012) Tanycytes of the hypothalamic median eminence form a diet-responsive neurogenic niche. Nat Neurosci 15:700–702. https://doi.org/10.1038/nn.3079

Lehtinen MK, Zappaterra MW, Chen X, Yang YJ, Hill AD, Lun M et al (2011) The cerebrospinal fluid provides a proliferative niche for neural progenitor cells. Neuron. https://doi.org/10.1016/j.neuron.2011.01.023

Lepelletier F-X, Mann DMA, Robinson AC, Pinteaux E, Boutin H (2017) Early changes in extracellular matrix in Alzheimer's disease. Neuropathol Appl Neurobiol 43:167–182. https://doi.org/10.1111/nan.12295

Li Z, Bao S, Wu Q, Wang H, Eyler C, Sathornsumetee S et al (2009) Hypoxia-inducible factors regulate tumorigenic capacity of glioma stem cells cancer stem cell specific molecules involved in neoangiogenesis, including HIF2α and its regulated factors. Cancer Cell 15:501–513. https://doi.org/10.1016/j.ccr.2009.03.018.

Li G, Fang L, Fernández G, Pleasure SJ (2013) The ventral hippocampus is the embryonic origin for adult neural stem cells in the dentate gyrus. Neuron 78:658–672. https://doi.org/10.1016/j.neuron.2013.03.019

Lim DA, Tramontin AD, Trevejo JM, Herrera DG, García-Verdugo JM, Alvarez-Buylla A (2000) Noggin antagonizes BMP signaling to create a niche for adult neurogenesis. Neuron 28:713–726

Limmer S, Weiler A, Volkenhoff A, Babatz F, Klämbt C (2014) The Drosophila blood-brain barrier: development and function of a glial endothelium. Front Neurosci 8:365. https://doi.org/10.3389/fnins.2014.00365

Lin X, Shi Y, Cao Y, Liu W (2016) Recent progress in stem cell differentiation directed by material and mechanical cues. Biomed Mater 11:14109. https://doi.org/10.1088/1748-6041/11/1/014109

Liu J, Eder PS, Brand AH (2014) Control of brain development and homeostasis by local and systemic insulin signalling. Diabetes Obes Metab 16:16–20. https://doi.org/10.1111/dom.12337

Long K, Moss L, Laursen L, Boulter L, Ffrench-Constant C (2016) Integrin signalling regulates the expansion of neuroepithelial progenitors and neurogenesis via Wnt7a and Decorin. Nat Commun 7:10354. https://doi.org/10.1038/ncomms10354

Louissaint A, Rao S, Leventhal C, Goldman SA (2002) Coordinated interaction of neurogenesis and angiogenesis in the adult songbird brain. Neuron 34:945–960

Loulier K, Lathia JD, Marthiens V, Relucio J, Mughal MR, Tang S-C et al (2009) beta1 integrin maintains integrity of the embryonic neocortical stem cell niche. PLoS Biol 7:e1000176. https://doi.org/10.1371/journal.pbio.1000176

Lovelace MD, Gu BJ, Eamegdool SS, Weible MW, Wiley JS, Allen DG et al (2015) P2X7 Receptors mediate innate phagocytosis by human neural precursor cells and neuroblasts. Stem Cells 33:526–541. https://doi.org/10.1002/stem.1864

Lun MP, Monuki ES, Lehtinen MK (2015) Development and functions of the choroid plexus–cerebrospinal fluid system. Nat Rev Neurosci 16:445–457. https://doi.org/10.1038/nrn3921

Luo J, Shook BA, Daniels SB, Conover JC (2008) Subventricular zone-mediated ependyma repair in the adult mammalian brain. J Neurosci 28:3804–3813. https://doi.org/10.1523/JNEUROSCI.0224-08.2008

Magnusson JP, Goritz C, Tatarishvili J, Dias DO, Smith EMK, Lindvall O et al (2014) A latent neurogenic program in astrocytes regulated by Notch signaling in the mouse. Science 346:237–241. https://doi.org/10.1126/science.346.6206.237

Mardones MD, Andaur GA, Varas-Godoy M, Henriquez JF, Salech F, Behrens MI et al (2016) Frizzled-1 receptor regulates adult hippocampal neurogenesis. Mol Brain 9:29. https://doi.org/10.1186/s13041-016-0209-3

Maria Galan-Moya E, Le Guelte A, Lima Fernandes E, Thirant C, Dwyer J, Bidere N et al (2011) Secreted factors from brain endothelial cells maintain glioblastoma stem-like cell expansion through the mTOR pathway. Nat Publ Gr 12:470–47639. https://doi.org/10.1038/embor.2011.39

Marques F, Sousa JC, Brito MA, Pahnke J, Santos C, Correia-Neves M et al (2016) The choroid plexus in health and in disease: dialogues into and out of the brain. Neurobiol Dis. https://doi.org/10.1016/j.nbd.2016.08.011

Marthiens V, Ffrench-Constant C (2009) Adherens junction domains are split by asymmetric division of embryonic neural stem cells. EMBO Rep 10:515–520. https://doi.org/10.1038/embor.2009.36

Marthiens V, Rujano MA, Pennetier C, Tessier S, Paul-Gilloteaux P, Basto R (2013) Centrosome amplification causes microcephaly. https://doi.org/10.1038/ncb2746

Mathew TC (1999) Association between supraependymal nerve fibres and the ependymal cilia of the mammalian brain. Anat Histol Embryol 28:193–197

Mazumdar J, O'Brien WT, Johnson RS, LaManna JC, Chavez JC, Klein PS et al (2010) O2 regulates stem cells through Wnt/β-catenin signalling. Nat Cell Biol 12:1007–1013. https://doi.org/10.1038/ncb2102

Mekki-Dauriac S, Agius E, Kan P, Cochard P (2002) Bone morphogenetic proteins negatively control oligodendrocyte precursor specification in the chick spinal cord. Development 129:5117–5130

Méndez O, Zavadil J, Esencay M, Lukyanov Y, Santovasi D, Wang S-C et al (2010) Knock down of HIF-1α in glioma cells reduces migration in vitro and invasion in vivo and impairs their ability to form tumor spheres. Mol Cancer 9. https://doi.org/10.1186/1476-4598-9-133

Mercier F, Douet V (2014) Bone morphogenetic protein-4 inhibits adult neurogenesis and is regulated by fractone-associated heparan sulfates in the subventricular zone. J Chem Neuroanat 57:54–61. https://doi.org/10.1016/j.jchemneu.2014.03.005

Mercier F, Kitasako JT, Hatton GI (2002) Anatomy of the brain neurogenic zones revisited: fractones and the fibroblast/macrophage network. J Comp Neurol 451:170–188. https://doi.org/10.1002/cne.10342

Mercier F, Cho Kwon Y, Kodama R (2011) Meningeal/vascular alterations and loss of extracellular matrix in the neurogenic zone of adult BTBR T+ tf/J mice, animal model for autism. Neurosci Lett 498:173–178. https://doi.org/10.1016/j.neulet.2011.05.014

Mich JK, Signer RA, Nakada D, Pineda A, Burgess RJ, Vue TY et al (2014) Prospective identification of functionally distinct stem cells and neurosphere-initiating cells in adult mouse forebrain. Elife 3:e02669. https://doi.org/10.7554/eLife.02669.

Migaud M, Butrille L, Batailler M (2015) Seasonal regulation of structural plasticity and neurogenesis in the adult mammalian brain: Focus on the sheep hypothalamus. Front Neuroendocrinol 37:146–157. https://doi.org/10.1016/j.yfrne.2014.11.004

Milošević NJ, Judaš M, Aronica E, Kostovic I (2014) Neural ECM in laminar organization and connectivity development in healthy and diseased human brain. Prog Brain Res 214:159–178. https://doi.org/10.1016/B978-0-444-63486-3.00007-4

Mirzadeh Z, Merkle FT, Soriano-Navarro M, Garcia-Verdugo JM, Alvarez-Buylla A (2008) Neural stem cells confer unique pinwheel architecture to the ventricular surface in neurogenic regions of the adult brain. Cell Stem Cell 3:265–278. https://doi.org/10.1016/j.stem.2008.07.004

Mobley AK, McCarty JH (2011) β8 integrin is essential for neuroblast migration in the rostral migratory stream. Glia. https://doi.org/10.1002/glia.21199

Morell M, Tsan Y, O'Shea KS (2015) Inducible expression of noggin selectively expands neural progenitors in the adult SVZ. Stem Cell Res 14:79–94. https://doi.org/10.1016/j.scr.2014.11.001

Motomiya M, Kobayashi M, Iwasaki N, Minami A, Matsuoka I (2007) Activity-dependent regulation of BRINP family genes. Biochem Biophys Res Commun 352:623–629. https://doi.org/10.1016/j.bbrc.2006.11.072

Murase S, Horwitz AF (2002) Deleted in colorectal carcinoma and differentially expressed integrins mediate the directional migration of neural precursors in the rostral migratory stream. J Neurosci 22:3568–3579

Nakada M, Nambu E, Furuyama N, Yoshida Y, Takino T, Hayashi Y, et al (2013) Integrin alpha3 is overexpressed in glioma stem-like cells and promotes invasion. doi:https://doi.org/10.1038/bjc.2013.218.

Nomura T, Göritz C, Catchpole T, Henkemeyer M, Frisén J (2010) EphB signaling controls lineage plasticity of adult neural stem cell niche cells. Cell Stem Cell 7:730–743. https://doi.org/10.1016/j.stem.2010.11.009

Noonan MA, Bulin SE, Fuller DC, Eisch AJ (2010) Reduction of adult hippocampal neurogenesis confers Vulnerability in an animal model of cocaine addiction. J Neurosci 30:304–315. https://doi.org/10.1523/JNEUROSCI.4256-09.2010

North HA, Pan L, McGuire TL, Brooker S, Kessler JA (2015) 1-Integrin alters ependymal stem cell bmp receptor localization and attenuates astrogliosis after spinal cord injury. J Neurosci 35:3725–3733. https://doi.org/10.1523/JNEUROSCI.4546-14.2015

O'Keeffe GC, Barker RA, Caldwell MA (2009a) Dopaminergic modulation of neurogenesis in the subventricular zone of the adult brain. Cell Cycle 8:2888–2894. https://doi.org/10.4161/cc.8.18.9512

O'Keeffe GC, Tyers P, Aarsland D, Dalley JW, Barker RA, Caldwell MA (2009b) Dopamine-induced proliferation of adult neural precursor cells in the mammalian subventricular zone is mediated through EGF. Proc Natl Acad Sci U S A 106:8754–8759. https://doi.org/10.1073/pnas.0803955106

O'Leary CJ, Bradford D, Chen M, White A, Blackmore DG, Cooper HM (2015) The netrin/RGM receptor, neogenin, controls adult neurogenesis by promoting neuroblast migration and cell cycle exit. Stem Cells 33:503–514. https://doi.org/10.1002/stem.1861

Oka N, Soeda A, Inagaki A, Onodera M, Maruyama H, Hara A et al (2007) VEGF promotes tumorigenesis and angiogenesis of human glioblastoma stem cells. Biochem Biophys Res Commun. https://doi.org/10.1016/j.bbrc.2007.06.094

Okano H, Temple S (2009) Cell types to order: temporal specification of CNS stem cells. Curr Opin Neurobiol 19:112–119. https://doi.org/10.1016/j.conb.2009.04.003

Ortega F, Gascón S, Masserdotti G, Deshpande A, Simon C, Fischer J et al (2013) Oligodendrogliogenic and neurogenic adult subependymal zone neural stem cells constitute distinct lineages and exhibit differential responsiveness to Wnt signalling. Nat Cell Biol 15:602–613. https://doi.org/10.1038/ncb2736

Ottone C, Krusche B, Whitby A, Clements M, Quadrato G, Pitulescu ME et al (2014) Direct cell-cell contact with the vascular niche maintains quiescent neural stem cells. Nat Cell Biol 16:1045–1056. https://doi.org/10.1038/ncb3045

Paez-Gonzalez P, Asrican B, Rodriguez E, Kuo CT (2014) Identification of distinct ChAT[+] neurons and activity-dependent control of postnatal SVZ neurogenesis. Nat Neurosci 17:934–942. https://doi.org/10.1038/nn.3734

Palmer TD, Willhoite AR, Gage FH (2000) Vascular niche for adult hippocampal neurogenesis. J Comp Neurol 425:479–494

Panchision DM (2009) The role of oxygen in regulating neural stem cells in development and disease. J Cell Physiol 220:562–568. https://doi.org/10.1002/jcp.21812

Pastrana E, Cheng LC, Doetsch F (2009) Simultaneous prospective purification of adult subventricular zone neural stem cells and their progeny. Proc Natl Acad Sci U S A 106:6387–6392

Pencea V, Bingaman KD, Wiegand SJ, Luskin MB (2001) Infusion of brain-derived neurotrophic factor into the lateral ventricle of the adult rat leads to new neurons in the parenchyma of the striatum, septum, thalamus, and hypothalamus. J Neurosci 21:6706–6717

Peñuelas S, Anido J, Prieto-Sánchez RM, Folch G, Barba I, Cuartas I et al (2009) TGF-β Increases glioma-initiating cell self-renewal through the induction of LIF in human glioblastoma. Cancer Cell. https://doi.org/10.1016/j.ccr.2009.02.011

Perera TD, Lu D, Thirumangalakudi L, Smith ELP, Yaretskiy A, Rosenblum LA et al (2011) Correlations between hippocampal neurogenesis and metabolic indices in adult nonhuman primates. Neural Plast 2011:1–6. https://doi.org/10.1155/2011/875307

Piccirillo SGM, Dietz S, Madhu B, Griffiths J, Price SJ, Collins VP et al (2012) Fluorescence-guided surgical sampling of glioblastoma identifies phenotypically distinct tumour-initiating cell populations in the tumour mass and margin. Br J Cancer 107:462–468. https://doi.org/10.1038/bjc.2012.271

Ping YF, Yao XH, Jiang JY, Zhao LT, Yu SC, Jiang T et al (2011) The chemokine CXCL12 and its receptor CXCR4 promote glioma stem cell-mediated VEGF production and tumour angiogenesis via PI3K/AKT signalling. J Pathol. https://doi.org/10.1002/path.2908

Pluchino S, Cossetti C (2013) How stem cells speak with host immune cells in inflammatory brain diseases. Glia 61:1379–1401. https://doi.org/10.1002/glia.22500

Pluchino S, Zanotti L, Rossi B, Brambilla E, Ottoboni L, Salani G et al (2005) Neurosphere-derived multipotent precursors promote neuroprotection by an immunomodulatory mechanism. Nature 436:266–271

Pollen AA, Nowakowski TJ, Chen J, Retallack H, Sandoval-Espinosa C, Nicholas CR et al (2015) Molecular identity of human outer radial glia during cortical development HHS public access. Cell 163:55–67. https://doi.org/10.1016/j.cell.2015.09.004.

Ponti G, Obernier K, Guinto C, Jose L, Bonfanti L, Alvarez-Buylla A (2013) Cell cycle and lineage progression of neural progenitors in the ventricular-subventricular zones of adult mice. Proc Natl Acad Sci U S A 110:E1045–E1054. https://doi.org/10.1073/pnas.1219563110

Qian X, Nguyen HN, Song MM, Hadiono C, Ogden SC, Hammack C et al (2016) Brain-region-specific organoids using mini-bioreactors for modeling ZIKV exposure. Cell. https://doi.org/10.1016/j.cell.2016.04.032

Qiang L, Wu T, Zhang H-W, Lu N, Hu R, Wang Y-J et al (2012) HIF-1alpha; is critical for hypoxia-mediated maintenance of glioblastoma stem cells by activating Notch signaling pathway. Cell Death Differ 19:284–29495. https://doi.org/10.1038/cdd.2011.95.

Radakovits R, Barros CS, Belvindrah R, Patton B, Muller U (2009) Regulation of radial glial survival by signals from the meninges. J Neurosci 29:7694–7705. https://doi.org/10.1523/JNEUROSCI.5537-08.2009

Rakic P (2003) Developmental and evolutionary adaptations of cortical radial glia. Cereb Cortex 13:541–549

Ramírez-Castillejo C, Sánchez-Sánchez F, Andreu-Agulló C, Ferrón SR, Aroca-Aguilar JD, Sánchez P et al (2006) Pigment epithelium-derived factor is a niche signal for neural stem cell renewal. Nat Neurosci 9:331–339. https://doi.org/10.1038/nn1657

Robins SC, Stewart I, McNay DE, Taylor V, Giachino C, Goetz M et al (2013) α-Tanycytes of the adult hypothalamic third ventricle include distinct populations of FGF-responsive neural progenitors. Nat Commun 4:2049. https://doi.org/10.1038/ncomms3049

Rosenberg GA (2017) Extracellular matrix inflammation in vascular cognitive impairment and dementia. Clin Sci (Lond) 131:425–437. https://doi.org/10.1042/CS20160604

Ruckh JM, Zhao JW, Shadrach JL, Van Wijngaarden P, Rao TN, Wagers AJ et al (2012) Rejuvenation of regeneration in the aging central nervous system. Cell Stem Cell 10:96–103

Sahay A, Scobie KN, Hill AS, O'Carroll CM, Kheirbek MA, Burghardt NS et al (2011) Increasing adult hippocampal neurogenesis is sufficient to improve pattern separation. Nature 472:466–470. https://doi.org/10.1038/nature09817

Sarkar S, Zemp FJ, Senger D, Robbins SM, Wee Yong V (2015) ADAM-9 is a novel mediator of tenascin-C-stimulated invasiveness of brain tumor-initiating cells. Neuro Oncol. https://doi.org/10.1093/neuonc/nou362

Sawamoto K, Wichterle H, Gonzalez-Perez O, Cholfin JA, Yamada M, Spassky N et al (2006) New neurons follow the flow of cerebrospinal fluid in the adult brain. Science 311:629–632. https://doi.org/10.1126/science.1119133

Schafer ST, Han J, Pena M, Bohlen Und v, Halbach O, Peters J, Gage FH (2015) The Wnt adaptor protein ATP6AP2 regulates multiple stages of adult hippocampal neurogenesis. J Neurosci 35:4983–4998. https://doi.org/10.1523/JNEUROSCI.4130-14.2015

Seidel S, Garvalov BK, Wirta V, Von Stechow L, Schänzer A, Meletis K et al (2010) A hypoxic niche regulates glioblastoma stem cells through hypoxia inducible factor 2α. Brain. https://doi.org/10.1093/brain/awq042

Semenza GL (2012) Hypoxia-inducible factors in physiology and medicine. Cell. https://doi.org/10.1016/j.cell.2012.01.021

Sen S, Dong M, Kumar S, Kreplak L (2009) Isoform-specific contributions of a-actinin to glioma cell mechanobiology. PLoS One 4. https://doi.org/10.1371/journal.pone.0008427

Sen S, Ng WP, Kumar S (2011) Contributions of talin-1 to glioma cell-matrix tensional homeostasis. J R Soc Interface 9:1311–1317. https://doi.org/10.1098/rsif.2011.0567

Shen Q (2004) Endothelial cells stimulate self-renewal and expand neurogenesis of neural stem cells. Science 304:1338–1340. https://doi.org/10.1126/science.1095505

Shen Q, Wang Y, Kokovay E, Lin G, Chuang S-M, Goderie SK et al (2008) Adult SVZ stem cells lie in a vascular niche: a quantitative analysis of niche cell-cell interactions. Cell Stem Cell 3:289–300. https://doi.org/10.1016/j.stem.2008.07.026

Shimogori T, VanSant J, Paik E, Grove EA (2004) Members of the Wnt, Fz, and Frp gene families expressed in postnatal mouse cerebral cortex. J Comp Neurol 473:496–510. https://doi.org/10.1002/cne.20135

Sibbe M, Kuner E, Althof D, Frotscher M (2015) Stem- and progenitor cell proliferation in the dentate gyrus of the reeler mouse. PLoS One 10:e0119643. https://doi.org/10.1371/journal.pone.0119643

Silva-Vargas V, Maldonado-Soto AR, Mizrak D, Codega P, Doetsch F (2016) Age-dependent niche signals from the choroid plexus regulate adult neural stem cells. Cell Stem Cell 19:643–652. https://doi.org/10.1016/j.stem.2016.06.013

Sirko S, von Holst A, Weber A, Wizenmann A, Theocharidis U, Götz M et al (2010) Chondroitin sulfates are required for fibroblast growth factor-2-dependent proliferation and maintenance in neural stem cells and for epidermal growth factor-dependent migration of their progeny. Stem Cells 28:775–787. https://doi.org/10.1002/stem.309

Sirko S, Behrendt G, Johansson PA, Tripathi P, Costa M, Bek S et al (2013) Reactive glia in the injured brain acquire stem cell properties in response to sonic hedgehog. [corrected]. Cell Stem Cell 12:426–439. https://doi.org/10.1016/j.stem.2013.01.019.

Smukler SR, Runciman SB, Xu S, Kooy DVD (2006) Embryonic stem cells assume a primitive neural stem cell fate in the absence of extrinsic influences. J Cell Biol 172:79. https://doi.org/10.1083/jcb.200508085

Soeda A, Park M, Lee D, Mintz A, Androutsellis-Theotokis A, Mckay R et al (2009) Hypoxia promotes expansion of the CD133-positive glioma stem cells through activation of HIF-1a. Oncogene 28:3949–3959. https://doi.org/10.1038/onc.2009.252

Sonego M, Gajendra S, Parsons M, Ma Y, Hobbs C, Zentar MP et al (2013) Fascin regulates the migration of subventricular zone-derived neuroblasts in the postnatal brain. J Neurosci 33:12171–12185. https://doi.org/10.1523/JNEUROSCI.0653-13.2013

Soumier A, Banasr M, Kerkerian-Le Goff L, Daszuta A (2010) Region- and phase-dependent effects of 5-HT(1A) and 5-HT(2C) receptor activation on adult neurogenesis. Eur Neuropsychopharmacol 20:336–345. https://doi.org/10.1016/j.euroneuro.2009.11.007

Stacpoole SRL, Bilican B, Webber DJ, Luzhynskaya A, He XL, Compston A et al (2011) Efficient derivation of NPCs, spinal motor neurons and midbrain dopaminergic neurons from hESCs at 3% oxygen. Nat Protoc 6:1229–1240. https://doi.org/10.1038/nprot.2011.380

Stahl R, Walcher T, De Juan Romero C, Pilz GA, Cappello S, Irmler M et al (2013) Trnp1 Regulates expansion and folding of the mammalian cerebral cortex by control of radial glial fate. Cell 153:535–549. https://doi.org/10.1016/j.cell.2013.03.027

Staquicini FI, Dias-neto E, Li J, Snyder EY, Sidman RL, Pasqualini R et al (2009) Discovery of a functional protein complex of netrin-4, laminin γ1 chain, and integrin α6β1 in mouse neural stem cells. Proc Natl Acad Sci U S A 106:2903–2908. https://doi.org/10.1073/pnas.0813286106

Stenzel D, Wilsch-Bräuninger M, Wong FK, Heuer H, Huttner WB (2014) Integrin αvβ3 and thyroid hormones promote expansion of progenitors in embryonic neocortex. Development 141:795–806. https://doi.org/10.1242/dev.101907

Syed MH, Mark B, Doe CQ (2017) Steroid hormone induction of temporal gene expression in Drosophila brain neuroblasts generates Summary: hormone induction of temporal gene expression in neural progenitors. Elife 6. https://doi.org/10.1101/121855.

Syková E, Nicholson C (2008) Diffusion in brain extracellular space. Physiol Rev 88:1277–1340. https://doi.org/10.1152/physrev.00027.2007

Tan X, Liu WA, Zhang X, Li Z, Brown KN, Shi S et al (2016) Vascular Influence on ventral telencephalic progenitors and neocortical interneuron production. Dev Cell 36:624–638. https://doi.org/10.1016/j.devcel.2016.02.023

Tavazoie M, Van der Veken L, Silva-Vargas V, Louissaint M, Colonna L, Zaidi B et al (2008) A specialized vascular niche for adult neural stem cells. Cell Stem Cell 3:279–288. https://doi.org/10.1016/j.stem.2008.07.025

Theocharidis U, Long K, Ffrench-Constant C, Faissner A (2014) Chapter 1—Regulation of the neural stem cell compartment by extracellular matrix constituents. Prog Brain Res 214:3–28. https://doi.org/10.1016/B978-0-444-63486-3.00001-3.

Thouvenot E, Lafon-Cazal M, Demettre E, Jouin P, Bockaert J, Marin P (2006) The proteomic analysis of mouse choroid plexus secretome reveals a high protein secretion capacity of choroidal epithelial cells. Proteomics 6:5941–5952. https://doi.org/10.1002/pmic.200600096

Tocharus C, Puriboriboon Y, Junmanee T, Tocharus J, Ekthuwapranee K, Govitrapong P (2014) Melatonin enhances adult rat hippocampal progenitor cell proliferation via ERK signaling pathway through melatonin receptor. Neuroscience. https://doi.org/10.1016/j.neuroscience.2014.06.026

Todd KL, Baker KL, Eastman MB, Kolling FW, Trausch AG, Nelson CE et al (2017) EphA4 regulates neuroblast and astrocyte organization in a neurogenic niche. J Neurosci 37:3738–3716. https://doi.org/10.1523/JNEUROSCI.3738-16.2017

Tong CK, Chen J, Cebrián-Silla A, Mirzadeh Z, Obernier K, Guinto CD et al (2014a) Axonal control of the adult neural stem cell niche. Cell Stem Cell 14:500–511. https://doi.org/10.1016/j.stem.2014.01.014

Tong CK, Han Y-G, Shah JK, Obernier K, Guinto CD, Alvarez-Buylla A (2014b) Primary cilia are required in a unique subpopulation of neural progenitors. Proc Natl Acad Sci 111:12438–12443. https://doi.org/10.1073/pnas.1321425111

Tong CK, Fuentealba LC, Shah JK, Lindquist RA, Ihrie RA, Guinto CD et al (2015) A dorsal SHH-dependent domain in the V-SVZ produces large numbers of oligodendroglial lineage cells in the postnatal brain. Stem Cell Rep. https://doi.org/10.1016/j.stemcr.2015.08.013

Torii M, Hashimoto-Torii K, Levitt P, Rakic P (2009) Integration of neuronal clones in the radial cortical columns by EphA and ephrin-A signalling. Nature 461:524–528. https://doi.org/10.1038/nature08362

Triviño-Paredes J, Patten AR, Gil-Mohapel J, Christie BR (2016) The effects of hormones and physical exercise on hippocampal structural plasticity. Front Neuroendocrinol 41:23–43. https://doi.org/10.1016/j.yfrne.2016.03.001

Umesh V, Rape AD, Ulrich TA, Kumar S (2014) Microenvironmental stiffness enhances glioma cell proliferation by stimulating epidermal growth factor receptor signaling. PLoS One. https://doi.org/10.1371/journal.pone.0101771

Vandeputte DAA, Troost D, Leenstra S, Ijlst-Keizers H, Ramkema M, Bosch DA et al (2002) Expression and distribution of Id helix-loop-helix proteins in human astrocytic tumors. Glia. https://doi.org/10.1002/glia.10076

Vasudevan A, Long JE, Crandall JE, Rubenstein JLR, Bhide PG (2008) Compartment-specific transcription factors orchestrate angiogenesis gradients in the embryonic brain. Nat Neurosci 11:429–439. https://doi.org/10.1038/nn2074

Velpula KK, Rehman AA, Chelluboina B, Dasari VR, Gondi CS, Rao JS et al (2012) Glioma stem cell invasion through regulation of the interconnected ERK, integrin α6 and N-cadherin signaling pathway. Cell Signal. https://doi.org/10.1016/j.cellsig.2012.07.002

Wang J, Wakeman TP, Lathia JD, Hjelmeland AB, Wang XF, White RR et al (2010) Notch promotes radioresistance of glioma stem cells. Stem Cells. https://doi.org/10.1002/stem.261

Wang H, Warner-Schmidt J, Varela S, Enikolopov G, Greengard P, Flajolet M (2015) Norbin ablation results in defective adult hippocampal neurogenesis and depressive-like behavior in mice. Proc Natl Acad Sci U S A 112:9745–9750. https://doi.org/10.1073/pnas.1510291112

Wang J, Fu X, Zhang D, Yu L, Li N, Lu Z et al (2017) ChAT-positive neurons participate in subventricular zone neurogenesis after middle cerebral artery occlusion in mice. Behav Brain Res 316:145–151. https://doi.org/10.1016/j.bbr.2016.09.007

Willaime-Morawek S, Seaberg RM, Batista C, Labbé E, Attisano L, Gorski JA et al (2006) Embryonic cortical neural stem cells migrate ventrally and persist as postnatal striatal stem cells. J Cell Biol 175:159–168. https://doi.org/10.1083/jcb.200604123

Wilsch-Bräuninger M, Florio M, Huttner WB (2016) Neocortex expansion in development and evolution—from cell biology to single genes. Curr Opin Neurobiol. https://doi.org/10.1016/j.conb.2016.05.004

Witt H, Mack SC, Ryzhova M, Bender S, Sill M, Isserlin R et al (2011) Delineation of two clinically and molecularly distinct subgroups of posterior fossa ependymoma. Cancer Cell. https://doi.org/10.1016/j.ccr.2011.07.007

Wong SY, Ulrich TA, Deleyrolle LP, MacKay JL, Lin JMG, Martuscello RT et al (2015) Constitutive activation of myosin-dependent contractility sensitizes glioma tumor-initiating cells to mechanical inputs and reduces tissue invasion. Cancer Res. https://doi.org/10.1158/0008-5472.CAN-13-3426

Xia S, Lal B, Tung B, Wang S, Goodwin CR, Laterra J (2016) Tumor microenvironment tenascin-C promotes glioblastoma invasion and negatively regulates tumor proliferation. Neuro-Oncology. https://doi.org/10.1093/neuonc/nov171

Yamamoto H, Mandai K, Konno D, Maruo T, Matsuzaki F, Takai Y (2015) Impairment of radial glial scaffold-dependent neuronal migration and formation of double cortex by genetic ablation of afadin. Brain Res 1620:139–152. https://doi.org/10.1016/j.brainres.2015.05.012

Yamashita T, Ninomiya M, Hernandez Acosta P, Garcia-Verdugo JM, Sunabori T, Sakaguchi M et al (2006) Subventricular zone-derived neuroblasts migrate and differentiate into mature neurons in the post-stroke adult striatum. J Neurosci 26:6627–6636. https://doi.org/10.1523/JNEUROSCI.0149-06.2006

Yan K, Wu Q, Yan DH, Lee CH, Rahim N, Tritschler I et al (2014) Glioma cancer stem cells secrete Gremlin1 to promote their maintenance within the tumor hierarchy. Genes Dev. https://doi.org/10.1101/gad.235515.113

Ying M, Tilghman J, Wei Y, Guerrero-Cazares H, Quinones-Hinojosa A, Ji H et al (2014) Kruppel-like factor-9 (KLF9) inhibits glioblastoma stemness through global transcription repression and integrin α6 inhibition. J Biol Chem 289:32742–32756. https://doi.org/10.1074/jbc.M114.588988

Yu JM, Kim JH, Song GS, Jung JS (2006) Increase in proliferation and differentiation of neural progenitor cells isolated from postnatal and adult mice brain by Wnt-3a and Wnt-5a. Mol Cell Biochem 288:17–28. https://doi.org/10.1007/s11010-005-9113-3

Zhang RL, Chopp M, Roberts C, Liu X, Wei M, Nejad-Davarani SP et al (2014) Stroke increases neural stem cells and angiogenesis in the neurogenic niche of the adult mouse. PLoS One 9:e113972. https://doi.org/10.1371/journal.pone.0113972

Zhu TS, Costello MA, Talsma CE, Flack CG, Crowley JG, Hamm LL et al (2011) Endothelial cells create a stem cell niche in glioblastoma by providing NOTCH ligands that nurture self-renewal of cancer stem-like cells. Cancer Res. https://doi.org/10.1158/0008-5472.CAN-10-4269

Ziu M, Schmidt NO, Cargioli TG, Aboody KS, Black PM, Carroll RS (2006) Glioma-produced extracellular matrix influences brain tumor tropism of human neural stem cells. J Neurooncol. https://doi.org/10.1007/s11060-006-9121-5

Chapter 7
Glioblastoma Stem Cells and Their Microenvironment

Anirudh Sattiraju, Kiran Kumar Solingapuram Sai, and Akiva Mintz

Abstract Glioblastoma (GBM) is the most common primary malignant astrocytoma associated with a poor patient survival. Apart from arising de novo, GBMs also occur due to progression of slower growing grade III astrocytomas. GBM is characterized by extensive hypoxia, angiogenesis, proliferation and invasion. Standard treatment options such as surgical resection, radiation therapy and chemotherapy have increased median patient survival to 14.6 months in adults but recurrent disease arising from treatment resistant cancer cells often results in patient mortality. These treatment resistant cancer cells have been found to exhibit stem cell like properties. Strategies to identify or target these Glioblastoma Stem Cells (GSC) have proven to be unsuccessful so far. Studies on cancer stem cells (CSC) within GBM and other cancers have highlighted the importance of paracrine signaling networks within their microenvironment on the growth and maintenance of CSCs. The study of GSCs and their communication with various cell populations within their microenvironment is therefore not only important to understand the biology of GBMs but also to predict response to therapies and to identify novel targets which could stymy support to treatment resistant cancer cells and prevent disease recurrence. The purpose of this chapter is to introduce the concept of GSCs and to detail the latest findings indicating the role of various cellular subtypes within their microenvironment on their survival, proliferation and differentiation.

Keywords GBM • Glioblastoma stem cells • Microenvironment • Cancer stem cells

A. Sattiraju • K.K.S. Sai • A. Mintz (✉)
Department of Radiology, Columbia University College of Physicians and Surgeons, New York, NY, USA
e-mail: asattira@wakehealth.edu; ksolinga@wakehealth.edu; am4754@cumc.columbia.edu

© Springer International Publishing AG 2017
A. Birbrair (ed.), *Stem Cell Microenvironments and Beyond*,
Advances in Experimental Medicine and Biology 1041,
DOI 10.1007/978-3-319-69194-7_7

7.1 Introduction

7.1.1 Disease Classification and Histopathology

Glioblastoma (GBM) is classified according to the 2007 WHO classification as a grade IV (high grade) astrocytoma (Louis et al. 2007). GBM is the most common and aggressive primary (arising *de novo*) malignant astrocytoma which is often characterized by extensive microvascular hyperplasia, hypercellularity, proliferation, diffuse infiltrating margins and necrotic foci, often surrounded by pseudopalisading cells, an ominous histopathological feature that distinguishes them from non-malignant "low-grade" gliomas (grade I and II) (Brat et al. 2004; Bissell and Radisky 2001; Rong et al. 2006; Van Meir et al. 2010). These pseudopalisades have previously been reported to be highly hypoxic and instigate microvascular hypercellularity through secretion of hypoxia inducible factors (HIFs), vascular endothelial growth factor A (VEGF-A) and interleukin 8 (IL-8). Using microarray analysis, The Cancer Genome Atlas (TCGA) researchers were able to identify genomic changes which drive GBM tumor development and classified the disease into classical (EGFR high, mutated *TP53* low), proneural (mutated *TP53* high, mutated *IDH1* high, mutated *PDGFA* high), mesenchymal (mutated *NF1* high, frequent mutations of *PTEN* and *TP53*) and neural subtypes (mutations in same genes as other subgroups; expression of neural genes) (Verhaak et al. 2010).

7.1.2 Therapeutic Challenges for Treating GBMs

Standard treatment strategies for GBM patients in the clinic include surgical resection, radiotherapy, chemotherapy with temozolomide (TMZ) and the recently FDA approved oscillating electric field therapy (Hottinger et al. 2014; Weller et al. 2012). Regardless of efforts to increase safety and efficacy of these treatments in the clinic, the median overall survival of patients has only extended to about 14.6 months. GBM is a highly infiltrative disease with cancer cells migrating extensively into surrounding normal neural tissue. It is therefore not possible to remove all tumor cells from patients through surgical resection, as they have been observed 2–3 cm from the original site of the tumor. Importantly, these invasive cells that are left behind after surgery close to the margins of the resection cavity give rise to even more aggressive tumors (Chaichana 2014; Eyupoglu et al. 2013; Yong and Lonser 2011). These cells that repopulate and maintain GBM tumors after gross surgical resection are thought to be undifferentiated glioblastoma stem cells (GSCs) (Jackson et al. 2015; Lathia et al. 2015; Ahmed et al. 2013; Cho et al. 2013). Seeding other normal regions of the brain with primary GBM cells during resection is another possible limitation of surgery.

The vasculature within the central nervous system protects neural tissue from harmful molecules by the formation of a blood-brain barrier (BBB). It comprises of astrocytes and pericytes which wrap around endothelial cell tight junctions (Abbott 2002; Abbott et al. 2010; Agarwal et al. 2013; Persidsky et al. 2006; Wolburg and Lippoldt 2002). The protective nature of the BBB causes several complications in delivering effective concentrations of therapies to tumor tissue as it regulates the extravasation of macromolecules and chemotherapy (Hawkins and Davis 2005; Pardridge 2005). GBMs exhibit high genetic heterogeneity due to clonal evolution of cancer cells making it difficult to target all GBM cells using a single biomarker targeted therapy that results in cancer cells surviving treatment and ultimate recurrence (Patel et al. 2014; Bonavia et al. 2011; Sottoriva et al. 2013; MDM et al. 2010). GSCs are thought to exist primarily in hypoxic or necrotic areas which are often inaccessible by chemotherapy and are thought to be resistant to chemotherapy due to overexpression of drug efflux pumps and their slow division rate (Seidel et al. 2010; Bar et al. 2010; Li et al. 2009; Heddleston et al. 2010; Heddleston et al. 2009; Carmeliet and Jain 2000; Dewhirst et al. 2008; Pistollato et al. 2010; Singh et al. 2004a; Murat et al. 2008a). Radiation therapy is another standard treatment option for GBM patients but has proven to be ineffective in completely eliminating the disease (Bao et al. 2006a; Bao et al. 2006b). Radiation causes the creation of free radicals in oxygenated areas that in turn cause DNA breaks in cells within the exposure field. GSCs are often resistant to radiation therapy as they are thought to exist primarily in hypoxic regions within the tumors where the creation of high amounts of free radicals is not possible (Bao et al. 2006a; Wang et al. 2010a). Radiation therapy also inflicts immediate and long term neuro-cognitive deficits due to damage to surrounding normal neural tissue (Tallet et al. 2012; Greene-Schloesser et al. 2012; Greene-Schloesser and Robbins 2012).

7.1.3 Attempts to Target GSCs and Role of Microenvironment

Strategies to bypass or to enhance the permeability of the BBB in order to deliver effective concentrations of chemotherapy have shown promise. Furthermore, engineering advances have allowed for limiting radiation exposure and protecting normal areas of brain (Konofagou et al. 2012; Baskar et al. 2012; Nhan et al. 2014; Burgess et al. 2014; Jordao et al. 2013). But strategies to identify GSC specific biomarkers and the targeting of therapies against them have not been fruitful thus far. Recent studies have highlighted the importance of the microenvironment including paracrine factors secreted within the perivascular niche, hypoxic areas and the extracellular matrix on the biology and maintenance of GSCs. It is therefore imperative to understand the important pathways within the GSC niche that are critical for their survival and maintenance. It is also important to identify the nature of origin of GSCs as it could allow elucidating their biology and understanding extracellular factors that are critical for their maintainance. In this chapter, we introduce you to recent finding regarding the GSC microenvironment and highlight critical steps that could be taken in the future.

7.2 Discovery of Cancer Cells Expressing Stem Cell Surface Markers

A stochastic model in which all cancer cells possessed similar proliferative capability was predicted to explain tumor progression. Cancers were therefore thought to arise upon spontaneous mutations that occur within terminally differentiated somatic cells leading to aberrant cellular proliferation (Hanahan and Weinberg 2011; Dalerba et al. 2007). Teratomas, which are benign masses arising from aberrantly proliferating undifferentiated stem cells were thought to occur only in rare cases due to genetic and developmental defects within the germline cell lineages. Researchers later discovered a subpopulation within acute myeloid leukemia (AML) patient samples associated with normal hematopoietic stem cells by sorting for expression of cell surface markers (Lapidot et al. 1994). Clonogenic assays showed that this small subset of cancer cells expressing normal stem markers, now termed cancer stem cells (CSCs), could give rise to *de novo* leukemia, histologically similar to parental disease, when engrafted in bone marrow of immunocompromised mice (Lapidot et al. 1994). Using ectopic xenograft assays, it was determined that CD38+/ CD34- immunophenotype were able to initiate de novo leukemia in immunocompromised mice (Lapidot et al. 1994). Subsequently, studies sorting for the presence normal stem cell markers in populations of cancer cells found similar CSC in solid cancers in locations such as the breast, liver, colon, prostrate, pancreas, skin (melanoma) and <u>brain</u> (Singh et al. 2003; Fang et al. 2005, 2010; Zhu et al. 2010; Tomuleasa et al. 2010).

The brain had long been thought to have limited differentiation and regenerative capabilities but the discovery of neural stem cells (NSCs) (prior to the discovery of CSCs) within distinct niches in the subventricular zone (SVZ) of the forebrain lateral ventricles and the subgranular zone (SGZ) in the dentate gyrus of the hippocampus indicated that neurogenesis within the brain could possibly occur throughout adult life (Riquelme et al. 2008; Vescovi et al. 2006; Eriksson et al. 1998). The discovery of CSCs in leukemia and other solid cancers led to studies aimed at discovering similar such cancer cells within tumors of the central nervous system. The initial discovery of neurosphere forming undifferentiated stem cells within human GBM biopsy samples (GSCs) and the identification of CD133+ as a marker for this cellular subpopulation within those samples alerted researchers to the possibility of cell populations within GBMs which might have greater differentiation capability than other cells within tumors (Ignatova et al. 2002; Uchida et al. 2000; Singh et al. 2004b). Xenograft studies found that the CD133+ subset of GBM cells could give rise to histologically identical orthotopic GBMs in immunocompromised rodents. Later studies also showed that these GSCs could be differentiated into cells showing decreased differentiation and proliferation capabilities (Laks et al. 2009; Kang et al. 2014; Bradshaw et al. 2016). These findings gave rise to a hierarchic model to predict tumor progression, in which undifferentiated cancer cells expressing stem cell surface markers were considered at the top of the hierarchy and cancer cells showing reduced differentiation capabilities, such as transiently amplifying progenitor cells and terminally differentiated cells were considered lower in hierarchy (Nguyen et al. 2012; Dick 2009; Cabrera et al. 2015; Kreso and Dick John 2014).

The existence of GSCs still remains a point of intense contention within scientific circles due to (i) the relatively small number of these cells within patient GBM samples, (ii) its uncertain cell of origin; and (iii) lack of a universal identification marker specific for GSC immunophenotype, as recent studies show that not all GSCs are CD133 + .

7.2.1 Identification and Isolation of GSCs Using Cell Specific Markers

A high variability in cell surface marker expression in observed within patient GBM samples and between samples from different patients due to the highly heterogeneous nature of the disease (MDM et al. 2010). Although initial studies had identified cell populations enriched for CD133 cell surface marker as GSCs, further reports identified GBM cells which did not show enrichment in their CD133 expression but formed neurospheres in culture and tumorigenesis in xenograft assays. Cellular heterogeneity within biopsy samples and inefficient antibody labeling during FACS could explain for the lack of expression of CD133 in these cells (Son et al. 2009; Ogden et al. 2008; Shackleton et al. 2009; Nishide et al. 2009; Kelly et al. 2009). GBMs consist of various clones expressing a diverse genetic profile and it is hypothesized that GSCs from one region of a tumor could express different cell surface markers when compared to GSCs in another region of the tumor. Therefore, the location within tumors from where GSCs are enriched may also play a vital role in there stem cell marker expression.

CD133+ GBM cells have been shown to be resistant to radiation and chemotherapy due to activation of cell cycle checkpoint pathways, enhancement of DNA repair and aberrant cell survival mechanisms. Gene expression profiling of treatment resistant GBM cells showed them to be enriched for CD133 expression (Bao et al. 2006a; Murat et al. 2008b). Tumors derived from CD133+ GBM cells have also been reported to be highly vascularized due to promotion of angiogenesis through secretion of VEGF and SDF-1. These tumors also show greater invasion into normal brain tissue when compared to other GBM cell lines (Bao et al. 2006b). Additionally, the stage of a patient's disease could also impact the cell surface marker expression of GSCs thereby showing variable results when comparing different patient samples. The particular therapy regiment along with its duration could also potentially impact cell surface marker expression within GBMs of patients. GSCs have also been reported to show high cellular plasticity. Recent reports have shown that not only can GSCs differentiate into progenitor and terminal cells, but their daughter cells can also de-differentiate back into GSCs depending on microenvironmental cues and adaptation to treatment (Lee et al. 2016; Safa et al. 2015). Terminal differentiated GBM cells could therefore express markers of different lineages simultaneously and have no counterparts in normal physiological lineages

(Heddleston et al. 2011). GSCs have also recently been shown to transdifferentiate into endothelial cells (Heddleston et al. 2011; Ricci-Vitiani et al. 2010; Wang et al. 2010b). These data indicate that CD133 could be considered as a biomarker of GSCs and provides survival advantages to tumors. However, evidence suggests that additional markers could more precisely identify GSC populations due their drivers and high plasticity.

Alternate cell surface markers such as integrin α6, SSEA-1, A2B5 and CD44 could also be considered to enrich for GSC population (Ogden et al. 2008). Integrin α6 high cells have been reported to be more tumorigenic than integrin α6 low cells. Integrin α6 co-segregates with CD133 but integrin α6 positive cells are enriched for neurosphere formation regardless of enrichment in CD133 expression. Knockdown of integrin α6 impedes neurosphere formation and tumor development in xenografts indicating that integrin α6 is probably expressed on a broader pool of GSCs (Lathia et al. 2010). SSEA-1 (stage specific embryonic antigen-1) is thought to be secreted by neural stem cells to modulate Wnt signaling and has been shown to be expressed on CD133+ tumors. SSEA-1+ GBM cells also show greater tumorigenicity when compared to SSEA-1- GBM cells (Son et al. 2009). A2B5+ GBM cells have also been shown to be highly tumorigenic compared to A2B5- GBM cells. This corroborates the fact that A2B5 has been shown to co-segregate with CD133 in flow cytometry studies (Ogden et al. 2008). But, similar to integrin α6, A2B5+/CD133- GBM cells showed tumor initiating capability in xenograft studies. This indicates that A2B5 could be used to enrich a broader pool of GSCs. Side population using Hoechst 33,342 was also shown to as a cell surface marker independent technique to enrich for GSCs. However, conflicting reports about the tumorigenic capability of isolated cells as well as viability issues after exposure to the dye have yet to be addressed (Bleau et al. 2009). Intracellular stem markers such as Olig2, Musashi, Bmi and Sox2 could also be used to verify enriched GSC populations (Bao et al. 2006a). These studies indicate that more than one biomarker may potentially have to be used in order to segregate a GSC population. Additionally, the set of biomarkers used to efficiently segregate GSCs from one patient's GBM may not efficiently segregate the entire pool of GSCs from another patient's GBM due to the highly heterogeneous nature of the disease.

7.2.2 Glioblastoma Cell of Origin

The stochastic and hierarchic models not only predict gross steps that occur in the process of tumor progression but also indicate possible cells of origin (COIs) in a variety of cancers. The identification and isolation of COIs in GBM and other cancers could not only indicate which of the two predicted models of tumor progression might be accurate but could also shed light on the emergence of CSCs and their role in tumors maintenance, which eventually could lead to identification of better targets for cancer therapy.

Currently, there are two lines of thought regarding COIs of cancers, one that assumes that somatic differentiated cancer cells undergo de-differentiation and

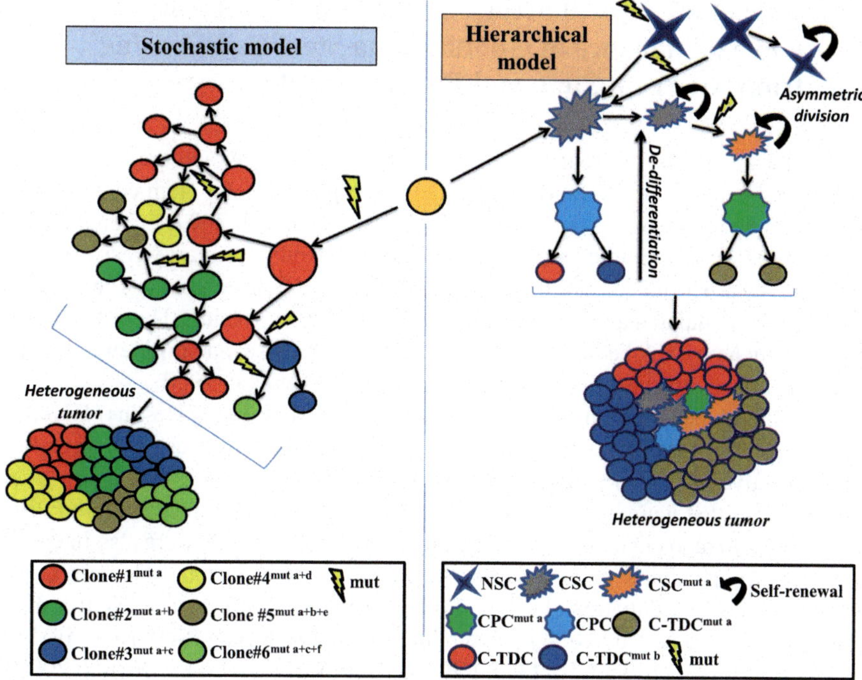

Fig. 7.1 Diagrammatic representation of stochastic and hierarchical models of tumor progression. *NSC* normal stem cells; *CSC* cancer stem cells; *CPC* cancer progenitor cells/transiently amplifying cells; *TDC* terminally differentiated cells; *mut* genetic mutation

transform upon genetic mutations and microenvironmental cues to give rise to CSCs (Hanahan and Weinberg 2011; Friedmann-Morvinski and Verma 2014). These CSCs could later give rise to progenitor cancer cells and terminally differentiated cancer cells. The other line of thought assumes that normal stem cells already present in various organs undergo genetic mutations and react to aberrant paracrine signaling within their microenvironment to give rise to altered undifferentiated malignant stem cells. These deviant stem cells can then give rise to aberrant progenitor cells with unregulated proliferative capabilities and later terminally differentiated cancer cells (Kreso and Dick John 2014; Reya et al. 2001; Tan et al. 2006) (Fig. 7.1).

7.2.3   Role of the Microenvironment in Maintenance of Glioblastomas Stem Cells

The site of origin of GSCs, either from a defined COIs or through de-differentiation from somatic GBM cells, is highly dependent on the microenvironment and the paracrine signaling networks. It is therefore important to understand the interactions of actors in the vicinity of established tumors.

7.3 Role of the Neural Stem Cell Niche in the Origin and Maintenance of Glioblastoma Stem Cells Arising from Aberrant Stem and Precursor Cells

Niches for stem cells are found in various organs in the body and are specific for each type of stem cell. These niches are not just repositories for stem cells but are also complex and dynamic ecosystems (Gilbertson and Rich 2007; Scheres 2007; Scadden 2006). In the brain, stem cells have been found to primarily reside in the SVZ and SGZ. The niches within SVZ and SGZ consist of heterogeneous cell populations, extracellular matrix proteins and other secreted proteins. The role of the stem cell niche within the brain is to regulate the self-renewal and differentiation of neural stem cells (Gage 2000; Merkle et al. 2004; Palmer et al. 2000).

SVZ, lining the later ventricles in the brain contains slow diving Nestin+/ GFAP+ type B astrocytes (NSCs), which are rapidly dividing type C astrocytes (progenitor cells) that give rise to type A astrocytes (neuroblasts). Type A astrocytes in turn give rise to committed neurons upon migrating anteriorly towards olfactory bulb (OB) (Alcantara Llaguno et al. 2015). The astrocytes within these niches are also in close contact with ependymal cells that line the cavity of the niches and play a role in preventing the differentiation of cell within the niche by (i) expressing CXCR4 (that binds to distally secreted SDF-1), (ii) binding to sonic hedgehog (SHH) and (iii) by secreting factors such as EGF, bFGF, IGF1, TGF-α, VEGF, ephrins (Doetsch 2003; Fidoamore et al. 2016). The other main component of SVZ and SGZ is a network of capillaries that are in close proximity to NSCs. These facilitate bi-directional communication between NSCs and endothelial cells through factors such as BDNF, VEGFC, PDGF, IL8, IGF-1 and bFGF (Riquelme et al. 2008; Leventhal et al. 1999; Ramirez-Castillejo et al. 2006). This close proximity to endothelial cells and the ensuing communication indicates a possibility that there may be amore permeable BBB within the stem cells niches allowing them access to systemic growth factors, nutrients and hormones (Fidoamore et al. 2016).

The extracellular matrix is yet another important player in regulation of neural stem cell fate. Studies show that tenascin C has a regulatory effect on NSC fate and number (Garcion et al. 2004; Tavazoie et al. 2008). Heparin sulfate proteoglycans (HSPs) have been shown to interact with BMP-2-2, HH, Wnts and other morphogens crucial in adult neurogenesis. HSPs have also been shown to interact with tenascin C, collagens, laminins, VEGF, EGF, FGFs, IGF-II, PDGF-AA, chemokines and cytokines (Doetsch 2003). NOTCH signaling which is an important paracrine signaling mechanism for regulating proliferation and differentiation of NSCs is also altered in GSCs, helping GSCs maintain an undifferentiated stem-like cell state. GSCs have been shown to lose oncogenic potential when NOTCH and its ligands Delta-like1 and Jagged-like-1 are downregulated (Stockhausen et al. 2009; Louvi and Artavanis-Tsakonas 2006; Fan et al. 2010). In addition, neuronal signals from ChAT+ (Choline acetyltransferase) neurons present within the SVZ, microglia and cerebrospinal fluid could also regulate NSC proliferation and differentiation (Paez-Gonzalez et al. 2014). These studies indicate the highly sophisticated and tightly regulated balance of various components of the neural stem cell niche that are responsible for maintaining NSCs. Dysregulation of any one of these components could result in a domino-effect which could cause de-regulated proliferation or differentiation of NSCs and their progenitor cells. Recent studies have also shown that deletion of *Nf1*, *Trp53* and *PTEN* in adult neural stem cells and their progenitors resulted in altered migration of aberrantly differentiated progenitors from SVZ. These altered cells gave rise to tumors in the brain instead of migrating and terminally differentiating into neurons. Aberrant mutations within cells of the NSC niche or NSCs themselves could results in GSCs, which may migrate elsewhere to form tumors or give rise to cells with similar differentiation potential through asymmetric division. These progeny can then in turn migrate to other regions of the brain to initiate tumors. Therefore, the site of origin of GSCs may not be the site of origin of GBMs (Alcantara Llaguno et al. 2015).

7.4 Role of Tumor Microenvironment in the Origin and Maintenance of De-Differentiated Somatic GBM Cells

7.4.1 Perivascular Niche

Extensive neovasculogenesis and abnormal morphology of vasculature is a common characteristic of GBMs. GBMs exhibit various forms of neovascularization which allow supply of essential nutrients and oxygen to tumor cells (Jhaveri et al. 2016).

The process of *de novo* formation of blood vessels *in situ*, termed as "Vasculogenesis" is thought to occur primarily during fetal development. However, recent research has shown that vessel formation can also result from circulating endothelial cells, tumor associated macrophages (TAMs), Tie-2+ monocytes and GSCs (De Palma et al. 2007; Venneri et al. 2007; Folkman and Shing 1992). Angiogenesis, a process of stimulating the sprouting of new blood vessels from preexisting vasculature is a critical step in tumor development and migration. GBMs are often characterized as having significantly increased angiogenesis and studies have indicated GSCs to play an important role in this process (Jhaveri et al. 2016). GSCs have been shown to overexpress factors that promote blood vessel formation such as VEGF and SDF-1α and their knockdown has been shown to significantly affect vessel formation in tumors *in vivo (Folkins et al.* 2009*)*. 20–90% of CD31+ endothelial cells were found to carry the same genetic mutations as tumor cells in human GBM samples (such as amplification of *EGFR* and chromosome 7) and GSCs cultured in endothelial cell culture conditions have been reported to transdifferentiate into CD31+ and Tie-2+ endothelial cells (Ricci-Vitiani et al. 2010). Studies also showed that co-implantation of GSCs and endothelial cells in immunocompromised mice resulted in accelerated initiation and growth of orthotopic

GBMs due to endothelial derived factors (Heddleston et al. 2011; Jhaveri et al. 2016). In addition, CD133+ GSC population has also been reported to contain a CD144+ (vascular endothelial-cadherin) cell population with the CD133+/CD144+ cell population showing an overexpression of pro-vasculogenesis markers such as CD31, CD105, CD 34 and VEGFR-2 (Wang et al. 2010b; Soda et al. 2011). In addition, circulating bone marrow derived cells have also been reported to be present within GBM vasculature but recent reports evaluating such transdifferentiating GSC or bone marrow derived cell populations have revealed that their presence within tumor vasculature is rare and might vary among tumors (Hardee and Zagzag 2012). GSCs have also been reported to transdifferentiate into pericytes but further investigation is needed to determine the role of such pericytes in tumor neovascularization (Cheng et al. 2013).

Nestin+/CD133+ stem-like cells within GBMs (GSCs) have been reported to reside in close proximity to capillary networks within niches in GBMs that resemble the neural stem cell niche. Endothelial and other vascular cells are thought to play a role in the maintainance of differentiation and proliferation of NSCs within the neural stem cell niche and the same is thought to occur in the case of GSCs. The dedifferentiation of GSCs into endothelial cells or the migration of endothelial cells that do not participate in the process of neovascularization towards GSCs could possibly serve as a mechanism to secrete factors to prevent maturation and proliferation of GSCs, which often impedes efficacy of radio- and chemo- therapies (Fidoamore et al. 2016; Calabrese et al. 2007). GBM cells have also been reported to infiltrate areas around brain vasculature and incorporate blood vessels into the tumor in a process termed as "Vascular Co-option". These tumor incorporated blood vessels are thought to undergo apoptosis inducing a state of hypoxia within these regions which could possibly maintain GSCs and stimulate the secretion of pro-angiogenic factors (Holash et al. 1999; Reiss et al. 2005; Liebelt et al. 2016).

SDF-1 or CXCL12 maintains NSCs in their niche and regulates their trafficking and homing within the brain. SDF-1 is a ligand for CXCR7 and is the only ligand for CXCR4, which is secreted either as an autocrine or paracrine factor in several cancers including GBM. Apart from promoting cancer cell proliferation, invasion, angiogenesis and cancer stem cell maintainance, SDF-1 has been shown to recruit bone marrow derived cells (BMDCs) that promote neovascularization, such as CXCR4 secreting endothelial precursor cells. GSCs express both CXCL12 and its receptors and endothelial cells within the brain have been reported to express CXCL12 to recruit GBM cells. The CXCR4/CXCL12 signaling axis is particularly high in pseudopalisading regions and invasive ends of GBM, indicating their importance in GSC maintainance and movement (Doetsch 2003; Sun et al. 2010). CXCL12 has also been reported to regulate tumor apoptosis by activating NF-κB, which reduces TNFα production. Radiation therapy, chemotherapy and treatment by VEGFR inhibitors activates CXCL12/CXCR4 pathway, resulting in the recruitment of myeloid BMDCs and promotion of angiogenesis and tumor invasion (Wang et al. 1996; Duda et al. 2011).

Notch signaling pathway is yet another way in which the perivascular niche plays an important role in regulating differentiation state of GSCs. *NOTCH1* silencing has been shown to interfere with the transdifferentiation of GSCs into endothelial cells and increased their sensitivity to radiation damage (Wang et al. 2010a; Fan et al. 2010). Nitric oxide (NO) has been shown to activate notch signaling pathway and maintains a stem-like phenotype in tumors. Along with *NOTCH1* overexpression, GSCs also express NO receptor, sGC and are often found in close proximity to endothelial cells that express endothelial nitric oxide synthase (eNOS) (Charles et al. 2010).

Tumor cells have also been reported to form a matrix embedded network capable of conducting fluids within the tumor through a process termed as "Vascular Mimicry". CD133+ GSCs have been reported to form tube networks in *in vitro* 3D matrigel experiments and GSCs have been reported to form tubular structures of vascular channels in tumors *in vivo* (El Hallani et al. 2010). Knockdown of VEGFR-2 in GSCs resulted in the loss of the ability to form these tubular structures and hypoxia is thought to play an important role in this process by upregulating CD144 expression in GSCs through HIF-1α and HIF-2α (Mao et al. 2013; Yao et al. 2013).

7.4.2 Hypoxic Niche

In healthy brain tissue, the normal physiological oxygen concentration ranges between 12.5 and 2.5%. GBM tissue however shows regions of mild hypoxia (2.5–0.5%) and severe hypoxia (0.5–0.1%) (Evans et al. 2004). It is hypothesized that the oxygen tension gradient within a tumor niche plays a vital role in differentiation of cells. The cells present in the periphery of tumor masses are thought to exhibit low proliferation rate, low levels of HIF1α and increased angiogenesis. The cells present at the tumor core are thought to exist in near anoxic conditions with very low proliferation rates and high levels of HIF1α. Cells present in the intermediate region of tumors are thought to high proliferation rate, form neurospheres in hypoxic conditions and show increased levels of expression of VEGF, Glut1 and carbonic anhydrase IX (CAIX) (Pistollato et al. 2010). Therefore, the presence of intratumoral hypoxia promotes the existence of a pool of stem-like cancer cells at the core of the tumor which are often resistant to radio- and chemo- therapies.

The importance of hypoxia in maintaining the differentiation state and proliferation of normal stem cells within their niches and its mechanism is well established. Within the bone marrow, hematopoietic stem cells (HSCs) migrate to hypoxic niches where they are maintained in a state of quiescence by the hypoxia induced protein, osteopontin (Stier et al. 2005). Severe hypoxia also prevents the differentiation of NSCs and embryonic stem cells without affecting their proliferation while also improving the generation of induced pluripotent stem cells (iPSCs) (Mathieu et al. 2014).

Neovascularization within GBM tissue often results in the formation of disorganized, chaotic and highly torturous blood vessels which are unable to effectively supply the entire tumor tissue with oxygen and nutrients. The lack of uniform oxygenation and the high proliferative rate of tumor cells results in the formation of regions of pseudopalisading necrosis that develop in order to protect the surrounding normal tissue from effects of hypoxia (Brat et al. 2004). Hypoxia and the activation of hypoxia response genes are thought to play a vital role in GBM progression, proliferation, aggressiveness and resistance to therapy. This was directly demonstrated in a recent multicenter trial that found hypoxia levels in GBM patients demonstrated by ^{18}F–FMISO PET/CT correlated with worse prognosis (Gerstner et al. 2016).

The effect of hypoxia on cells is mediated through intracellular family of proteins called hypoxia inducible factors (HIFs) which form transcriptional complexes consisting of HIF-β subunit (ARNT- aryl hydrocarbon nuclear translocator) which is constitutively expressed and oxygen regulated HIF-α subunits which belong to the basic helix-loop-helix-Per-Arnt-Sims (PAS) family of transcriptional activators. HIF1α (ubiquitously expressed), HIF 2α and HIF3α (tissue specific expression) are the three mammalian HIF-1 subunits. Even though the *HIF-1α* is highly transcribed and translated in normoxic conditions, it is rapidly hydroxylated on two conserved proline residues (P402 and P564) on the oxygen dependent degradation domain (ODD) by HIF specific prolyl hydroxylases PHD1, PHD2 and PHD3. Hydroxylated HIF-1α is then recognized by the von Hippel-Lindau tumor suppressor (pVHL), a subunit of E3 ubiquitin ligase which ubiquitinates HIF-1α for degradation by 26S proteosome (Nath and Szabo 2012).

Under hypoxic conditions, the hydroxylation of ODD of HIF-1α and its subsequent recognition by pVHL is inhibited, resulting in the accumulation of HIF-1α protein within the cytoplasm. In such conditions, HIF-1α translocated into the nucleus and dimerizes with HIF-1β to form HIF-1α/β dimer complex. HIF-1α/β dimer binds to HIF response elements (HRE) which contain the core consensus sequence 5'RCGTG-3' (R = purine residue) along with coactivators p300 and CBP. HREs are present within promoters, introns and 3' enhanced regions of many stress response gene families which facilitate adaptations to hypoxic conditions such as angiogenesis, hematopoietic growth factors, glucose transporters and glycolytic enzymes thereby affecting cell proliferation, survival and movement (Nath and Szabo 2012; Semenza 2010, 2013).

HIF-1α levels also increase due to metabolic and genetic changes within tumors such as increased production of H_2O_2 (which stabilizes HIF-1α). Increase in the levels of HIF-1α in response to low oxygen pressure leads to reprogramming of tumor metabolism towards glycolysis, thereby increasing the expression of glucose uptake receptors, glycolytic enzymes, lactate productions and reducing conversion of pyruvate to acetyl coenzyme A. HIF-1α also increases the conversion of glucose to glycogen by activating expression of hexokinases (HK1 and HK2), glycogen synthase (GYS1), UDP- glucose pyrophosphorylase (UGP2), phosphoglucomutase 1 (PGM1), glycogen branching enzyme (GBE1) and PPP1R3C. PPP1R3C activates GYS1 and also inhibits expression of liver-type glycogen phosphorylase (PYGL) which breaks down glycogen. The reduced oxygen availability in GBM results in increased oxidative phosphorylation and causes increased ROS generation, which can lead to additional mutations. HIF-1α is increased through PI3K/AKT pathways upon downregulation of PTEN. Additionally, PTEN mutations and its altered degradation also increase HIF-1α levels within tumor cells (Fidoamore et al. 2016; Nath and Szabo 2012; Semenza 2010, 2013). HIF-1α is also thought to be involved in increasing mutational rate in tumors by reducing levels of mismatch repair protein (mutS) and promoting glucose flux through a non-oxidative arm of the pentose phosphate pathway (Zhong et al. 1999). HIF-1α also increases the expression of VEGF and VEGF receptors (FLT-1 and FLK-1), plasminogen activator inhibitor-1 (PAI-1), angiopoietins (Ang-1 and Ang-2) and matrix metalloproteinases (MMP-2 and MMP-9) thereby promoting angiogenesis and invasion (Semenza 2013; Mendez et al. 2010). HIF-1α also reduced the sensitivity of GBM cells towards pro-differentiation and pro-apoptotic signals such as bone morphogenic proteins (BMPs) (Pistollato et al. 2009; Persano et al. 2012). HIF-1α also activates the expression of multidrug resistance 1 (MDR1) gene which encodes for P-glycoprotein (P-gp), belonging to a family of ATP binding cassette (ABC) transporters which acts as a drug efflux pump, thereby reducing intracellular concentration of various chemotherapeutics (Chou et al. 2012; Chen et al. 2014). HIF-1α also stabilizes NF-κB and contributes to suppressing hypoxia related apoptosis through expression of NF-κB target genes such as Bax, Bcl-2, Bcl-xL (Gorlach and Bonello 2008).

Hypoxia has also been reported to induce expression of stem cell and GSC markers such as CD133, *Oct4* and *Sox2*. Studies also report differentiation of CD133+ cell population upon exposure of GSCs to normoxic conditions, indicating that reduced levels of HIF-1α affect their differentiation state (McCord et al. 2009).

HIF-1α and HIF-2α share 75% homology but have distinct functions within cells. HIF-2α is also regulated by PHD hydroxylation at the transcriptional level as opposed to HIF-1α which is only regulated at the translational level. Genes such as *OCT4, Nanog, Sox2, Serpin B9,* and *TGF-α* are specifically regulated by HIF-2α. Recent studies have shown that HIF-2α is preferentially expressed in GSCs and knockdown of HIF-2α reduces self-renewal of GSCs indicating an important role for HIF-2α in maintaining the stem-like differentiation state of GSCs. Studies have also reported that the expression of HIF-2α in non-stem GBM cells induce expression of stem cell markers such as *Oct4*, *myc* and *Nanog* and neurosphere formation. Expression of non-degradable HIF-2α also increases tumorigenic potential of non-stem GBM cells *in vivo* and increases ratio of GSCs to non-stem GBM cells (Li et al. 2009; Heddleston et al. 2009). Future studies into the specific role of HIFs in the maintainance and self-renewal of GSCs and their effect on the GSC microenvironment such as promoting angiogenesis and invasion could help us better understand the GSC biology and lead to identification of better targets within GSC microenvironment.

7.4.3   *ECM and Paracrine Factors*

The components of the extracellular matrix that form a complex of macromolecules within the tumor cell niche are essential for the survival and migration of GBM cells. Remodeling of ECM to facilitate processes such as angiogenesis and pro-survival signals through integrin mediated signaling cascades promotes GBM growth and progression. Overexpression of basement membrane protein laminin 8 and its cellular receptor integrin $\alpha6\beta1$ in GBM cells has been reported to promote tumor progression (Huang et al. 2012). Laminin is also a critical component in adherent GSC cultures, upon which GSCs tend to form tumorspheres. Heparin sulphate binds to basic Fibroblast Growth Factor (bFGF) and stimulates GBM cell growth and prevents radiation induced cell death (Bao et al. 2006a; Folkman et al. 1988). Binding of GBM cells to ECM components allows for intracellular signal transduction through formation of multimeric complexes termed focal adhesions with other proteins such as focal adhesion kinase (FAK) (Fidoamore et al. 2016; Gilmore and Romer 1996).

Integrin play a vital role in the interactions of GBM cells with components of the ECM and vascular cells such as endothelial cells and pericytes. Integrin $\alpha6$ is highly expressed in NSCs where it heterodimerizes with integrin$\beta1$ and integrin $\beta4$ and binds to ECM protein laminin. Through its binding to laminin, integrin $\alpha6\beta1$ regulates self-renewal and differentiation by favoring adhesion to ventricular zone (Fortunel et al. 2003). GBM cells that overexpress integrin $\alpha6$ were reported to show self-renewal and ability to differentiate into CNS lineage indicating that integrin $\alpha6$ expression confers stem-like cellular state (Lathia et al. 2010). Integrin $\beta1$ plays an important role in perivascular niche where it promotes GBM invasion and functions together with CXCR4 to regulate critical stem cell pathways such as SHH, Wnt, and Notch. Integrin $\alpha3$ is overexpressed on invasive GBM and on GBM cells in close proximity to endothelial cells and is thought to regulate invasiveness of GSCs through ERK1/2 pathway (Nakada et al. 2013). Overexpression of integrin $\alpha v\beta5$ and integrin $\alpha v\beta3$ is associated with heightened invasiveness of GBMs and overexpression of integrin $\alpha v\beta8$ is associated with a more infiltrative phenotype in GBMs. Paracrine factors such as TGF-$\beta1$ and TGF-$\beta2$ are associated with an increase in the integrins $\alpha v\beta3$, $\alpha v\beta5$ and $\alpha v\beta8$ and thus result in aggressive GBMs (Fidoamore et al. 2016). Interaction of vitronectin (VN) with integrins $\alpha v\beta3$ and $\alpha v\beta5$ has been reported to enhance expression of Bcl-2 and Bcl-X_L and confer chemoresistance at invasive ends of GBM. Overexpression of integrins $\alpha v\beta3$, $\alpha v\beta5$ and $\beta1$ along with an increase in synthesis of basement membrane components such as fibronectin and matrigel have been reported to confer radio-resistence to GBM cells (Uhm et al. 1999).

Cadherins stimulate intracellular signaling upon intercellular adhesion between GBM cells and regulators of their cell fate such as cdc42, protein kinase C (PKC), β-catenin and Numb. N-cadherin plays a vital role in the NSC niche where it maintains the stem cell differentiation state while overexpression of E-cadherin in GBM samples is associated with poor patient outcomes. E-cadherin expressing CD133+ GSCs have also been reported to transdifferentiate into endothelial cells. Cadherin 11 plays a vital role by enhancing migration of GBM cells in tumors. Expression of cadherins is regulated by interleukin 8 (IL-8) and transcriptional activators such as FoxP2, FoxP4, Twist and Snail (Fidoamore et al. 2016; Brooks et al. 2013). Tenascin C has been reported to be highly expressed after radiation therapy in GBMs and its expression is correlated with poor patient survival as it promotes tumor cell growth and preserves differentiation state of GSCs (Leins et al. 2003; Mannino and Chalmers 2011) (Fig. 7.2).

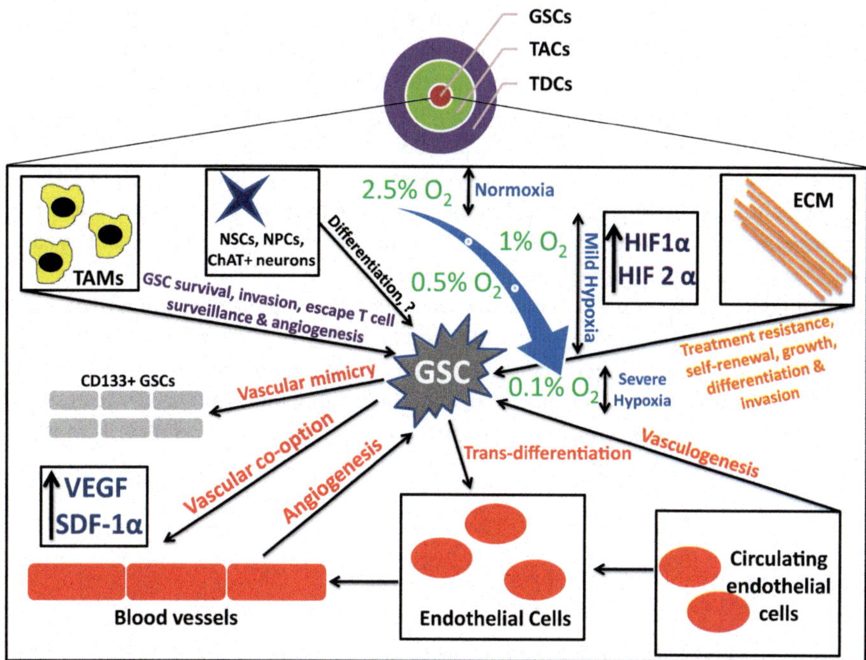

Fig. 7.2 Summary of interactions between GSCs and components of their microenvironment. *GSC* glioblastoma stem cells; *TAC* transiently amplifying cells; *TDC* terminally differentiated cells; *NSC* neural stem cells; *NPC* neural progenitor cells; *ChAT* choline acetyltransferase; *TAM* tumor associated macrophages; *ECM* extracellular matrix; *VEGF* vascular endothelial growth factor; *SDF-1* stromal derived factor-1; O_2 oxygen; *HIF* hypoxia inducible factor

7.5 Future Directions

Standard forms of therapy have been ineffective in significantly increasing the overall survival of GBM patients and the failures of these therapies have been pinned on the highly heterogeneous nature of GBMs. As summarized above, a wide variety of studies that have investigated the existence of a stem-like GBM population within tumors and their resistance to chemo- and radiation therapy. The expansion and re-establishment of aggressive GBMs post therapy by these stem-like cells is thought to be one of the major reasons for disease recurrence which is a major cause of GBM patient mortality. It is therefore imperative to design therapeutics which can effectively target GSCs which exist within hypoxic regions and invasive ends of GBMs.

As mentioned above, recent reports have brought to light the gaps in knowledge regarding cellular plasticity and the role played by the GBM microenvironment. Studies have shown that non-stem GBM cells de-differentiate into GSCs along with transdifferentiation of GSCs into cells of the perivascular niche. Therapies designed to target GSC specific biomarkers and angiogenic blood vessels might therefore prove to be ineffective in future. The recent development of transgenic models which can recapitulate, to a certain extent, the heterogenous nature of GBMs could allow for the better identification of microenvironmental components which have a relevant impact on GSC proliferation and survival. Knowledge about the interactions between GSCs and their microenvironment, cellular plasticity within GBMs and the cellular signaling cascades that maintain stemness of GSCs would allow for the targeting of multiple critical events within tumors. Delivering GSC targeted lethal therapeutics using genetically engineered cellular components of GSC microenvironment could potentially reduce disease recurrence (Birbrair et al. 2016).

Investigations of the biology of GSCs and how they become resistant to therapy can also extend to greater understanding of the role of the microenvironment in their maintenance and survival. It is therefore essential to appreciate the highly integrated nature of the GSC niche consisting of various cell populations with regards to their cellular functions and identity.

References

Abbott NJ (2002) Astrocyte-endothelial interactions and blood-brain barrier permeability. J Anat 200(6):629–638

Abbott NJ, Patabendige AAK, Dolman DEM, Yusof SR, Begley DJ (2010) Structure and function of the blood–brain barrier. Neurobiol Dis 37(1):13–25

Agarwal S, Manchanda P, Vogelbaum MA, Ohlfest JR, Elmquist WF (2013) Function of the blood-brain barrier and restriction of drug delivery to invasive glioma cells: findings in an orthotopic rat xenograft model of glioma. Drug Metab Dispos 41(1):33–39

Ahmed AU, Auffinger B, Lesniak MS (2013) Understanding glioma stem cells: rationale, clinical relevance and therapeutic strategies. Expert Rev Neurother 13(5):545–555

Alcantara Llaguno SR, Wang Z, Sun D et al (2015) Adult lineage restricted CNS progenitors specify distinct glioblastoma subtypes. Cancer Cell 28(4):429–440

Bao S, Wu Q, McLendon RE et al (2006a) Glioma stem cells promote radioresistance by preferential activation of the DNA damage response. Nature 444(7120):756–760

Bao S, Wu Q, Sathornsumetee S et al (2006b) Stem cell-like glioma cells promote tumor angiogenesis through vascular endothelial growth factor. Cancer Res 66(16):7843–7848

Bar EE, Lin A, Mahairaki V, Matsui W, Eberhart CG (2010) Hypoxia increases the expression of stem-cell markers and promotes clonogenicity in glioblastoma neurospheres. Am J Pathol 177(3):1491–1502

Baskar R, Lee KA, Yeo R, Yeoh K-W (2012) Cancer and radiation therapy: current advances and future directions. Int J Med Sci 9(3):193–199

Birbrair A, Sattiraju A, Zhu D et al (2016) Novel peripherally derived neural-like stem cells as therapeutic carriers for treating glioblastomas. Stem Cells Transl Med 14:2016

Bissell MJ, Radisky D (2001) Putting tumours in context. Nat Rev Cancer 1(1):46–54

Bleau AM, Hambardzumyan D, Ozawa T et al (2009) PTEN/PI3K/Akt pathway regulates the side population phenotype and ABCG2 activity in glioma tumor stem-like cells. Cell Stem Cell 4(3):226–235

Bonavia R, Inda M-d-M, Cavenee WK, Furnari FB (2011) Heterogeneity maintenance in glioblastoma: a social network. Cancer Res 71(12):4055–4060

Bradshaw A, Wickremsekera A, Tan ST, Peng L, Davis PF, Itinteang T (2016) Cancer stem cell hierarchy in glioblastoma multiforme. Front Surg 3:21

Brat DJ, Castellano-Sanchez AA, Hunter SB et al (2004) Pseudopalisades in glioblastoma are hypoxic, express extracellular matrix proteases, and are formed by an actively migrating cell population. Cancer Res 64(3):920–927

Brooks MD, Sengupta R, Snyder SC, Rubin JB (2013) Hitting them where they live: targeting the glioblastoma perivascular stem cell niche. Curr Pathobiol Rep 1(2):101–110

Burgess A, Nhan T, Moffatt C, Klibanov AL, Hynynen K (2014) Analysis of focused ultrasound-induced blood-brain barrier permeability in a mouse model of Alzheimer's disease using two-photon microscopy. J Control Release 192:243–248

Cabrera MC, Hollingsworth RE, Hurt EM (2015) Cancer stem cell plasticity and tumor hierarchy. World J Stem Cells 7(1):27–36

Calabrese C, Poppleton H, Kocak M et al (2007) A perivascular niche for brain tumor stem cells. Cancer Cell 11(1):69–82

Carmeliet P, Jain RK (2000) Angiogenesis in cancer and other diseases. Nature 407(6801):249–257

Chaichana KL (2014) The need to continually redefine the goals of surgery for glioblastoma. Neurooncology 30:2014

Charles N, Ozawa T, Squatrito M et al (2010) Perivascular nitric oxide activates notch signaling and promotes stem-like character in PDGF-induced glioma cells. Cell Stem Cell 6(2):141–152

Chen J, Ding Z, Peng Y et al (2014) HIF-1α inhibition reverses multidrug resistance in colon cancer cells via downregulation of MDR1/P-glycoprotein. PLoS One 9(6):e98882

Cheng L, Huang Z, Zhou W et al (2013) Glioblastoma stem cells generate vascular pericytes to support vessel function and tumor growth. Cell 153(1):139–152

Cho DY, Lin SZ, Yang WK et al (2013) Targeting cancer stem cells for treatment of glioblastoma multiforme. Cell Transplant 22(4):731–739

Chou C-W, Wang C-C, Wu C-P et al (2012) Tumor cycling hypoxia induces chemoresistance in glioblastoma multiforme by upregulating the expression and function of ABCB1. Neurooncology 14(10):1227–1238

Dalerba P, Cho RW, Clarke MF (2007) Cancer stem cells: models and concepts. Annu Rev Med 58:267–284

De Palma M, Murdoch C, Venneri MA, Naldini L, Lewis CE (2007) Tie2-expressing monocytes: regulation of tumor angiogenesis and therapeutic implications. Trends Immunol 28(12):519–524

Dewhirst MW, Cao Y, Moeller B (2008) Cycling hypoxia and free radicals regulate angiogenesis and radiotherapy response. Nat Rev Cancer 8(6):425–437

Dick JE (2009) Looking ahead in cancer stem cell research. Nat Biotechnol 27(1):44–46

Doetsch F (2003) A niche for adult neural stem cells. Curr Opin Genet Dev 13(5):543–550

Duda DG, Kozin SV, Kirkpatrick ND, Xu L, Fukumura D, Jain RK (2011) CXCL12 (SDF1alpha)-CXCR4/CXCR7 pathway inhibition: an emerging sensitizer for anticancer therapies? Clin Cancer Res 17(8):2074–2080

El Hallani S, Boisselier B, Peglion F et al (2010) A new alternative mechanism in glioblastoma vascularization: tubular vasculogenic mimicry. Brain 133(Pt 4):973–982

Eriksson PS, Perfilieva E, Bjork-Eriksson T et al (1998) Neurogenesis in the adult human hippocampus. Nat Med 4(11):1313–1317

Evans SM, Judy KD, Dunphy I et al (2004) Comparative measurements of hypoxia in human brain tumors using needle electrodes and EF5 binding. Cancer Res 64(5):1886–1892

Eyupoglu IY, Buchfelder M, Savaskan NE (2013) Surgical resection of malignant gliomas—role in optimizing patient outcome. Nat Rev Neurol 9(3):141–151

Fan X, Khaki L, Zhu TS et al (2010) NOTCH pathway blockade depletes CD133-positive glioblastoma cells and inhibits growth of tumor neurospheres and xenografts. Stem Cells 28(1):5–16

Fang D, Nguyen TK, Leishear K et al (2005) A tumorigenic subpopulation with stem cell properties in melanomas. Cancer Res 65(20):9328–9337

Fang DD, Kim YJ, Lee CN et al (2010) Expansion of CD133(+) colon cancer cultures retaining stem cell properties to enable cancer stem cell target discovery. Br J Cancer 102(8):1265–1275

Fidoamore A, Cristiano L, Antonosante A et al (2016) Glioblastoma stem cells microenvironment: the paracrine roles of the niche in drug and radioresistance. Stem Cells Int 2016:6809105

Folkins C, Shaked Y, Man S et al (2009) Glioma tumor stem-like cells promote tumor angiogenesis and vasculogenesis via vascular endothelial growth factor and stromal-derived factor 1. Cancer Res 69(18):7243–7251

Folkman J, Shing Y (1992) Angiogenesis. J Biol Chem 267(16):10931–10934

Folkman J, Klagsbrun M, Sasse J, Wadzinski M, Ingber D, Vlodavsky I (1988) A heparin-binding angiogenic protein—basic fibroblast growth factor—is stored within basement membrane. Am J Pathol 130(2):393–400

Fortunel NO, Otu HH, Ng HH et al (2003) Comment on " 'Stemness': transcriptional profiling of embryonic and adult stem cells" and "a stem cell molecular signature". Science 302(5644):393. author reply 393

Friedmann-Morvinski D, Verma IM (2014) Dedifferentiation and reprogramming: origins of cancer stem cells. EMBO Rep 15(3):244–253

Gage FH (2000) Mammalian neural stem cells. Science 287(5457):1433–1438

Garcion E, Halilagic A, Faissner A, French-Constant C (2004) Generation of an environmental niche for neural stem cell development by the extracellular matrix molecule tenascin C. Development 131(14):3423–3432

Gerstner E, Zhang Z, Fink J et al (2016) ACRIN 6684: assessment of tumor hypoxia in newly diagnosed GBM using 18F-FMISO PET and MRI. Clin Cancer Res 22:5079–5086

Gilbertson RJ, Rich JN (2007) Making a tumour's bed: glioblastoma stem cells and the vascular niche. Nat Rev Cancer 7(10):733–736

Gilmore AP, Romer LH (1996) Inhibition of focal adhesion kinase (FAK) signaling in focal adhesions decreases cell motility and proliferation. Mol Biol Cell 7(8):1209–1224

Gorlach A, Bonello S (2008) The cross-talk between NF-kappaB and HIF-1: further evidence for a significant liaison. Biochem J 412(3):e17–e19

Greene-Schloesser D, Robbins ME (2012) Radiation-induced cognitive impairment-from bench to bedside. Neurooncology 14(suppl 4):iv37–iv44

Greene-Schloesser D, Robbins ME, Peiffer AM, Shaw EG, Wheeler KT, Chan MD (2012) Radiation-induced brain injury: a review. Front Oncol 2:73

Hanahan D, Weinberg RA (2011) Hallmarks of cancer: the next generation. Cell 144(5):646–674

Hardee ME, Zagzag D (2012) Mechanisms of glioma-associated neovascularization. Am J Pathol 181(4):1126–1141

Hawkins BT, Davis TP (2005) The blood-brain barrier/neurovascular unit in health and disease. Pharmacol Rev 57(2):173–185

Heddleston JM, Li Z, Hjelmeland AB, Rich JN (2009) The hypoxic microenvironment maintains glioblastoma stem cells and promotes reprogramming towards a cancer stem cell phenotype. Cell Cycle 8(20):3274–3284

Heddleston JM, Li Z, Lathia JD, Bao S, Hjelmeland AB, Rich JN (2010) Hypoxia inducible factors in cancer stem cells. Br J Cancer 102(5):789–795

Heddleston JM, Hitomi M, Venere M et al (2011) Glioma stem cell maintenance: the role of the microenvironment. Curr Pharm Des 17(23):2386–2401

Holash J, Maisonpierre PC, Compton D et al (1999) Vessel cooption, regression, and growth in tumors mediated by angiopoietins and VEGF. Science 284(5422):1994–1998

Hottinger AF, Stupp R, Homicsko K (2014) Standards of care and novel approaches in the management of glioblastoma multiforme. Chin J Cancer 33(1):32–39

Huang P, Rani MR, Ahluwalia MS et al (2012) Endothelial expression of TNF receptor-1 generates a proapoptotic signal inhibited by integrin alpha6beta1 in glioblastoma. Cancer Res 72(6):1428–1437

Ignatova TN, Kukekov VG, Laywell ED, Suslov ON, Vrionis FD, Steindler DA (2002) Human cortical glial tumors contain neural stem-like cells expressing astroglial and neuronal markers in vitro. Glia 39(3):193–206

Jackson M, Hassiotou F, Nowak A (2015) Glioblastoma stem-like cells: at the root of tumor recurrence and a therapeutic target. Carcinogenesis 36(2):177–185

Jhaveri N, Chen TC, Hofman FM (2016) Tumor vasculature and glioma stem cells: contributions to glioma progression. Cancer Lett 380(2):545–551

Jordao JF, Thevenot E, Markham-Coultes K et al (2013) Amyloid-beta plaque reduction, endogenous antibody delivery and glial activation by brain-targeted, transcranial focused ultrasound. Exp Neurol 248:16–29

Kang TW, Choi SW, Yang SR et al (2014) Growth arrest and forced differentiation of human primary glioblastoma multiformE by a novel small molecule. Sci Rep 4:5546

Kelly JJ, Stechishin O, Chojnacki A et al (2009) Proliferation of human glioblastoma stem cells occurs independently of exogenous mitogens. Stem Cells 27(8):1722–1733

Konofagou EE, Tung YS, Choi J, Deffieux T, Baseri B, Vlachos F (2012) Ultrasound-induced blood-brain barrier opening. Curr Pharm Biotechnol 13(7):1332–1345

Kreso A, Dick John E (2014) Evolution of the cancer stem cell model. Cell Stem Cell 14(3):275–291

Laks DR, Masterman-Smith M, Visnyei K et al (2009) Neurosphere formation is an independent predictor of clinical outcome in malignant glioma. Stem Cells 27(4):980–987

Lapidot T, Sirard C, Vormoor J et al (1994) A cell initiating human acute myeloid leukaemia after transplantation into SCID mice. Nature 367(6464):645–648

Lathia JD, Gallagher J, Heddleston JM et al (2010) Integrin alpha 6 regulates glioblastoma stem cells. Cell Stem Cell 6(5):421–432

Lathia JD, Mack SC, Mulkearns-Hubert EE, Valentim CLL, Rich JN (2015) Cancer stem cells in glioblastoma. Genes Dev 29(12):1203–1217

Lee G, Hall RR, Ahmed AU (2016) Cancer stem cells: cellular plasticity, niche, and its clinical relevance. J Stem Cell Res Ther 6(10):363

Leins A, Riva P, Lindstedt R, Davidoff MS, Mehraein P, Weis S (2003) Expression of tenascin-C in various human brain tumors and its relevance for survival in patients with astrocytoma. Cancer 98(11):2430–2439

Leventhal C, Rafii S, Rafii D, Shahar A, Goldman SA (1999) Endothelial trophic support of neuronal production and recruitment from the adult mammalian subependyma. Mol Cell Neurosci 13(6):450–464

Li Z, Bao S, Wu Q et al (2009) Hypoxia-inducible factors regulate tumorigenic capacity of glioma stem cells. Cancer Cell 15(6):501–513

Liebelt BD, Shingu T, Zhou X, Ren J, Shin SA, Hu J (2016) Glioma stem cells: signaling, microenvironment, and therapy. Stem Cells Int 2016:7849890

Louis DN, Ohgaki H, Wiestler OD et al (2007) The 2007 WHO classification of tumours of the central nervous system. Acta Neuropathol 114(2):97–109

Louvi A, Artavanis-Tsakonas S (2006) Notch signalling in vertebrate neural development. Nat Rev Neurosci 7(2):93–102

Mannino M, Chalmers AJ (2011) Radioresistance of glioma stem cells: Intrinsic characteristic or property of the 'microenvironment-stem cell unit'? Mol Oncol 5(4):374–386

Mao XG, Xue XY, Wang L et al (2013) CDH5 is specifically activated in glioblastoma stem-like cells and contributes to vasculogenic mimicry induced by hypoxia. Neurooncology 15(7):865–879

Mathieu J, Zhou W, Xing Y et al (2014) Hypoxia inducible factors have distinct and stage-specific roles during reprogramming of human cells to pluripotency. Cell Stem Cell 14(5):592–605

McCord AM, Jamal M, Shankavaram UT, Lang FF, Camphausen K, Tofilon PJ (2009) Physiologic oxygen concentration enhances the stem-like properties of CD133+ human glioblastoma cells in vitro. Mol Cancer Res 7(4):489–497

MDM I, Bonavia R, Mukasa A et al (2010) Tumor heterogeneity is an active process maintained by a mutant EGFR-induced cytokine circuit in glioblastoma. Genes Dev 24(16):1731–1745

Mendez O, Zavadil J, Esencay M et al (2010) Knock down of HIF-1alpha in glioma cells reduces migration in vitro and invasion in vivo and impairs their ability to form tumor spheres. Mol Cancer 9:133

Merkle FT, Tramontin AD, Garcia-Verdugo JM, Alvarez-Buylla A (2004) Radial glia give rise to adult neural stem cells in the subventricular zone. Proc Natl Acad Sci U S A 101(50):17528–17532

Murat A, Migliavacca E, Gorlia T et al (2008a) Stem cell–related "Self-Renewal" signature and high epidermal growth factor receptor expression associated with resistance to concomitant chemoradiotherapy in glioblastoma. J Clin Oncol 26(18):3015–3024

Murat A, Migliavacca E, Gorlia T et al (2008b) Stem cell-related "self-renewal" signature and high epidermal growth factor receptor expression associated with resistance to concomitant chemoradiotherapy in glioblastoma. J Clin Oncol 26(18):3015–3024

Nakada M, Nambu E, Furuyama N et al (2013) Integrin [alpha]3 is overexpressed in glioma stem-like cells and promotes invasion. Br J Cancer 108(12):2516–2524

Nath B, Szabo G (2012) Hypoxia and hypoxia inducible factors: diverse roles in liver diseases. Hepatology 55(2):622–633

Nguyen LV, Vanner R, Dirks P, Eaves CJ (2012) Cancer stem cells: an evolving concept. Nat Rev Cancer 12(2):133–143

Nhan T, Burgess A, Lilge L, Hynynen K (2014) Modeling localized delivery of Doxorubicin to the brain following focused ultrasound enhanced blood-brain barrier permeability. Phys Med Biol 59(20):5987–6004

Nishide K, Nakatani Y, Kiyonari H, Kondo T (2009) Glioblastoma formation from cell population depleted of Prominin1-expressing cells. PLoS One 4(8):e6869

Ogden AT, Waziri AE, Lochhead RA et al (2008) Identification of A2B5+CD133- tumor-initiating cells in adult human gliomas. Neurosurgery 62(2):505–514. discussion 514-505

Paez-Gonzalez P, Asrican B, Rodriguez E, Kuo CT (2014) Identification of distinct ChAT+ neurons and activity-dependent control of postnatal SVZ neurogenesis. Nat Neurosci 17(7):934–942

Palmer TD, Willhoite AR, Gage FH (2000) Vascular niche for adult hippocampal neurogenesis. J Comp Neurol 425(4):479–494

Pardridge WM (2005) The blood-brain barrier: bottleneck in brain drug development. NeuroRx 2(1):3–14

Patel AP, Tirosh I, Trombetta JJ et al (2014) Single-cell RNA-seq highlights intratumoral heterogeneity in primary glioblastoma. Science 344(6190):1396–1401

Persano L, Pistollato F, Rampazzo E et al (2012) BMP2 sensitizes glioblastoma stem-like cells to Temozolomide by affecting HIF-1[alpha] stability and MGMT expression. Cell Death Dis e412:3

Persidsky Y, Ramirez SH, Haorah J, Kanmogne GD (2006) Blood–brain barrier: structural components and function under physiologic and pathologic conditions. J Neuroimmune Pharmacol 1(3):223–236

Pistollato F, Chen H-L, Rood BR et al (2009) Hypoxia and HIF1α repress the differentiative effects of BMPs in high-grade glioma. Stem Cells 27(1):7–17

Pistollato F, Abbadi S, Rampazzo E et al (2010) Intratumoral hypoxic gradient drives stem cells distribution and MGMT expression in glioblastoma. Stem Cells 28(5):851–862

Ramirez-Castillejo C, Sanchez-Sanchez F, Andreu-Agullo C et al (2006) Pigment epithelium-derived factor is a niche signal for neural stem cell renewal. Nat Neurosci 9(3):331–339

Reiss Y, Machein MR, Plate KH (2005) The role of angiopoietins during angiogenesis in gliomas. Brain Pathol 15(4):311–317

Reya T, Morrison SJ, Clarke MF, Weissman IL (2001) Stem cells, cancer, and cancer stem cells. Nature 414(6859):105–111

Ricci-Vitiani L, Pallini R, Biffoni M et al (2010) Tumour vascularization via endothelial differentiation of glioblastoma stem-like cells. Nature 468(7325):824–828

Riquelme PA, Drapeau E, Doetsch F (2008) Brain micro-ecologies: neural stem cell niches in the adult mammalian brain. Philos Trans R Soc Lond Ser B Biol Sci 363(1489):123–137

Rong Y, Durden DL, Van Meir EG, Brat DJ (2006) 'Pseudopalisading' necrosis in glioblastoma: a familiar morphologic feature that links vascular pathology, hypoxia, and angiogenesis. J Neuropathol Exp Neurol 65(6):529

Safa AR, Saadatzadeh MR, Cohen-Gadol AA, Pollok KE, Bijangi-Vishehsaraei K (2015) Glioblastoma stem cells (GSCs) epigenetic plasticity and interconversion between differentiated non-GSCs and GSCs. Genes Diseases 2(2):152–163

Scadden DT (2006) The stem-cell niche as an entity of action. Nature 441(7097):1075–1079

Scheres B (2007) Stem-cell niches: nursery rhymes across kingdoms. Nat Rev Mol Cell Biol 8(5):345–354

Seidel S, Garvalov BK, Wirta V et al (2010) A hypoxic niche regulates glioblastoma stem cells through hypoxia inducible factor 2 alpha. Brain 133(Pt 4):983–995

Semenza GL (2010) Defining the role of hypoxia-inducible factor 1 in cancer biology and therapeutics. Oncogene 29(5):625–634

Semenza GL (2013) HIF-1 mediates metabolic responses to intratumoral hypoxia and oncogenic mutations. J Clin Invest 123(9):3664–3671

Shackleton M, Quintana E, Fearon ER, Morrison SJ (2009) Heterogeneity in cancer: cancer stem cells versus clonal evolution. Cell 138(5):822–829

Singh SK, Clarke ID, Terasaki M et al (2003) Identification of a cancer stem cell in human brain tumors. Cancer Res 63(18):5821–5828

Singh SK, Hawkins C, Clarke ID et al (2004a) Identification of human brain tumour initiating cells. Nature 432

Singh SK, Clarke ID, Hide T, Dirks PB (2004b) Cancer stem cells in nervous system tumors. Oncogene 23(43):7267–7273

Soda Y, Marumoto T, Friedmann-Morvinski D et al (2011) Transdifferentiation of glioblastoma cells into vascular endothelial cells. Proc Natl Acad Sci U S A 108(11):4274–4280

Son MJ, Woolard K, Nam DH, Lee J, Fine HA (2009) SSEA-1 is an enrichment marker for tumor-initiating cells in human glioblastoma. Cell Stem Cell 4(5):440–452

Sottoriva A, Spiteri I, Piccirillo SGM et al (2013) Intratumor heterogeneity in human glioblastoma reflects cancer evolutionary dynamics. Proc Natl Acad Sci 110(10):4009–4014

Stier S, Ko Y, Forkert R et al (2005) Osteopontin is a hematopoietic stem cell niche component that negatively regulates stem cell pool size. J Exp Med 201(11):1781–1791

Stockhausen M-T, Kristoffersen K, Poulsen HS (2009) The functional role of Notch signaling in human gliomas. Neurooncology 12(2):199–211

Sun X, Cheng G, Hao M et al (2010) CXCL12/CXCR4/CXCR7 chemokine axis and cancer progression. Cancer Metastasis Rev 29(4):709–722

Tallet AV, Azria D, Barlesi F et al (2012) Neurocognitive function impairment after whole brain radiotherapy for brain metastases: actual assessment. Radiat Oncol 7(1):77

Tan BT, Park CY, Ailles LE, Weissman IL (2006) The cancer stem cell hypothesis: a work in progress. Lab Investig 86(12):1203–1207

Tavazoie M, Van der Veken L, Silva-Vargas V et al (2008) A specialized vascular niche for adult neural stem cells. Cell Stem Cell 3(3):279–288

Tomuleasa C, Soritau O, Rus-Ciuca D et al (2010) Isolation and characterization of hepatic cancer cells with stem-like properties from hepatocellular carcinoma. J Gastrointestin Liver Dis 19(1):61–67

Uchida N, Buck DW, He D et al (2000) Direct isolation of human central nervous system stem cells. Proc Natl Acad Sci 97(26):14720–14725

Uhm JH, Dooley NP, Kyritsis AP, Rao JS, Gladson CL (1999) Vitronectin, a glioma-derived extracellular matrix protein, protects tumor cells from apoptotic death. Clin Cancer Res 5(6):1587–1594

Van Meir EG, Hadjipanayis CG, Norden AD, Shu HK, Wen PY, Olson JJ (2010) Exciting new advances in neuro-oncology: the avenue to a cure for malignant glioma. CA Cancer J Clin 60(3):166–193

Venneri MA, De Palma M, Ponzoni M et al (2007) Identification of proangiogenic TIE2-expressing monocytes (TEMs) in human peripheral blood and cancer. Blood 109(12):5276–5285

Verhaak RG, Hoadley KA, Purdom E et al (2010) Integrated genomic analysis identifies clinically relevant subtypes of glioblastoma characterized by abnormalities in PDGFRA, IDH1, EGFR, and NF1. Cancer Cell 17(1):98–110

Vescovi AL, Galli R, Reynolds BA (2006) Brain tumour stem cells. Nat Rev Cancer 6(6):425–436

Wang CY, Mayo MW, Baldwin AS Jr (1996) TNF- and cancer therapy-induced apoptosis: potentiation by inhibition of NF-kappaB. Science 274(5288):784–787

Wang J, Wakeman TP, Latha JD et al (2010a) Notch promotes radioresistance of glioma stem cells. Stem Cells 28(1):17–28

Wang R, Chadalavada K, Wilshire J et al (2010b) Glioblastoma stem-like cells give rise to tumour endothelium. Nature 468(7325):829–833

Weller M, Cloughesy T, Perry JR, Wick W (2012) Standards of care for treatment of recurrent glioblastoma—are we there yet? Neurooncology 2012

Wolburg H, Lippoldt A (2002) Tight junctions of the blood–brain barrier: development, composition and regulation. Vasc Pharmacol 38(6):323–337

Yao X, Ping Y, Liu Y et al (2013) Vascular endothelial growth factor receptor 2 (VEGFR-2) plays a key role in vasculogenic mimicry formation, neovascularization and tumor initiation by Glioma stem-like cells. PLoS One 8(3):e57188

Yong RL, Lonser RR (2011) Surgery for glioblastoma multiforme: striking a balance. World Neurosurg 76(6):528–530

Zhong H, De Marzo AM, Laughner E et al (1999) Overexpression of hypoxia-inducible factor 1alpha in common human cancers and their metastases. Cancer Res 59(22):5830–5835

Zhu Z, Hao X, Yan M et al (2010) Cancer stem/progenitor cells are highly enriched in CD133+CD44+ population in hepatocellular carcinoma. Int J Cancer 126(9):2067–2078

Chapter 8
Plasticity of the Muscle Stem Cell Microenvironment

Ivana Dinulovic, Regula Furrer, and Christoph Handschin

Abstract Satellite cells (SCs) are adult muscle stem cells capable of repairing damaged and creating new muscle tissue throughout life. Their functionality is tightly controlled by a microenvironment composed of a wide variety of factors, such as numerous secreted molecules and different cell types, including blood vessels, oxygen, hormones, motor neurons, immune cells, cytokines, fibroblasts, growth factors, myofibers, myofiber metabolism, the extracellular matrix and tissue stiffness. This complex niche controls SC biology—quiescence, activation, proliferation, differentiation or renewal and return to quiescence. In this review, we attempt to give a brief overview of the most important players in the niche and their mutual interaction with SCs. We address the importance of the niche to SC behavior under physiological and pathological conditions, and finally survey the significance of an artificial niche both for basic and translational research purposes.

Keywords Skeletal muscle • Muscle regeneration • Satellite cells • Stem cell-niche • Muscular dystrophies • Extracellular matrix

8.1 Satellite Cells

Over the past half a century, the focus of research on muscle regeneration has shifted from other myogenic cells of muscle tissue to satellite cells (SCs), from developmental myogenesis to adult muscle regeneration, from cell-intrinsic properties of SCs to the relevance of extrinsic factors delivered by their niche. SCs, small, inactive cells wedged between the myofiber and the surrounding extracellular matrix (ECM), have attracted the attention of scientists since their discovery 56 years ago (Mauro 1961). The astonishing translational potential of SCs continues to fascinate, and the ever expanding knowledge of SCs and their microenvironment paves the

I. Dinulovic • R. Furrer • C. Handschin (✉)
Biozentrum, University of Basel, Basel, Switzerland
e-mail: christoph.handschin@unibas.ch

© Springer International Publishing AG 2017
A. Birbrair (ed.), *Stem Cell Microenvironments and Beyond*,
Advances in Experimental Medicine and Biology 1041,
DOI 10.1007/978-3-319-69194-7_8

way for the development of novel cell and gene therapies, in vitro disease models and preclinical drug testing paradigms. Here, we discuss different aspects of SC biology and the niche in health and disease. For a more detailed assessment of the particularities of SCs and the SC niche, we direct readers to several recent reviews focusing on the extracellular matrix (Thomas et al. 2015), blood vessels (Mounier et al. 2011), bioengineering (Bursac et al. 2015), SC function from a cell-intrinsic perspective (Almada and Wagers 2016) and an extensive review on SC biology (Yin et al. 2013).

8.1.1 Skeletal Muscle Regeneration and Muscle Stem Cells

Comprising approximately 40% of body weight, skeletal muscle can be considered the largest organ in the human body (Janssen et al. 2000). Muscle not only supports breathing and movement, but is also a very important metabolic and endocrine organ. It comes as no surprise that skeletal muscle has a remarkable capability to repair damage caused by injuries or simple everyday wear-and-tear. As numerous animal studies demonstrate, skeletal muscle is able to regain near original morphology and functionality after several weeks of serious damage caused by injection of myotoxic agents (e.g. cardiotoxin, bupivacaine, barium chloride or notexin), freezing, crushing, or complete mincing and re-transplantation (Rosenblatt 1992; Carlson and Gutmann 1972; Fink et al. 2003; Dinulovic et al. 2016a; Warren et al. 2004). However, aging, traumatic injuries in humans resulting in volumetric muscle loss and various myopathies result in impaired functionality and inability of the tissue to regain homeostatic conditions.

SCs are the main cells responsible for sustaining skeletal muscle morphology and functionality throughout the lifetime of an individual. They are largely lineage-committed adult stem cells located at the periphery of muscle fibers, situated between the sarcolemma (the myofiber membrane) and basal lamina (BL) (Mauro 1961), in close proximity to blood vessels (Mounier et al. 2011) and the neuromuscular junction (Kelly 1978). This specific environment surrounding SCs is known as the SC niche.

Under resting conditions, SCs are in the G0 phase (non-cycling state) and quiescent, with a heterochromatic nucleus and a thin rim of cytoplasm containing scarce organelles. These cells are most commonly distinguished by the expression of the paired box transcription factor Pax7. SCs have a tremendous myogenic potential and self-renewal capabilities, as demonstrated by single-fiber (Collins et al. 2005) as well as single cell (Sacco et al. 2008) implantation in irradiated muscles of immunodeficient mice.

The classical cascade of regeneration resembles that of prenatal skeletal muscle development (Bentzinger et al. 2012). In response to injury or other stimuli, SCs become activated, increase in size and begin to proliferate. The majority of the progeny reduces Pax7 and induces MyoD expression. After several rounds of proliferation, these myoblasts start to express myogenin and exit the cell cycle as myocytes.

The myocytes subsequently fuse in order to form new or repair existing myofibers (depending on the severity of injury). The myofibers then express myogenic regulatory factor 4 (MRF-4) and grow, supported by hypertrophy, until reaching their pre-injury size. At the same time, a part of the SC progeny reacquires high Pax7 levels and returns to quiescence, thereby replenishing the SC pool and maintaining sufficient reserves for future rounds of regeneration.

Besides SCs, several other cell types, such as muscle side-population (SP) cells, muscle-derived stem cells (MDSCs), bone marrow stem cells, PW1[+] interstitial cells, CD133[+] cells, mesoangioblasts (MABs) and pericytes, can successfully regenerate muscles and some can even reconstitute the niche upon transplantation into damaged muscle (Péault et al. 2007). However, the contribution of these cells seems to be very low under physiological conditions and dependent on SCs, which are essential for skeletal muscle regeneration and therefore represent the true stem cells of muscle tissue (Lepper et al. 2011; Sambasivan et al. 2011; McCarthy et al. 2011; Murphy et al. 2011).

According to their gene expression profiles and their characteristics in vitro SCs stemming from different muscle groups (e.g. head vs. limb muscles) are heterogeneous. Nevertheless, SCs from the masseter muscle (head) are able to regenerate the extensor digitorum longus (EDL) muscle (limb) as efficiently as SCs from the EDL muscle (Ono et al. 2010), attesting to the enormous influence of the in vivo microenvironment on the behavior and functionality of SCs, which in some cases can overcome the intrinsic differences between SCs.

8.1.2 *The Heterogeneity of Satellite Cells and Its Dependence on the Niche*

Several studies have addressed the heterogeneity of SC populations in regard to their renewal potential. Interestingly, SC heterogeneity was not only reported between different muscle beds, but also observed between SCs on the same muscle fibers, thereby implicating additional factors besides ontogeny and composition of the fiber type as possible causes. According to these studies, only a small proportion of SCs are *bona fide* stem cells, whereas the vast majority are committed progenitors with limited stemness. For example, Chakkalakal et al. discovered heterogeneity among SCs based on their proliferative history, suggesting that cells that cycle less frequently have higher self-renewal potential (Chakkalakal et al. 2012). On a related note, Rocheteau et al. evaluated differential DNA strand segregation, where one daughter cell retains the template strands, stays in the niche and returns to quiescence, while the other daughter cell receives newly synthesized DNA strands, continues to proliferate and finally differentiates (Rocheteau et al. 2012). It was suggested that such DNA strand segregation would prevent accumulation of proliferation-associated mutations in the stem cell, and therefore provide a lifelong supply of progenitors. Similarly, in a lineage tracing experiment with Myf5-Cre/

ROSA-YFP mice, Kuang et al. found that the majority of SCs are Pax7[+]/Myf5[+], and only small subset are Pax7[+]/Myf5[−] cells (Kuang et al. 2007). Upon isolation and transplantation, both cell populations are capable of proliferating and differentiating, but only Myf5[−] SCs occupy the niche in the transplanted muscle. In addition, after in vivo activation, Pax7[+]/Myf5[+] (committed progenitors) are exclusively prone to symmetrical division, giving rise to more committed progenitors, whereas Pax7[+]/Myf5[−] (true stem cells) on the other hand can divide both symmetrically and asymmetrically, producing uncommitted and committed daughter cells. Mechanistically, the asymmetrical distribution of the Par complex results in p38α/β MAPK activation and MyoD expression only in the committed daughter (Troy et al. 2012). Importantly, the capability to control the orientation of the cell division is tightly coupled to the SC niche. Following asymmetric division, the uncommitted progenitor remains in the niche in contact with the BL, whereas the committed progenitor is pushed towards the muscle fiber, thus losing contact with the niche. In contrast, both daughter cells retain contact with the BL and the myofiber during a stem cell pool expansion through symmetric division of Pax7[+]/Myf5[−] cells.

8.2 The Satellite Cell Niche in Quiescence and Regeneration

SC quiescence, activation, proliferation, differentiation and renewal are intricately connected to the niche. There is a plethora of cell-cell and cell-matrix interactions, numerous paracrine and endocrine molecules (e.g. growth factors and cytokines), as well as biophysical properties of muscle that have a direct effect on the SC. However, this communication is bidirectional, as the SCs themselves also influence their local environment. The most important factors governing the niche in quiescence and activation are depicted in Figs. 8.1 and 8.2, respectively.

8.2.1 The Extracellular Matrix

In homeostatic conditions, SCs are situated just outside the muscle fiber, in direct contact with the sarcolemma and the ECM. The ECM surrounding muscle fibers is called the basal membrane (BM) and it consists of two parts—the reticular lamina (RL) and the BL, the latter being in direct contact with the fiber. The BM is a mesh composed of various glycoproteins and proteoglycans with sequestered growth factors. The main components of the RL are fibrillar collagens, whereas the main components of the BL are laminin-2 ($\alpha2\beta1\gamma1$) and non-fibrillar collagen IV (Sanes 2003). The laminins and collagen of the BL self-assemble into networks that are cross-linked by the glycoprotein nidogen. This network provides binding sites for components of the RL on one, and the sarcolemma and SC membrane on the other side. In addition, proteoglycans such as perlecan are anchored to the main BL mesh and bind polypeptidic growth factors with their glycosaminoglycan chains.

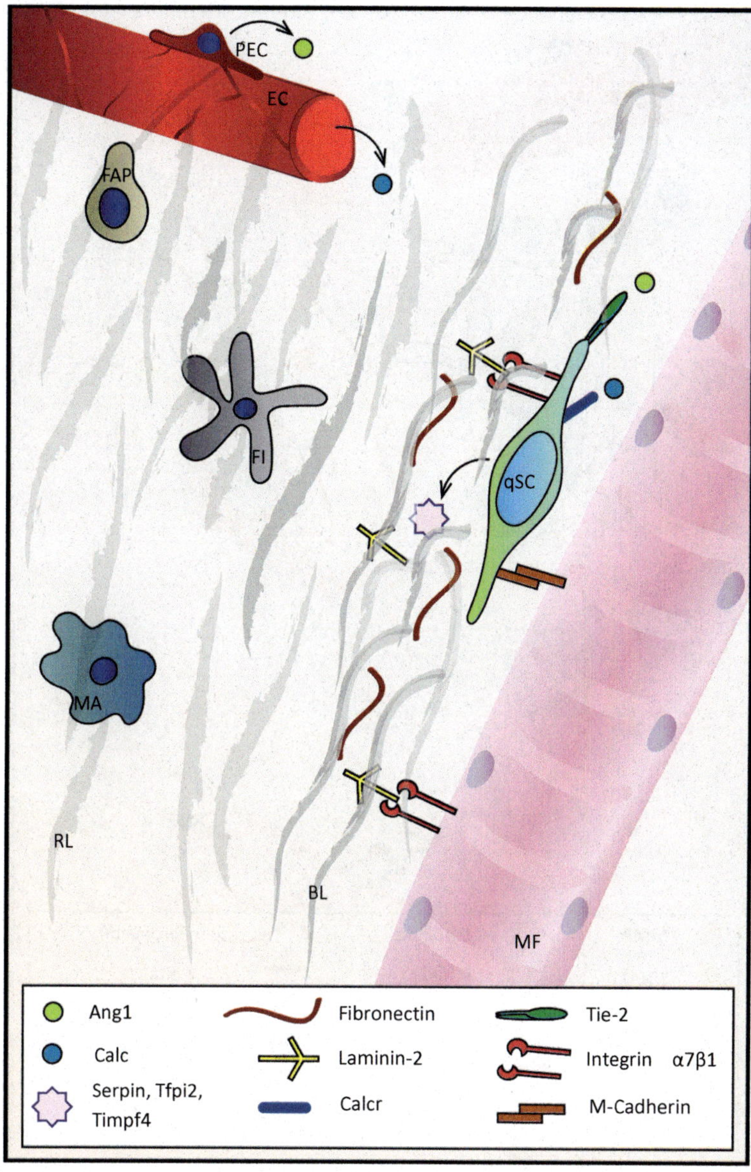

Fig. 8.1 The quiescent niche. Under homeostatic conditions, the SC and its niche are in a quiescent (q) state. The qSC receives signals that keep it dormant, such as angiopoietin 1 (Ang1) from periendothelial cells (PECs) and calcitonin (Calc) from the bloodstream, but at the same time actively contributes to the niche state by secreting protease inhibitors (Serpin, Tfpi2, Timpf4). The SC is attached with M-cadherin to the myofiber (MF) and with integrin to the basal lamina (BL). Fibro/adipogenic progenitors (FAPs) and macrophages (MAs) are present in low numbers and are mainly inactive. Further abbreviations: *Calcr* calcitonin receptor; *EC* endothelial cell; *FI* fibroblast; *RL* reticular lamina; *Tie-2* receptor for Ang1

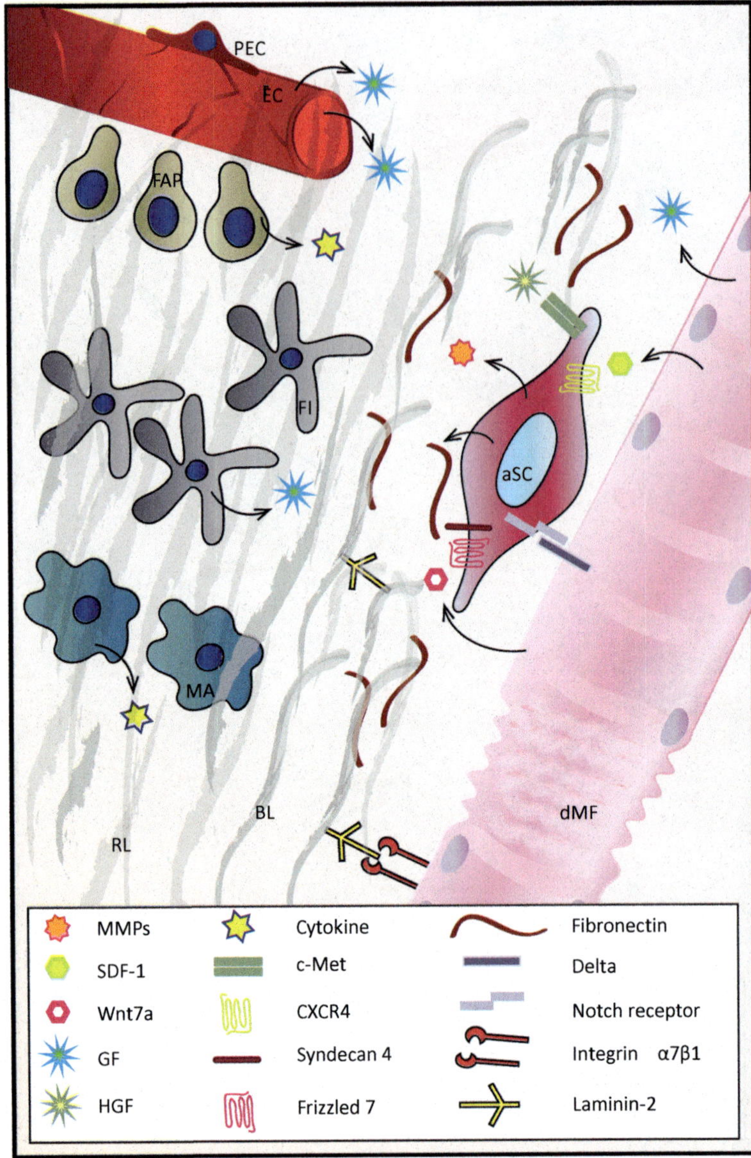

Fig. 8.2 The activated niche. The damaged myofiber (dMF) changes the signaling balance in the niche and causes activation (a) of the SC by liberating hepatocyte growth factor (HGF). The aSC increases in size and begin secreting matrix metalloproteases (MMPs), as well as fibronectin, which together with Wnt7a supports SC proliferation through the Frizzled 7 and syndecan 4 receptors. The dMF expresses Delta and secretes Wnt7a, stromal cell-derived factor 1 (SDF-1) and growth factors (GFs) that further regulate SC behavior. GFs are also secreted by fibroblasts (FIs), endothelial cells (ECs) and are delivered by the bloodstream to the niche. In addition, cytokine secreting fibro/adipogenic progenitors (FAPs) and macrophages (MAs) increase in number. The result is higher density of the basal lamina (BL) and reticular lamina (RL). Further abbreviations: *c-Met* receptor for HGF; *CXCR4* receptor for SDF-1; *Notch* receptor for Delta; *PEC* periendothelial cell

These growth factors, including fibroblast growth factors (FGFs), epidermal growth factor (EGF), insulin-like growth factors (IGFs) and hepatocyte growth factor (HGF), are secreted by various components of the niche, such as muscle fibers, interstitial cells and SCs, or can be delivered to the niche by blood vessels.

Integrins on the SC membrane and the sarcolemma bind to laminins in the BL, forming focal adhesions and contributing to mechanical stability between the ECM and intracellular cytoskeleton. However, these interactions also have important signaling functions. The main integrin isoforms on SCs are $\alpha7$ and $\beta1$, which bind to laminin-2 on the BL side (Blanco-Bose et al. 2001). After SC activation, the expression of integrins on the SC membrane changes, along with the preference for binding partners in the BL. For example, activated, but not quiescent SCs express the $\beta3$ integrin isoform, which probably binds to fibronectin (FN) in a complex with the αv chain (Liu et al. 2011). Both quiescent and activated SCs also express the transmembrane heparin sulfate proteoglycans syndecan-3 and syndecan-4. These proteins form complexes with different tyrosine kinases such as c-Met and FGF receptor (FGFR) on the SC membrane and are consequently important not only for cell adhesion to the BL, but also for SC activation (Cornelison et al. 2001).

Expression profiles of quiescent and activated SCs suggest that SCs actively contribute to maintaining niche quiescence while remaining highly sensitive to activating stimuli (Pallafacchina et al. 2010). Quiescent SCs express the protease inhibitors Serpin and Tfpi2 as well as metalloprotease inhibitor Timpf4. Upon activation, however, these genes become downregulated, and instead, SCs start expressing the matrix metalloproteases MMP-2 and MMP-9 (Guérin and Holland 1995). MMPs are major enzymes responsible for ECM degradation.

Activated SCs also produce FN, an ECM glycoprotein whose role in SC maintenance by enabling their attachment to the niche has recently been demonstrated (Lukjanenko et al. 2016). SC-produced FN potentiates Wnt7a signaling through the receptor complex syndecan-4/Frizzled-7, thereby supporting symmetric division of SCs and expansion of the stem cell pool (Bentzinger et al. 2013). Specific knockdown of FN in SCs leads to a drastic reduction in symmetric division, in particular in the Pax7$^+$/Myf5$^-$ population, leading to a drop in SC numbers during regeneration.

Collagen VI is another BL component essential for preserving the SC pool. Fibroblasts are the prime producers of this protein as well as many other BL components. Collagen VI knock-out mice exhibit reduced regeneration and an inability to maintain SC numbers following injury. This defect is, however, rescued by transplanting wild-type fibroblasts, demonstrating the critical importance of non-SC-autonomous ECM factors in SC maintenance (Urciuolo et al. 2013).

8.2.2 The Muscle Fiber

On the apical side, SCs are bound to a muscle fiber, and M-cadherin is the main adhesion protein supporting the connection between these two cell types. Myofibers are important regulators of SC state: for example, myofiber damage or stretch induces nitric oxide (NO) synthesis in the BL, which is able to activate MMPs, and through that action liberate ECM-bound HGF, allowing its binding to the c-Met receptor on SCs. This HGF signaling through c-Met has been proposed as an initial activation signal for SCs (Tatsumi et al. 2006).

SCs are furthermore affected by the Notch and Wnt signaling pathways in regard to quiescence, activation, proliferation and differentiation (Yin et al. 2013). Proof-of-concept was provided in different studies, e.g. by ablation of RBP-Jκ, a downstream mediator of Notch. This ablation leads to spontaneous activation and differentiation of SCs without a proliferative phase, precipitating depletion of the SC pool and thus indicating that Notch signaling is essential for SC quiescence (Bjornson et al. 2012; Mourikis et al. 2012). Upon injury, damaged fibers express Delta, a ligand of the Notch receptor, which stimulates SC proliferation. In addition, regenerating fibers synthesize Wnt7a, which induces SC symmetrical cell division by binding to the Frizzled7 receptor (Polesskaya et al. 2003).

In regeneration, myofibers secrete stromal cell-derived factor-1 (SDF-1), which binds to the C-X-C motif chemokine receptor 4 (CXCR4) on SCs and induces SC chemoattraction (Ratajczak et al. 2003). Injured fibers and other cells of the niche also secrete FGFs, EGF and IGFs, which further regulate SC proliferation and differentiation. For instance, FGF-2 induces proliferation and represses differentiation of progenitor cells by binding to the tyrosine kinase FGFR and activating the Ras/MAPK pathway (Fedorov et al. 2001). Likewise, IGF-II supports proliferation, while the pleiotropic functions of IGF-I include stimulation of SC proliferation, differentiation, migration and anti-inflammatory effects on the niche (reviewed in (Philippou et al. 2007)). These effects of IGF-I are mediated through several signal transduction pathways, all initiated by IGF-I binding to the tyrosine receptor kinase IGF1R. The situation is further complicated by the existence of multiple IGF-I isoforms, as well as IGF binding proteins (IGFBPs) secreted by activated SCs, whose function is to transport IGFs and modulate their half-life (reviewed in (Jones and Clemmons 1995)). On the other hand, myofibers also secrete myostatin (Mstn), a member of the transforming growth factor β (TGF-β) family and negative regulator of muscle growth that has been implicated in reducing SC activation and self-renewal (McCroskery et al. 2003).

Much attention has been given to metabolic reprograming of SCs, that is, the effects of the metabolism of a SC on its fate (Tang and Rando 2014). Some research proposes that in quiescence, SCs primarily rely on fatty acid oxidation (Ryall et al. 2015), whereas upon activation, they increase substrate utilization through glycolysis, and finally switch to oxidative phosphorylation during differentiation (Wagatsuma and Sakuma 2013). Other studies suggest that activated SCs depend more on oxidative phosphorylation (Tang and Rando 2014; Rodgers et al. 2014; Cerletti et al. 2012). It also remains unclear how metabolic substrate utilization in skeletal fibers (the SC niche) influences the SC state. Experiments with caloric restriction have suggested that the increased fatty acid oxidation and mitochondrial activity in the fiber in this context probably induce SC activation through increased oxidative phosphorylation (Cerletti et al. 2012).

Effects of fiber metabolism on SCs are furthermore implied by the observation that resting SC numbers are considerably higher in oxidative slow-twitch compared to glycolytic fast-twitch myofibers (Gibson and Schultz 1982). Moreover, a similar difference in SC numbers can be achieved by endurance exercise, which promotes a switch from glycolytic to oxidative fibers (Shefer et al. 2010; Wilson et al. 2012). Although a conclusive explanation for the correlation between SC numbers and the oxidative fiber type remains elusive, the metabolic properties and the vascularization have been linked to this observation. The existence of a denser blood vessel network in slow fibers is of particular interest given close vicinity of the majority of SCs to blood vessels (Mounier et al. 2011). However, this simple view has recently been challenged. Namely, mice with myofiber-specific overexpression of peroxisome proliferator-activated receptor γ coactivator 1α (PGC-1α), a nodal modulator of oxidative metabolism, exhibit both a switch to oxidative fibers and increased capillarity (Lin et al. 2002), but nevertheless have fewer SCs, albeit with an increased myogenic capacity (Dinulovic et al. 2016b). In fact, in regard to most metabolic and contractile traits, PGC-1α transgenic, *bona fide* oxidative and endurance-trained muscles are indistinguishable. Interestingly, the muscle fiber PGC-1α transgene affects expression of BM components FN and tenascin-C (Dinulovic et al. 2016b), which might account for the increased myogenic potential of the SCs. However, a possible influence of other differences in the microenvironment, for instance the increased percentage of M2 macrophages in resting conditions in these animals, should not be overlooked (Dinulovic et al. 2016a; Furrer et al. 2017). Therefore, an alternative explanation for the correlation between SC number and fiber type could be a difference in ECM organization. For example, the slow soleus muscle has double the amount of collagen IV and half the amount of laminin-2 compared to the fast rectus femoris in rats (Kovanen et al. 1988). However, the link between SC number and fiber type-specific ECM composition is still poorly understood and thus awaits further research.

8.2.3 *Blood Vessels, Oxygen, (Peri)Endothelial Cells and Secreted Systemic Factors*

The close proximity of SCs and capillaries suggests that blood vessels are an important part of the niche. Indeed, the close correlation between a well-developed capillary network and successful skeletal muscle regeneration has been demonstrated (Arsic et al. 2004; Ochoa et al. 2007). This is not surprising given the fact that a myriad of factors and cells that modify the SC niche, such as hormones and monocytes, are delivered by blood vessels. In addition, endothelial cells can secrete growth factors (EGF, IGF-I, bFGF) including vascular endothelial growth factor (VEGF) and platelet-derived growth factor BB (PDGF-BB), which promote SC proliferation (Christov et al. 2007). Conversely, differentiating myogenic cells also secrete VEGF, thereby stimulating angiogenesis (Rhoads et al. 2009). Interestingly, peri-endothelial cells, such as smooth muscle cells, secrete angiopoietin 1 (Ang1), which regulates the SC state by binding to the Tie-2 receptor that is highly expressed in resting SCs. This interaction in turn induces the expression of quiescence markers and blocks the expression of differentiation markers in SCs through ERK1/2 signaling (Abou-Khalil et al. 2009), resulting in a return to quiescence at the end of regeneration.

A reduction in partial oxygen pressure has also emerged as an essential factor in SC biology. Hypoxia is a critical factor for many stem cells, with a strong link between low oxygen levels and the undifferentiated cell state (Mohyeldin et al. 2010). Myoblasts cultured under hypoxic conditions show increased quiescence and higher self-renewal efficiency upon transplantation in vivo (Liu et al. 2012).

Finally, systemic, circulating factors facilitate the adjustment of SCs to distal processes away from the niche. For example, calcitonin, a thyroid hormone that is secreted in response to high blood calcium levels, is important for SC dormancy and sub-laminar location. It exerts its effects by binding to the calcitonin receptor (Calcr), which is expressed by resting, but not by activated SCs (Fukada et al. 2007). A specific knock-out of Calcr in SCs results in their relocation from the niche and loss by apoptosis (Yamaguchi et al. 2015). Likewise, SC-specific knock-out of the androgen receptor, which is expressed in this cell population (Sinha-Hikim et al. 2004), leads to induction of Mstn expression, a fiber-type switch and a reduction in muscle mass and strength (Dubois et al. 2014).

8.2.4 Motor Neurons, Fibroblasts, Fibro/Adipogenic Progenitors and Immune Cells

In slow-twitch muscles, SCs are located in close proximity to the neuromuscular junction (NMJ) (Kelly 1978), and the difference in SC numbers between slow- and fast-twitch fibers is correlated with the pattern of neuron firing (Gibson and Schultz 1982). When denervated, skeletal muscle fibers undergo atrophy, to which SCs initially respond with activation and proliferation similar to what is observed in damaged muscle, but after several weeks of denervation, SC number declines due to loss of proliferative capacity and apoptosis (Kuschel et al. 1999; Jejurikar et al. 2002). Conversely, it has been shown that developing muscle produces neurotrophins, which function as retrograde survival factors for the motor neuron (Griesbeck et al. 1995), and SCs secrete the axonal guidance factor semaphorin 3A with possible implications in muscle regeneration (Tatsumi et al. 2009). Although initially found to have a role in neuron survival, neurotrophins are emerging as important modulatory factors for various cell populations and tissues including skeletal muscle. For example, nerve growth factor (NGF) is expressed by regenerating fibers, which implies its involvement in muscle regeneration. Similarly, SC expression of brain-derived neurotrophic factor (BDNF) is important for SC maintenance, and consequently affects muscle regeneration (Clow and Jasmin 2010; Menetrey et al. 2000).

Fibroblasts contribute to the niche by secreting growth factors and structural components of the BL. Temporary thickening of the ECM coupled with an increase in the number of muscle tissue fibroblasts is a hallmark of muscle regeneration (Serrano et al. 2011). Furthermore, interactions between Tcf4[+] fibroblasts and SCs are necessary for successful regeneration. Selective, conditional ablation of SCs in Pax7[CreERT2/+];R26R[DTA/+] mice leads to insufficient proliferation of fibroblasts in the initial phases of regeneration and fibrosis at the later stages, whereas the partial ablation of fibroblasts in Tcf4[CreERT2/+];R26R[DTA/+] mice causes reduced proliferation and precocious differentiation of SCs, resulting in a decreased diameter of regenerated muscles and depletion of the SC pool (Murphy et al. 2011).

Skeletal muscle-resident mesenchymal progenitors expressing PDGFRα are known as fibro/adipogenic progenitors (FAPs) due to their ability to differentiate into adipocytes and fibroblasts (Uezumi et al. 2010). In homeostatic conditions, these cells are in close proximity to blood vessels (Pretheeban et al. 2012), and their number quickly rises in the event of muscle damage. FAPs facilitate myofiber formation and myoblast differentiation by secreting specific ECM components and cytokines, respectively (Joe et al. 2010). These cells also display the ability to remove necrotic tissue (Heredia et al. 2013), thereby supporting muscle regeneration. Interestingly, proper signaling from myotubes and eosinophils prevents FAP differentiation into adipocytes (Uezumi et al. 2010).

Immune cells are additional important players in defining the SC niche in regeneration. Some of these cells, like tissue macrophages and mast cells, are permanent members of the niche, but their importance in modulating the SC microenvironment in quiescence is likely limited. However, they take on an active role upon sterile injury, which induces muscle fiber damage and necrosis. Resident immune cells react by secreting cytokines and chemokines including tumor necrosis factor α (TNF-α), interleukin 6 (IL-6) and macrophage inflammatory protein 2 (MIP-2), which primarily drive the extravasation of neutrophils (Wang and Thorlacius 2005; Brigitte et al. 2010). Next, neutrophils secrete MIP-1α, monocyte chemoattractant protein-1 (MCP-1) and other cytokines attracting monocytes from blood vessels, which rapidly become the most abundant inflammatory cell type in the damaged tissue (Scapini et al. 2000). Depending on the milieu of inflammatory signals and immune cells present in the niche, the macrophages derived from the monocytes can acquire the M1 or M2 type. M1 macrophages secrete proinflammatory cytokines (TNF-α, IL-1β) and are characteristic of the early post-injury stages. They are essential for the removal of necrotic tissue and promote SC proliferation. Upon clearance of cellular debris, the altered conditions in the niche promote an increase in the number of M2 macrophages, which secrete anti-inflammatory cytokines (IL-4, IGF-I, TGF-β) and support the differentiation stages of regeneration (Ceafalan et al. 2014; Arnold et al. 2007). Temporal regulation of the inflammatory cascade is crucial in the process. For example, suppression of M1 macrophages leads to reduced SC proliferation, persistence of necrosis and results in fat and fibrotic tissue accumulation. Likewise, suppression of the switch from the M1 to the M2 type negatively affects myogenesis and myofiber growth (Segawa et al. 2008; Deng et al. 2012; Summan et al. 2006). In addition to paracrine signaling, macrophages establish direct contact with myoblasts and myotubes through cell adhesion interactions (e.g. via VCAM-1-VLA-4, ICAM-1-LFA-1, PECAM-1-PECAM-1 and CX3CL1-

CX3CR1), which prevent apoptosis of myogenic cells (Sonnet et al. 2006). Apart from innate immunity, cells of the adaptive immune system are also central to regulating SC behavior during sterile injury. An instrumental role of T regulatory cells in proper SC expansion and muscle regeneration, as well as in the M1 to M2 macrophage switch after injury has been described (Castiglioni et al. 2015; Burzyn et al. 2013).

8.2.5 *Other Stem Cells with Myogenic Capacity*

Apart from the true stem cells of muscle tissue and the battalion of auxiliary cells that participate in skeletal muscle regeneration, additional multipotent progenitors from muscle and other organs can contribute to this process. The function and ontogeny of many of these heterogeneous groups of cells are unclear, but interestingly, they are all found in close proximity to blood vessels (Péault et al. 2007). In addition, they have not been fully characterized and therefore, their interaction remains obscure.

Some of those myogenic progenitors, such as SP cells, CD133$^+$ cells and MDSCs, can be found in skeletal muscle tissue, while others originate from the blood vessel wall, for example MABs, pericytes, endothelial as well as myo-endothelial cells (Péault et al. 2007). Interestingly, these blood vessel-derived progenitors have a myogenic potential even when isolated from organs other than skeletal muscle (e.g. pericytes and endothelial cells isolated from adult human pancreas or adipose tissue) (Péault et al. 2007).

Unlike SCs, all these myogenic progenitors are able to cross the blood wall and home in on muscle tissue when administered via the bloodstream. They can engraft skeletal muscle, albeit often to a low extent, as demonstrated for SP cells (Péault et al. 2007). Although systemic delivery represents an enormous advantage in cell therapy for muscle disorders, some problems do exist. For example, the majority of intravenously injected cells end up trapped in filter organs (liver, lung, spleen) instead of muscle. In addition, systemic delivery of cells can cause blood flow obstruction, e.g. pulmonary embolism and myocardial infarction, resulting in ischemia and tissue damage (Berry 2015). And finally, their potential to form several cell populations poses a danger of e.g. ectopic formation of bone tissue in muscle (Birbrair et al. 2014).

Due to lack of specific markers, these progenitors are heterogeneous in nature and possibly consist of several groups of cells with different functions. For instance, in skeletal muscle tissue, type 2 pericytes (nestin$^+$) have myogenic and angiogenic potential, while type 1 pericytes (nestin$^-$) have adipogenic and fibrogenic potential. Importantly, other resulting cell populations depend on the pericyte microenvironment, which is perturbed in dystrophic conditions and aging (Birbrair et al. 2015).

Some of these multipotent progenitors offer several potential advantages over SCs in terms of systemic delivery, better survival and proliferation potential, leading to increased regenerative capacity, as demonstrated for MDSCs (Qu-Petersen et al.

2002). Intra-arterial delivery of MABs in a mouse model for limb-girdle muscular dystrophy (LGMD) and a dog model of Duchenne muscular dystrophy (DMD) also resulted in amelioration of the dystrophic phenotype (Sampaolesi et al. 2003, 2006). However, the outcome of transplantation can largely depend on the local microenvironment, e.g. caused by tissue damage, or the presence or absence of other cell populations and signaling molecules (Birbrair et al. 2014, 2015). Importantly, the significance of these cell populations to contribute to tissue maintenance upon skeletal muscle damage or in skeletal muscle formation under physiological conditions is largely undetermined. In any case however, even if the contribution of these populations to skeletal muscle formation independent of cell therapy is minor or non-existent, these cells could still significantly contribute to regeneration by secreting paracrine factors, such as growth factors, as suggested for pericytes and SPs (Birbrair et al. 2014; Péault et al. 2007).

8.2.6 The Biophysical Properties of Muscle

Aside from other factors of the niche, rigidity of the microenvironment can profoundly affect SC behavior. The elastic stiffness of uninjured skeletal muscle is ~12 kPa, and ECM deposition during regeneration increases this value (Engler et al. 2004). SCs can sense and react to this biophysical property of the environment through focal adhesions (Geiger et al. 2009). When cultured on rigid plastic dishes (~10^6kPa), SCs quickly lose their quiescence and stemness. Myoblasts cultured on hydrogels prefer a substrate stiffness of ~21 kPa, while softer (~ 3 kPa) and stiffer (~ 80 kPa) gels reduce their proliferative rate (Boonen et al. 2009). In line, SCs cultured on soft hydrogels that mimic the stiffness of natural muscle (12 kPa) are able to self-renew and significantly improve their contribution to muscle regeneration upon transplantation (Gilbert et al. 2010).

8.3 The Satellite Cell Niche in Pathological Contexts

Aging, muscle dystrophies and related pathologies invariably lead to perturbed conditions of the SC niche. These changes can cause a reduction or an expansion in the SC pool, irresponsiveness to stimuli and therefore a reduced SC activation rate, aberrant proliferation and precocious or reduced differentiation, or SC senescence and apoptosis upon activation. For example, a disproportion of symmetric and asymmetric SC division might tip the balance towards SC loss in aging and a pathological SC expansion with a reduced number of myogenic progenitors in dystrophic conditions (Chang et al. 2016). Irrespective of the dysregulation, the outcome is diminished SC regenerative capacity in both contexts.

Although some of the pathological changes are SC intrinsic, altering the niche can alleviate the underlying condition in many cases. Nevertheless, it is difficult to precisely discriminate between intrinsic and extrinsic origins of the SC pathology due to the bidirectional signaling between SCs and their microenvironment. Importantly, the niche can induce modifications in SC properties that can persist even after removal of SCs from the niche, and are hence perceived as "intrinsic".

8.3.1 *The Satellite Cell Niche in Aging*

With advanced age, skeletal muscle mass and neuromuscular performance diminishes, a condition termed sacropenia. Decreased fiber and motor neuron numbers, reduced fiber size, a myofiber switch towards the oxidative type and loss of myonuclei resulting in an increase in myonuclear domain size are all common observations in aging, collectively resulting in a marked decrease of the efficiency of muscle regeneration (Larsson and Ansved 1995; Faulkner et al. 2007). The reduction of the SC pool has been proposed as an explanation for the underlying condition (Shefer et al. 2010). However, based on conflicting results in different studies (Conboy et al. 2003), the prevailing opinion is that a drop in the myogenic potential of SCs might be the causative factor of the impaired regenerative capacity. Some characteristics of the aged niche that lead to reduced potency of SCs are illustrated in Fig. 8.3.

Some changes in the aged niche are precipitated by aberrant signaling. For instance, lack of Delta upregulation by injured aged muscles leads to reduced Notch signaling in SCs and hence reduced SC proliferation — a phenotype that can be overcome by alternative Notch activation (Conboy et al. 2003). Interestingly, experiments with heterochronic, parabiotic pairings (a shared circulatory system between a young and an old animal) demonstrated that systemic factors at least partially account for the perturbed SC biology, as the exposure to young blood restored otherwise reduced Notch signaling and improved SC proliferation in old mice (Conboy et al. 2005). The subsequent search for rejuvenating humoral factors led to the implication of the hormone oxytocin (Elabd et al. 2014) and growth factor GDF11 (Sinha et al. 2014; Walker et al. 2016) as systemic factors that decline with age and whose induction is able to revert aging-related SC pathology. However, the function of GDF11 in promoting muscle and cardiac health in aging has been largely discredited in more recent studies (Schafer et al. 2016; Egerman et al. 2015; Harper et al. 2016). Exacerbated canonical Wnt signaling due to elevated circulating Wnt activators in aged mice was also suggested as being responsible for aging-related tissue fibrosis and conversion of myoblasts into fibroblasts, a process that can be curbed by Wnt inhibitors (Brack et al. 2007). Increased NF-κB and TGF-β signaling in aged muscles are additional examples of how the immediate niche can negatively impact the regenerative potential of SCs (Oh et al. 2016; Carlson et al. 2008).

ECM deposition in the aged niche in general is thought to act as a damper and therefore exert a negative influence on the activation potential of SCs, e.g. by increasing tissue stiffness. For example, slow muscles boost the expression of

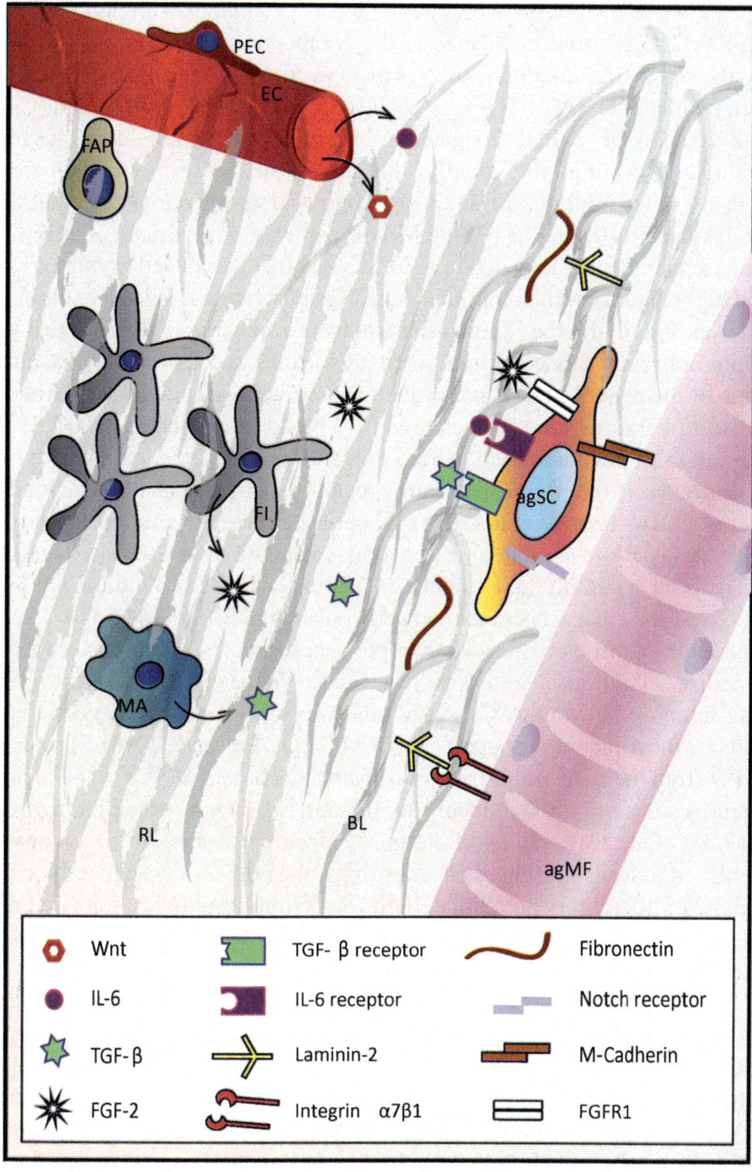

Fig. 8.3 The aged niche. The aged myofiber (agMF) has a smaller diameter, fewer myonuclei and reduced levels of Delta, a ligand of Notch, which is crucial to SC proliferation. Further changes to the niche include increased fibroblast (FI) numbers and a denser basal and reticular lamina (BL, RL) with reduced fibronectin levels, active macrophages (MAs) secreting TGF-β, as well as increased levels of FGF-2, IL-6 and Wnt signaling molecules. Further abbreviations: *EC* endothelial cell; *FAP*, fibro/adipogenic progenitor; *PEC*, periendothelial cell

collagen IV while fast muscles elevate the levels of laminin with aging (Kovanen et al. 1988). The ensuing imbalance in the components of the BL in old muscle disturbs the signal transduction pathways that govern SCs in the niche, such as those triggered by higher levels of TGF-β, a negative regular of SC proliferation (Carlson et al. 2008), and FGF-2. FGF-2 signaling through FGFR1 results in SC loss based on unmitigated cycling. Importantly, this effect can be reverted by increasing Spry1 in SCs, an inhibitor of FGF signaling and preserver of SC quiescence (Chakkalakal et al. 2012). The p38α/β MAPK pathway, downstream of FGF signaling, is consequently overactivated in aged SCs, leading to reduced asymmetric division and higher numbers of committed daughter cells, hence resulting in diminished self-renewal. Improving the SC environment by transplanting old SCs into a young host could not revert this condition, in contrast to the successful pharmacological inhibition of the p38α/β MAPK pathway in SCs (Cosgrove et al. 2014; Bernet et al. 2014). Most likely triggered by increased IL-6 blood levels, the JAK/STAT signaling pathway is also overactivated in aged SCs and results in a reduction of symmetric division and self-renewal, which can be reverted with pharmacological inhibitors (Price et al. 2014; Tierney et al. 2014). In geriatric mice (30 months of age), SCs lose their ability for reversible quiescence by switching to pre-senescence. At that age, the respective stimuli fail to induce SC activation and proliferation, but instead prompt senescence in a process termed geroconversion. Silencing of p16^{INK4a}, a cell cycle inhibitor that triggers the switch to pre-senescence, is able to restore the activation and proliferation potential of SCs (Sousa-Victor et al. 2014). Intriguingly, blocking autophagy in young SCs causes senescence, while its restoration in old age reestablishes the regenerative potential of SCs (Garcia-Prat et al. 2016). Furthermore, loss of FN from the aged BL prevents sufficient attachment of SCs to the niche and thus disturbs signaling through focal adhesion kinase, thereby precipitating SC loss (Lukjanenko et al. 2016). In addition, mislocalization of integrin β1 on aged and dystrophic SCs leads to impaired sensitivity to FGF-2, consequently causing reduced SC proliferation and ultimately SC depletion, resulting in impaired regeneration. In both models, activation of β1-integrin reverts the impairment of SC function (Rozo et al. 2016).

Hormonal and pharmacological interventions, calorie restriction as well as cell therapy have been proposed for the prevention and treatment of sarcopenia. However, to date, physical activity remains the most efficacious approach to combating this disease (Jang et al. 2011), e.g. by boosting the number and myogenic capacity of SCs (Shefer et al. 2010; Snijders et al. 2009). Although an SC pathology is most likely not the only driving force for development of sarcopenia, SC dysfunction contributes to impaired muscle regeneration and increased fibrosis (Brack et al. 2007). Recent advances in understanding aberrant signal transduction pathways and communication between aged SCs and their niche will potentially offer new pharmacological avenues in the treatment of sarcopenia that could circumvent the inherent problems of exercise interventions in geriatric patients.

8.3.2   The Satellite Cell Niche in Dystrophic Conditions

Muscular dystrophies are a heterogeneous group of sporadic and inherited disorders that lead to progressive muscle wasting and weakness. Fiber size variation, fiber necrosis followed by inflammation, and muscle tissue replacement by fat and scar tissue are often hallmarks of these pathologies, depending on the severity of the dystrophy in question (Emery 2002). Many dystrophies are caused by a mutation in structural proteins of the cytoskeleton, membrane or ECM, which comprise a part of the SC niche.

One of the most common and extensively studied dystrophy is DMD, which arises due to a genetic mutation in the structural protein dystrophin. Lack of dystrophin, a member of the membrane-bound protein complex, leads to the improper connection of the cytoskeleton to the ECM, rendering fibers more prone to mechanical damage. As a consequence, recurring rounds of degeneration and regeneration form a vicious cycle and impose proliferative pressure on SCs. It has been proposed that progressive worsening of the disease over time is at least partially due to telomere shortening and ultimately loss of the regenerative potential of SCs (Sacco et al. 2010). The most important signaling molecules implicated in the dystrophic niche are presented in Fig. 8.4.

Infiltrating macrophages and T cells induce fibrosis through secretion of profibrotic cytokines, which in chronic diseases such as muscular dystrophies result in fibrotic tissue formation at the expense of functional muscle tissue (Mann et al. 2011). For instance, in acute injury, a wave of TNF-α-secreting M1 macrophages induces a reduction of the preceding FAP expansion, thereby limiting ECM accumulation. Under chronic conditions, however, loss of proper control of macrophage polarization results in exacerbated TGF-β secretion that in turn causes FAP persistence and fibrosis (Lemos et al. 2015). Therefore, anti-inflammatory drugs like corticosteroids, despite their potential pro-atrophic side effects, are the current standard of care for DMD. A big portion of current DMD therapy-related research focuses on intercepting the pathways implicated in fibrotic tissue formation, namely those triggered by TGF-β and Mstn (Bentzinger et al. 2010).

Interestingly, SC fate conversion from the myogenic to the fibrogenic lineage can contribute to fibrosis development in DMD. Thus, increased Wnt signaling in dystrophic muscle triggers TGF-β2 secretion, which in turn induces pro-fibrotic gene expression in SCs, thereby limiting their myogenic potential (Biressi et al. 2014). Besides progressive fibrosis, the SC niche in DMD is affected by other events, such as alterations in the BL with differential expression of laminin α2, laminin β1 and collagen IV, which are implicated in the direct interactions with SCs (Hayashi et al. 1993), as well as that of decorin and biglycan, proteoglycans linked to TGF-β sequestration (Fadic et al. 2006). These changes presumably also contribute to alterations in muscle stiffness, which further affects SC behavior. In addition, perturbed conditions can alter the differentiation of several multipotent progenitor populations in the muscle, including FAPs, resulting in extracellular fat accumulation (Uezumi et al. 2010). Of note, these alterations to the SC niche can be extrapolated

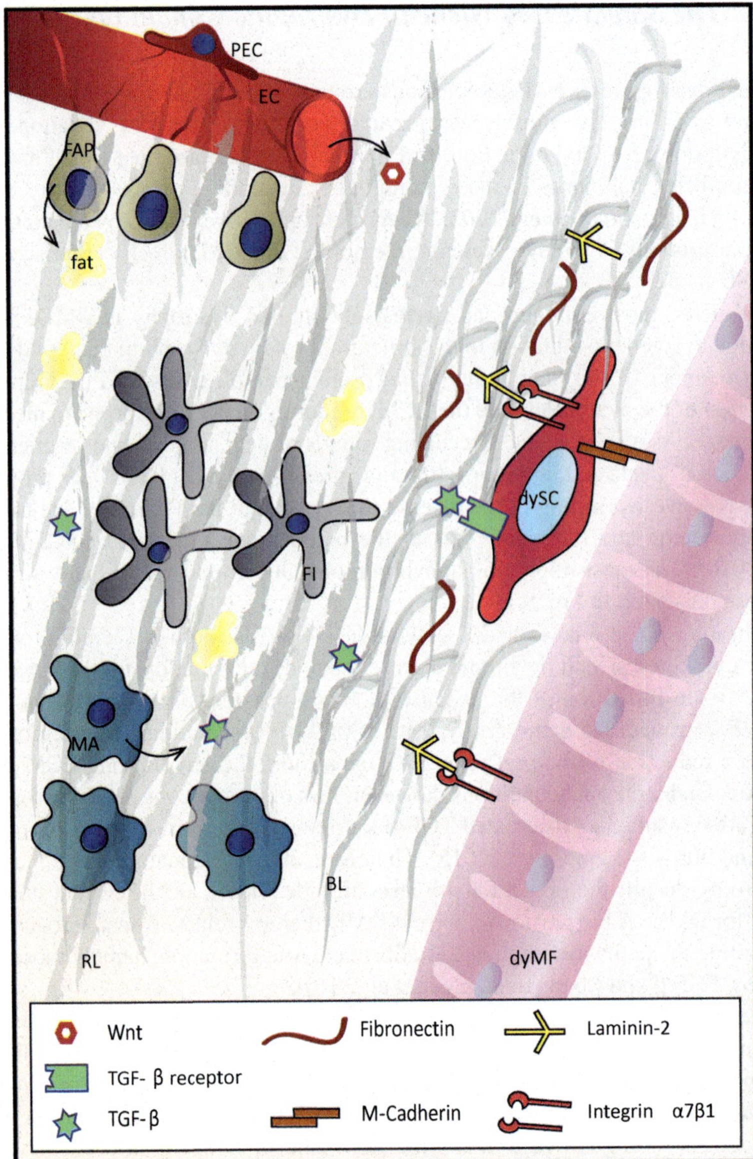

Fig. 8.4 The dystrophic niche. The dystrophic fiber (dyMF) is smaller in diameter with centrally located nuclei, a hallmark of ongoing damage and regeneration. As a result, dySCs are under high proliferative pressure. Similar to the aged niche, elevated levels of TGF-β and Wnt, as well as increased numbers of fibroblasts (FIs), are common features of the dystrophic niche. In addition, inflammation caused by high numbers of fibro/adipogenic progenitors (FAPs) and macrophages (MAs) contributes to the formation of fat and scar tissue (a denser basal and reticular lamina—BL, RL) at the expense of muscle tissue. Further abbreviations: *EC* endothelial cell; *PEC* periendothelial cell

to other dystrophies and muscle pathologies with prominent fibrosis and fat accumulation, even diseases such as type 2 diabetes (Berria et al. 2006; Goodpaster and Wolf 2004).

The niche has been the primary focus of research on SC dysfunction in DMD, mainly due to a body of literature suggesting that dystrophin expression is limited to differentiated myofibers. However, recent findings suggest a direct role of SCs in the pathology based on the discovery that dystrophin is also expressed in activated SCs and is important for establishing cell polarity, thus enabling asymmetric SC division (Dumont et al. 2015). Lack of SC dystrophin therefore results in reduced numbers of committed progenitors and differentiated myocytes, as well as a higher numbers of $Myf5^-$ progenitors. However, both increased and decreased SC numbers have been reported in DMD, a discrepancy that could be due to the difference in age of the subjects in the studies in question (Kottlors and Kirschner 2010; Jiang et al. 2014). Given the reciprocal regulation between SCs and fibroblasts (Murphy et al. 2011), it will be interesting to further explore the role of SC dystrophin in fibrotic tissue accumulation and other DMD symptoms.

Dysferlinopathy is another example of a muscular dystrophy (LGMD) with a complex etiology. In this disease, a mutation in the structural protein dysferlin primarily prevents myotubes from patching contraction-induced small ruptures in the sarcolemma. However, dysferlinopathy also affects proper muscle regeneration, where impairment in the release of cytokines upon injury results in reduced neutrophil recruitment and leads to a prolonged inflammatory phase, creating a suboptimal environment for successful regeneration by SCs (Chiu et al. 2009).

Despite extensive efforts, no treatment for most of these debilitating diseases has been found so far. Therapies are mainly symptomatic and palliative, relying on corticosteroids as well as pulmonary and cardiac management in the case of DMD (Wagner et al. 2007). Experimental treatments centered on stem cell therapy (e.g. SC transplantation), gene therapy (e.g. antisense oligonucleotide exon skipping, viral delivery of mini-dystrophin, CRISPR/Cas9-mediated deletion) and pharmacology (e.g. Mstn blockade) might, however, result in therapeutic breakthroughs in the future (Chakkalakal et al. 2005; Fairclough et al. 2013; Young et al. 2016; Mendell and Rodino-Klapac 2016).

8.4 Future Directions: An Artificial Niche

Autologous SC therapy represents one of the most promising treatments both for dystrophies and sacropenia. In sacropenia, enhancement of the myogenic potential of SCs and expansion of *bona fide* SCs in vitro prior to their transplantation in order to boost regeneration would most likely be sufficient, while in dystrophic conditions, the approach would comprise stem cell and gene therapy, including correction of a relevant genetic mutation in vitro. However, several hurdles impede the success of such trials. For example, the inability of SCs to home in on muscle with systemic delivery (Elster et al. 2013), poor migration when delivered intramuscularly

(Bentzinger et al. 2014), as well as reduced proliferation, immediate differentiation and apoptosis of injected cells have been reported. These effects are further compounded by the rapid and irreversible loss of SC stemness in culture, resulting in reduced myogenic potential upon transplantation (Montarras et al. 2005). Thus, as expanding the stem cell population is a necessary step prior to implantation, improving the intrinsic myogenic potential of SCs, e.g. by overexpressing PGC-1α, can help to lead to enhanced early muscle tissue formation after transplantation (Haralampieva et al. 2017). Furthermore, attempts have been made to mimic the SC niche in vitro to circumvent some of the aforementioned problems.

Bioengineering efforts have made progress in creating 3D biomimetics as acellular or cellular scaffolds for use in regenerative therapy (Handschin et al. 2015). From cylindrically shaped, collagen I-based gels to various natural hydrogels and finally fibrin gels, conditions conductive to increasing cell survival, fusion and maturation are constantly improving (Bursac et al. 2015). For example, in the case of trauma-induced volumetric muscle loss, acellular biodegradable materials filled with anti-fibrotic and pro-myogenic factors on one, and angiogenic and neurotrophic factors on the other hand, would possibly provide optimal conditions to tip the balance towards functional muscle tissue instead of scar tissue formation when transplanted in a timely manner (Sicari et al. 2014; Shvartsman et al. 2014). These scaffolds would provide not only fast infiltration and proper activation of the myogenic cells of the host, but also support fast establishment of the vascular and neural network necessary to support the newly formed muscle tissue. Other conditions such as aging and dystrophies require, however, more intricate cellular approaches, with biomaterials that closely resemble the SC niche in terms of stiffness and composition, enabling the cell-matrix interactions that are crucial for proper SC function. In that regard, polyethylene glycol hydrogels cross-linked with laminin have been used successfully in improving SC self-renewal in vitro and engraftment in vivo (Gilbert et al. 2010). This substrate, in combination with pharmacological inhibition of the p38α/β MAPK pathway, was also able to reverse the age-related SC pathology (Cosgrove et al. 2014).

Besides identification of ECM proteins as crucial components of an artificial niche, the search for extrinsic factors that would enable SC expansion in vitro without loss of cell stemness has led to the discovery of a cocktail of four cytokines. Intrigued by the role of CD4[+] and CD8[+] T cells in regeneration, Fu and colleagues identified T cell-derived factors that are responsible for increased SC proliferation. They defined a pro-inflammatory cytokine combination composed of IL-1α, IL-13, TNF-α and INF-γ that is sufficient and necessary to maintain SC potency in vitro (Fu et al. 2015). This combination of cytokines promoted proliferation and limited differentiation of SCs for 20 passages. The gene expression profile of cells expanded in this way suggests that these cells retain at least some of the features of freshly isolated SCs, such as high Pax7 and low MyoD expression. SCs expanded under such conditions were not only able to engraft efficiently and occupy the niche upon transplantation into muscle, but also to respond to secondary injury by undergoing activation and self-renewal (Fu et al. 2015). In addition, the transplantation efficiency of such expanded cells in vitro was comparable to freshly isolated SCs. Since the cocktail in question has been optimized for murine SCs, efforts will have to be made to find proper conditions and factors for human SCs.

Recently, Quarta and colleagues successfully mimicked the in vivo microenvironment of SCs by using a defined serum-free quiescence medium and artificial muscle fibers. A 3D microscaffold with an elasticity between 1 and 2 kPa based on collagen, recombinant laminin and $\alpha 4\beta 1$ integrin provided optimal stiffness and enabled signaling pathways to keep the cells in reversible quiescence (Quarta et al. 2016). This method proved effective in keeping both murine and human SCs in a quiescent state for up to a week. With this system, the engraftment potential and self-renewal of cultured cells upon transplantation surpassed that of freshly isolated SCs and was comparable to SCs associated with their native fibers. These results confirm the importance of the niche and mimicking the in vivo microenvironment for maintaining SC stemness in vitro (Quarta et al. 2016).

These studies provide crucial insights into the optimal conditions for keeping SCs in a quiescent state in vitro, SC propagation, and preservation of the stemness for subsequent in vivo transplantation. Importantly, an artificial niche not only enables disease modeling and gene therapy, but also provides an amenable experimental system for toxicology screenings of novel drugs, thereby reducing the burden of animal studies (Madden et al. 2015; Huh et al. 2012). Together with novel imaging and cell tracking techniques (Haralampieva et al. 2016), the increasing knowledge about SC biology, the importance of the niche, and the interplay of SCs with myofibers and other cell types will hopefully result in novel therapeutic approaches to treating sarcopenia, muscular dystrophies and other skeletal muscle-associated pathologies.

References

Abou-Khalil R, Le Grand F, Pallafacchina G, Valable S, Authier FJ, Rudnicki MA, Gherardi RK, Germain S, Chretien F, Sotiropoulos A, Lafuste P, Montarras D, Chazaud B (2009) Autocrine and paracrine angiopoietin 1/Tie-2 signaling promotes muscle satellite cell self-renewal. Cell Stem Cell 5(3):298–309

Almada AE, Wagers AJ (2016) Molecular circuitry of stem cell fate in skeletal muscle regeneration, ageing and disease. Nat Rev Mol Cell Biol 17(5):267–279

Arnold L, Henry A, Poron F, Baba-Amer Y, van Rooijen N, Plonquet A, Gherardi RK, Chazaud B (2007) Inflammatory monocytes recruited after skeletal muscle injury switch into antiinflammatory macrophages to support myogenesis. J Exp Med 204(5):1057–1069

Arsic N, Zacchigna S, Zentilin L, Ramirez-Correa G, Pattarini L, Salvi A, Sinagra G, Giacca M (2004) Vascular endothelial growth factor stimulates skeletal muscle regeneration in vivo. Mol Ther 10(5):844–854

Bentzinger CF, von Maltzahn J, Rudnicki MA (2010) Extrinsic regulation of satellite cell specification. Stem Cell Res Ther 1(3):27

Bentzinger CF, Wang YX, Rudnicki M (2012) Building muscle: molecular regulation of myogenesis. Cold Spring Harb Perspect Biol 4(2)

Bentzinger CF, Wang YX, von Maltzahn J, Soleimani VD, Yin H, Ma R (2013) Fibronectin regulates Wnt7a signaling and satellite cell expansion. Cell Stem Cell 12(1):75–87

Bentzinger CF, von Maltzahn J, Dumont NA, Stark DA, Wang YX, Nhan K, Frenette J, Cornelison DD, Rudnicki MA (2014) Wnt7a stimulates myogenic stem cell motility and engraftment resulting in improved muscle strength. J Cell Biol 205(1):97–111

Bernet JD, Doles JD, Hall JK, Kelly Tanaka K, Ta C, Olwin BB (2014) p38 MAPK signaling underlies a cell-autonomous loss of stem cell self-renewal in skeletal muscle of aged mice. Nat Med 20(3):265–271

Berria R, Wang L, Richardson DK, Finlayson J, Belfort R, Pratipanawatr T, De Filippis EA, Kashyap S, Mandarino LJ (2006) Increased collagen content in insulin-resistant skeletal muscle. Am J Physiol Endocrinol Metab 290(3):E560–E565

Berry SE (2015) Concise review: mesoangioblast and mesenchymal stem cell therapy for muscular dystrophy: progress, challenges, and future directions. Stem Cells Transl Med 4(1):91–98

Birbrair A, Zhang T, Wang ZM, Messi ML, Mintz A, Delbono O (2014) Pericytes: multitasking cells in the regeneration of injured, diseased, and aged skeletal muscle. Front Aging Neurosci 6:245

Birbrair A, Zhang T, Wang ZM, Messi ML, Mintz A, Delbono O (2015) Pericytes at the intersection between tissue regeneration and pathology. Clin Sci (Lond) 128(2):81–93

Biressi S, Miyabara EH, Gopinath SD, Carlig PM, Rando TA (2014) A Wnt-TGFbeta2 axis induces a fibrogenic program in muscle stem cells from dystrophic mice. Sci Transl Med 6(267):267ra176

Bjornson CRR, Cheung TH, Liu L, Tripathi PV, Steeper KM, Ta R (2012) Notch signaling is necessary to maintain quiescence in adult muscle stem cells. Stem Cells 30(2):232–242

Blanco-Bose WE, Yao CC, Kramer RH, Blau HM (2001) Purification of mouse primary myoblasts based on alpha 7 integrin expression. Exp Cell Res 265(2):212–220

Boonen KJ, Rosaria-Chak KY, Baaijens FP, van der Schaft DW, Post MJ (2009) Essential environmental cues from the satellite cell niche: optimizing proliferation and differentiation. Am J Physiol Cell Physiol 296(6):C1338–C1345

Brack AS, Conboy MJ, Roy S, Lee M, Kuo CJ, Keller C, Ta R (2007) Increased Wnt signaling during aging alters muscle stem cell fate and increases fibrosis. Science 317(5839):807–810

Brigitte M, Schilte C, Plonquet A, Baba-Amer Y, Henri A, Charlier C, Tajbakhsh S, Albert M, Gherardi RK, Chrétien F (2010) Muscle resident macrophages control the immune cell reaction in a mouse model of notexin-induced myoinjury. Arthritis Rheum 62(1):268–279

Bursac N, Juhas M, Rando TA (2015) Synergizing engineering and biology to treat and model skeletal muscle injury and disease. Annu Rev Biomed Eng 17:217–242

Burzyn D, Kuswanto W, Kolodin D, Shadrach JL, Cerletti M, Jang Y, Sefik E, Tan TG, Wagers AJ, Benoist C, Mathis D (2013) A special population of regulatory T cells potentiates muscle repair. Cell 155(6):1282–1295

Carlson BM, Gutmann E (1972) Development of contractile properties of minced muscle regenerates in the rat. Exp Neurol 36(2):239–249

Carlson ME, Hsu M, Conboy IM (2008) Imbalance between pSmad3 and Notch induces CDK inhibitors in old muscle stem cells. Nature 454(7203):528–532

Castiglioni A, Corna G, Rigamonti E, Basso V, Vezzoli M, Monno A, Almada AE, Mondino A, Wagers AJ, Manfredi AA, Rovere-Querini P (2015) FOXP3+ T Cells recruited to sites of sterile skeletal muscle injury regulate the fate of satellite cells and guide effective tissue regeneration. PLoS One 10(6):e0128094

Ceafalan LC, Popescu BO, Hinescu ME (2014) Cellular players in skeletal muscle regeneration. Biomed Res Int 2014:957014

Cerletti M, Jang YC, Finley LW, Haigis MC, Wagers AJ (2012) Short-term calorie restriction enhances skeletal muscle stem cell function. Cell Stem Cell 10(5):515–519

Chakkalakal JV, Thompson J, Parks RJ, Jasmin BJ (2005) Molecular, cellular, and pharmacological therapies for Duchenne/Becker muscular dystrophies. FASEB J 19(8):880–891

Chakkalakal JV, Jones KM, Basson MA, Brack AS (2012) The aged niche disrupts muscle stem cell quiescence. Nature 490(7420):355–360

Chang NC, Chevalier FP, Rudnicki MA (2016) Satellite cells in muscular dystrophy—lost in polarity. Trends Mol Med 22(6):479–496

Chiu YH, Hornsey MA, Klinge L, Jorgensen LH, Laval SH, Charlton R, Barresi R, Straub V, Lochmuller H, Bushby K (2009) Attenuated muscle regeneration is a key factor in dysferlin-deficient muscular dystrophy. Hum Mol Genet 18(11):1976–1989

Christov C, Chretien F, Abou-Khalil R, Bassez G, Vallet G, Authier FJ, Bassaglia Y, Shinin V, Tajbakhsh S, Chazaud B, Gherardi RK (2007) Muscle satellite cells and endothelial cells: close neighbors and privileged partners. Mol Biol Cell 18(4):1397–1409

Clow C, Jasmin BJ (2010) Brain-derived neurotrophic factor regulates satellite cell differentiation and skeletal muscle regeneration. Mol Biol Cell 21(13):2182–2190

Collins CA, Olsen I, Zammit PS, Heslop L, Petrie A, Partridge TA, Morgan JE (2005) Stem cell function, self-renewal, and behavioral heterogeneity of cells from the adult muscle satellite cell niche. Cell 122(2):289–301

Conboy IM, Conboy MJ, Smythe GM, Ta R (2003) Notch-mediated restoration of regenerative potential to aged muscle. Science 302(5650):1575–1577

Conboy IM, Conboy MJ, Wagers AJ, Girma ER, Weissman IL, Ta R (2005) Rejuvenation of aged progenitor cells by exposure to a young systemic environment. Nature 433(7027):760–764

Cornelison DD, Filla MS, Stanley HM, Rapraeger AC, Olwin BB (2001) Syndecan-3 and syndecan-4 specifically mark skeletal muscle satellite cells and are implicated in satellite cell maintenance and muscle regeneration. Dev Biol 239(1):79–94

Cosgrove BD, Gilbert PM, Porpiglia E, Mourkioti F, Lee SP, Corbel SY, Llewellyn ME, Delp SL, Blau HM (2014) Rejuvenation of the muscle stem cell population restores strength to injured aged muscles. Nat Med 20(3):255–264

Deng B, Wehling-Henricks M, Villalta SA, Wang Y, Tidball JG (2012) IL-10 triggers changes in macrophage phenotype that promote muscle growth and regeneration. J Immunol 189(7):3669–3680

Dinulovic I, Furrer R, Di Fulvio S, Ferry A, Beer M, Handschin C (2016a) PGC-1alpha modulates necrosis, inflammatory response, and fibrotic tissue formation in injured skeletal muscle. Skelet Muscle 6:38

Dinulovic I, Furrer R, Beer M, Ferry A, Cardel B, Handschin C (2016b) Muscle PGC-1alpha modulates satellite cell number and proliferation by remodeling the stem cell niche. Skelet Muscle 6(1):39

Dubois V, Laurent MR, Sinnesael M, Cielen N, Helsen C, Clinckemalie L, Spans L, Gayan-Ramirez G, Deldicque L, Hespel P, Carmeliet G, Vanderschueren D, Claessens F (2014) A satellite cell-specific knockout of the androgen receptor reveals myostatin as a direct androgen target in skeletal muscle. FASEB J 28(7):2979–2994

Dumont NA, Wang YX, von Maltzahn J, Pasut A, Bentzinger CF, Brun CE, Rudnicki MA (2015) Dystrophin expression in muscle stem cells regulates their polarity and asymmetric division. Nat Med 21(12):1455–1463

Egerman MA, Cadena SM, Gilbert JA, Meyer A, Nelson HN, Swalley SE, Mallozzi C, Jacobi C, Jennings LL, Clay I, Laurent G, Ma S, Brachat S, Lach-Trifilieff E, Shavlakadze T, Trendelenburg AU, Brack AS, Glass DJ (2015) GDF11 increases with age and inhibits skeletal muscle regeneration. Cell Metab 22(1):164–174

Elabd C, Cousin W, Upadhyayula P, Chen RY, Chooljian MS, Li J, Kung S, Jiang KP, Conboy IM (2014) Oxytocin is an age-specific circulating hormone that is necessary for muscle maintenance and regeneration. Nat Commun 5:4082

Elster JL, Rathbone CR, Liu Z, Liu X, Barrett HH, Rhoads RP, Allen RE (2013) Skeletal muscle satellite cell migration to injured tissue measured with 111In-oxine and high-resolution SPECT imaging. J Muscle Res Cell Motil 34(5–6):417–427

Emery AE (2002) The muscular dystrophies. Lancet 359(9307):687–695

Engler AJ, Griffin MA, Sen S, Bonnemann CG, Sweeney HL, Discher DE (2004) Myotubes differentiate optimally on substrates with tissue-like stiffness: pathological implications for soft or stiff microenvironments. J Cell Biol 166(6):877–887

Fadic R, Mezzano V, Alvarez K, Cabrera D, Holmgren J, Brandan E (2006) Increase in decorin and biglycan in Duchenne muscular dystrophy: role of fibroblasts as cell source of these proteoglycans in the disease. J Cell Mol Med 10(3):758–769

Fairclough RJ, Wood MJ, Davies KE (2013) Therapy for Duchenne muscular dystrophy: renewed optimism from genetic approaches. Nat Rev Genet 14(6):373–378

Faulkner JA, Larkin LM, Claflin DR, Brooks SV (2007) Age-related changes in the structure and function of skeletal muscles. Clin Exp Pharmacol Physiol 34(11):1091–1096

Fedorov YV, Rosenthal RS, Olwin BB (2001) Oncogenic Ras-induced proliferation requires autocrine fibroblast growth factor 2 signaling in skeletal muscle cells. J Cell Biol 152(6):1301–1305

Fink E, Fortin D, Serrurier B, Ventura-Clapier R, Bigard AX (2003) Recovery of contractile and metabolic phenotypes in regenerating slow muscle after notexin-induced or crush injury. J Muscle Res Cell Motil 24(7):421–429

Fu X, Xiao J, Wei Y, Li S, Liu Y, Yin J, Sun K, Sun H, Wang H, Zhang Z, Zhang BT, Sheng C, Wang H, Hu P (2015) Combination of inflammation-related cytokines promotes long-term muscle stem cell expansion. Cell Res 25(9):1082–1083

Fukada S, Uezumi A, Ikemoto M, Masuda S, Segawa M, Tanimura N, Yamamoto H, Miyagoe-Suzuki Y, Takeda S (2007) Molecular signature of quiescent satellite cells in adult skeletal muscle. Stem Cells 25(10):2448–2459

Furrer R, Eisele PS, Schmidt A, Beer M, Handschin C (2017) Paracrine cross-talk between skeletal muscle and macrophages in exercise by PGC-1alpha-controlled BNP. Sci Rep 7:40789

Garcia-Prat L, Martinez-Vicente M, Perdiguero E, Ortet L, Rodriguez-Ubreva J, Rebollo E, Ruiz-Bonilla V, Gutarra S, Ballestar E, Serrano AL, Sandri M, Munoz-Canoves P (2016) Autophagy maintains stemness by preventing senescence. Nature 529(7584):37–42

Geiger B, Spatz JP, Bershadsky AD (2009) Environmental sensing through focal adhesions. Nat Rev Mol Cell Biol 10(1):21–33

Gibson MC, Schultz E (1982) The distribution of satellite cells and their relationship to specific fiber types in soleus and extensor digitorum longus muscles. Anat Rec 202(3):329–337

Gilbert PM, Havenstrite KL, Magnusson KE, Sacco A, Leonardi NA, Kraft P, Nguyen NK, Thrun S, Lutolf MP, Blau HM (2010) Substrate elasticity regulates skeletal muscle stem cell self-renewal in culture. Science 329(5995):1078–1081

Goodpaster BH, Wolf D (2004) Skeletal muscle lipid accumulation in obesity, insulin resistance, and type 2 diabetes. Pediatr Diabetes 5(4):219–226

Griesbeck O, Parsadanian AS, Sendtner M, Thoenen H (1995) Expression of neurotrophins in skeletal muscle: quantitative comparison and significance for motoneuron survival and maintenance of function. J Neurosci Res 42(1):21–33

Guérin CW, Holland PC (1995) Synthesis and secretion of matrix-degrading metalloproteases by human skeletal muscle satellite cells. Dev Dyn 202(1):91–99

Handschin C, Mortezavi A, Plock J, Eberli D (2015) External physical and biochemical stimulation to enhance skeletal muscle bioengineering. Adv Drug Deliv Rev 82-83:168–175

Haralampieva D, Betzel T, Dinulovic I, Salemi S, Stoelting M, Kramer SD, Schibli R, Sulser T, Handschin C, Eberli D, Ametamey SM (2016) Noninvasive PET imaging and tracking of engineered human muscle precursor cells for skeletal muscle tissue engineering. J Nucl Med 57(9):1467–1473

Haralampieva D, Salemi S, Dinulovic I, Sulser T, MA S, Handschin C, Eberli D (2017) Human muscle precursor cells overexpressing PGC-1alpha enhance early skeletal muscle tissue formation. Cell Transplant 26(6):1103–1114

Harper SC, Brack A, MacDonnell S, Franti M, Olwin BB, Bailey BA, Rudnicki MA, Houser SR (2016) Is growth differentiation factor 11 a realistic therapeutic for aging-dependent muscle defects? Circ Res 118(7):1143–1150. discussion 1150

Hayashi YK, Engvall E, Arikawa-Hirasawa E, Goto K, Koga R, Nonaka I, Sugita H, Arahata K (1993) Abnormal localization of laminin subunits in muscular dystrophies. J Neurol Sci 119(1):53–64

Heredia JE, Mukundan L, Chen FM, Mueller AA, Deo RC, Locksley RM, Rando TA, Chawla A (2013) Type 2 innate signals stimulate fibro/adipogenic progenitors to facilitate muscle regeneration. Cell 153(2):376–388

Huh D, Leslie DC, Matthews BD, Fraser JP, Jurek S, Hamilton GA, Thorneloe KS, McAlexander MA, Ingber DE (2012) A human disease model of drug toxicity-induced pulmonary edema in a lung-on-a-chip microdevice. Sci Transl Med 4(159):159ra147

Jang YC, Sinha M, Cerletti M, Dall'Osso C, Wagers aJ (2011) Skeletal muscle stem cells: effects of aging and metabolism on muscle regenerative function. Cold Spring Harb Symp Quant Biol 76:101–111

Janssen I, Heymsfield SB, Wang ZM, Ross R (2000) Skeletal muscle mass and distribution in 468 men and women aged 18-88 yr. J Appl Physiol 89(1):81–88

Jejurikar SS, Marcelo CL, Kuzon WM Jr (2002) Skeletal muscle denervation increases satellite cell susceptibility to apoptosis. Plast Reconstr Surg 110(1):160–168

Jiang C, Wen Y, Kuroda K, Hannon K, Rudnicki MA, Kuang S (2014) Notch signaling deficiency underlies age-dependent depletion of satellite cells in muscular dystrophy. Dis Model Mech 7(8):997–1004

Joe AW, Yi L, Natarajan A, Le Grand F, So L, Wang J, Rudnicki MA, Rossi FM (2010) Muscle injury activates resident fibro/adipogenic progenitors that facilitate myogenesis. Nat Cell Biol 12(2):153–163

Jones JI, Clemmons DR (1995) Insulin-like growth factors and their binding proteins: biological actions. Endocr Rev 16(1):3–34

Kelly AM (1978) Perisynaptic satellite cells in the developing and mature rat soleus muscle. Anat Rec 190(4):891–903

Kottlors M, Kirschner J (2010) Elevated satellite cell number in Duchenne muscular dystrophy. Cell Tissue Res 340(3):541–548

Kovanen V, Suominen H, Risteli J, Risteli L (1988) Type IV collagen and laminin in slow and fast skeletal muscle in rats--effects of age and life-time endurance training. Coll Relat Res 8(2):145–153

Kuang S, Kuroda K, Le Grand F, Rudnicki MA (2007) Asymmetric self-renewal and commitment of satellite stem cells in muscle. Cell 129(5):999–1010

Kuschel R, Yablonka-Reuveni Z, Bornemann A (1999) Satellite cells on isolated myofibers from normal and denervated adult rat muscle. J Histochem Cytochem 47(11):1375–1384

Larsson L, Ansved T (1995) Effects of ageing on the motor unit. Prog Neurobiol 45(5):397–458

Lemos DR, Babaeijandaghi F, Low M, Chang CK, Lee ST, Fiore D, Zhang RH, Natarajan A, Nedospasov SA, Rossi FM (2015) Nilotinib reduces muscle fibrosis in chronic muscle injury by promoting TNF-mediated apoptosis of fibro/adipogenic progenitors. Nat Med 21(7):786–794

Lepper C, Ta P, Fan C-M (2011) An absolute requirement for Pax7-positive satellite cells in acute injury-induced skeletal muscle regeneration. Development 138(17):3639–3646

Lin J, Wu H, Tarr PT, Zhang C-Y, Wu Z, Boss O, Michael LF, Puigserver P, Isotani E, Olson EN, Lowell BB, Bassel-Duby R, Spiegelman BM (2002) Transcriptional co-activator PGC-1 alpha drives the formation of slow-twitch muscle fibres. Nature 418(6899):797–801

Liu H, Niu A, Chen S-E, Li Y-P (2011) Beta3-integrin mediates satellite cell differentiation in regenerating mouse muscle. FASEB J 25(6):1914–1921

Liu W, Wen Y, Bi P, Lai X, Liu XS, Liu X, Kuang S (2012) Hypoxia promotes satellite cell self-renewal and enhances the efficiency of myoblast transplantation. Development 139(16):2857–2865

Lukjanenko L, Jung MJ, Hegde N, Perruisseau-Carrier C, Migliavacca E, Rozo M, Karaz S, Jacot G, Schmidt M, Li L, Metairon S, Raymond F, Lee U, Sizzano F, Wilson DH, Dumont NA, Palini A, Fassler R, Steiner P, Descombes P, Rudnicki MA, Fan CM, von Maltzahn J, Feige JN, Bentzinger CF (2016) Loss of fibronectin from the aged stem cell niche affects the regenerative capacity of skeletal muscle in mice. Nat Med 22(8):897–905

Madden L, Juhas M, Kraus WE, Truskey GA, Bursac N (2015) Bioengineered human myobundles mimic clinical responses of skeletal muscle to drugs. elife 4:e04885

Mann CJ, Perdiguero E, Kharraz Y, Aguilar S, Pessina P, Serrano AL, Muñoz-Cánoves P (2011) Aberrant repair and fibrosis development in skeletal muscle. Skelet Muscle 1(1):21–21

Mauro A (1961) Satellite cell of skeletal muscle fibers. J Biophys Biochem Cytol 9:493–495

McCarthy JJ, Mula J, Miyazaki M, Erfani R, Garrison K, Farooqui AB, Srikuea R, Ba L, Grimes B, Keller C, Van Zant G, Campbell KS, Ka E, Dupont-Versteegden EE, Ca P (2011) Effective fiber hypertrophy in satellite cell-depleted skeletal muscle. Development 138(17):3657–3666

McCroskery S, Thomas M, Maxwell L, Sharma M, Kambadur R (2003) Myostatin negatively regulates satellite cell activation and self-renewal. J Cell Biol 162(6):1135–1147

Mendell JR, Rodino-Klapac LR (2016) Duchenne muscular dystrophy: CRISPR/Cas9 treatment. Cell Res 26(5):513–514

Menetrey J, Kasemkijwattana C, Day CS, Bosch P, Vogt M, FH F, Moreland MS, Huard J (2000) Growth factors improve muscle healing in vivo. J Bone Joint Surg Br 82(1):131–137

Mohyeldin A, Garzon-Muvdi T, Quinones-Hinojosa A (2010) Oxygen in stem cell biology: a critical component of the stem cell niche. Cell Stem Cell 7(2):150–161

Montarras D, Morgan J, Collins C, Relaix F, Zaffran S, Cumano A, Partridge T, Buckingham M (2005) Direct isolation of satellite cells for skeletal muscle regeneration. Science 309(5743):2064–2067

Mounier R, Chretien F, Chazaud B (2011) Blood vessels and the satellite cell niche. Curr Top Dev Biol 96:121–138

Mourikis P, Sambasivan R, Castel D, Rocheteau P, Bizzarro V, Tajbakhsh S (2012) A critical requirement for notch signaling in maintenance of the quiescent skeletal muscle stem cell state. Stem Cells 30(2):243–252

Murphy MM, Ja L, Mathew SJ, Da H, Kardon G (2011) Satellite cells, connective tissue fibroblasts and their interactions are crucial for muscle regeneration. Development 138(17):3625–3637

Ochoa O, Sun D, Reyes-Reyna SM, Waite LL, Michalek JE, McManus LM, Shireman PK (2007) Delayed angiogenesis and VEGF production in CCR2−/− mice during impaired skeletal muscle regeneration. Am J Physiol Regul Integr Comp Physiol 293(2):R651–R661

Oh J, Sinha I, Tan KY, Rosner B, Dreyfuss JM, Gjata O, Tran P, Shoelson SE, Wagers AJ (2016) Age-associated NF-kappaB signaling in myofibers alters the satellite cell niche and re-strains muscle stem cell function. Aging 8(11):2871–2896

Ono Y, Boldrin L, Knopp P, Morgan JE, Zammit PS (2010) Muscle satellite cells are a functionally heterogeneous population in both somite-derived and branchiomeric muscles. Dev Biol 337(1):29–41

Pallafacchina G, François S, Regnault B, Czarny B, Dive V, Cumano A, Montarras D, Buckingham M (2010) An adult tissue-specific stem cell in its niche: a gene profiling analysis of in vivo quiescent and activated muscle satellite cells. Stem Cell Res 4(2):77–91

Péault B, Rudnicki M, Torrente Y, Cossu G, Tremblay JP, Partridge T, Gussoni E, Kunkel LM, Huard J (2007) Stem and progenitor cells in skeletal muscle development, maintenance, and therapy. Mol Ther 15(5):867–877

Philippou A, Halapas A, Maridaki M, Koutsilieris M (2007) Type I insulin-like growth factor receptor signaling in skeletal muscle regeneration and hypertrophy. J Musculoskelet Neuronal Interact 7(3):208–218

Polesskaya A, Seale P, Ma R (2003) Wnt signaling induces the myogenic specification of resident CD45+ adult stem cells during muscle regeneration. Cell 113(7):841–852

Pretheeban T, Lemos DR, Paylor B, Zhang RH, Rossi FM (2012) Role of stem/progenitor cells in reparative disorders. Fibrogenesis Tissue Repair 5(1):20

Price FD, von Maltzahn J, Bentzinger CF, Dumont NA, Yin H, Chang NC, Wilson DH, Frenette J, Rudnicki MA (2014) Inhibition of JAK-STAT signaling stimulates adult satellite cell function. Nat Med 20(10):1174–1181

Quarta M, Brett JO, DiMarco R, De Morree A, Boutet SC, Chacon R, Gibbons MC, Garcia VA, Su J, Shrager JB, Heilshorn S, Rando TA (2016) An artificial niche preserves the quiescence of muscle stem cells and enhances their therapeutic efficacy. Nat Biotechnol 34(7):752–759

Qu-Petersen Z, Deasy B, Jankowski R, Ikezawa M, Cummins J, Pruchnic R, Mytinger J, Cao B, Gates C, Wernig A, Huard J (2002) Identification of a novel population of muscle stem cells in mice: potential for muscle regeneration. J Cell Biol 157(5):851–864

Ratajczak MZ, Majka M, Kucia M, Drukala J, Pietrzkowski Z, Peiper S, Janowska-Wieczorek A (2003) Expression of functional CXCR4 by muscle satellite cells and secretion of SDF-1 by muscle-derived fibroblasts is associated with the presence of both muscle progenitors in bone marrow and hematopoietic stem/progenitor cells in muscles. Stem Cells 21(3):363–371

Rhoads RP, Johnson RM, Rathbone CR, Liu X, Temm-Grove C, Sheehan SM, Hoying JB, Allen RE (2009) Satellite cell-mediated angiogenesis in vitro coincides with a functional hypoxia-inducible factor pathway. Am J Physiol Cell Physiol 296(6):C1321–C1328

Rocheteau P, Gayraud-Morel B, Siegl-Cachedenier I, Blasco MA, Tajbakhsh S (2012) A subpopulation of adult skeletal muscle stem cells retains all template DNA strands after cell division. Cell 148(12):112–125

Rodgers JT, King KY, Brett JO, Cromie MJ, Charville GW, Maguire KK, Brunson C, Mastey N, Liu L, Tsai CR, Goodell MA, Rando TA (2014) mTORC1 controls the adaptive transition of quiescent stem cells from G0 to G(Alert). Nature 510(7505):393–396

Rosenblatt JD (1992) A time course study of the isometric contractile properties of rat extensor digitorum longus muscle injected with bupivacaine. Comp Biochem Physiol Comp Physiol 101(2):361–367

Rozo M, Li L, Fan CM (2016) Targeting beta1-integrin signaling enhances regeneration in aged and dystrophic muscle in mice. Nat Med 22(8):889–896

Ryall JG, Dell'Orso S, Derfoul A, Juan A, Zare H, Feng X, Clermont D, Koulnis M, Gutierrez-Cruz G, Fulco M, Sartorelli V (2015) The NAD(+)-dependent SIRT1 deacetylase translates a metabolic switch into regulatory epigenetics in skeletal muscle stem cells. Cell Stem Cell 16(2):171–183

Sacco A, Doyonnas R, Kraft P, Vitorovic S, Blau HM (2008) Self-renewal and expansion of single transplanted muscle stem cells. Nature 456(7221):502–506

Sacco A, Mourkioti F, Tran R, Choi J, Llewellyn M, Kraft P, Shkreli M, Delp S, Pomerantz JH, Artandi SE, Blau HM (2010) Short telomeres and stem cell exhaustion model Duchenne muscular dystrophy in mdx/mTR mice. Cell 143(7):1059–1071

Sambasivan R, Yao R, Kissenpfennig A, Van Wittenberghe L, Paldi A, Gayraud-Morel B, Guenou H, Malissen B, Tajbakhsh S, Galy A (2011) Pax7-expressing satellite cells are indispensable for adult skeletal muscle regeneration. Development 138(17):3647–3656

Sampaolesi M, Torrente Y, Innocenzi A, Tonlorenzi R, D'Antona G, Pellegrino MA, Barresi R, Bresolin N, De Angelis MG, Campbell KP, Bottinelli R, Cossu G (2003) Cell therapy of alpha-sarcoglycan null dystrophic mice through intra-arterial delivery of mesoangioblasts. Science 301(5632):487–492

Sampaolesi M, Blot S, D'Antona G, Granger N, Tonlorenzi R, Innocenzi A, Mognol P, Thibaud JL, Galvez BG, Barthelemy I, Perani L, Mantero S, Guttinger M, Pansarasa O, Rinaldi C, Cusella De Angelis MG, Torrente Y, Bordignon C, Bottinelli R, Cossu G (2006) Mesoangioblast stem cells ameliorate muscle function in dystrophic dogs. Nature 444(7119):574–579

Sanes JR (2003) The basement membrane/basal lamina of skeletal muscle. J Biol Chem 278(15):12601–12604

Scapini P, Lapinet-Vera JA, Gasperini S, Calzetti F, Bazzoni F, Cassatella MA (2000) The neutrophil as a cellular source of chemokines. Immunol Rev 177:195–203

Schafer MJ, Atkinson EJ, Vanderboom PM, Kotajarvi B, White TA, Moore MM, Bruce CJ, Greason KL, Suri RM, Khosla S, Miller JD, Bergen HR III, LeBrasseur NK (2016) Quantification of GDF11 and myostatin in human aging and cardiovascular disease. Cell Metab 23(6):1207–1215

Segawa M, Fukada S, Yamamoto Y, Yahagi H, Kanematsu M, Sato M, Ito T, Uezumi A, Hayashi S, Miyagoe-Suzuki Y, Takeda S, Tsujikawa K, Yamamoto H (2008) Suppression of macrophage functions impairs skeletal muscle regeneration with severe fibrosis. Exp Cell Res 314(17):3232–3244

Serrano AL, Mann CJ, Vidal B, Ardite E, Perdiguero E, Munoz-Canoves P (2011) Cellular and molecular mechanisms regulating fibrosis in skeletal muscle repair and disease. Curr Top Dev Biol 96:167–201

Shefer G, Rauner G, Yablonka-Reuveni Z, Benayahu D (2010) Reduced satellite cell numbers and myogenic capacity in aging can be alleviated by endurance exercise. PLoS One 5(10):e13307

Shvartsman D, Storrie-White H, Lee K, Kearney C, Brudno Y, Ho N, Cezar C, McCann C, Anderson E, Koullias J, Tapia JC, Vandenburgh H, Lichtman JW, Mooney DJ (2014) Sustained delivery of VEGF maintains innervation and promotes reperfusion in ischemic skeletal muscles via NGF/GDNF signaling. Mol Ther 22(7):1243–1253

Sicari BM, Rubin JP, Dearth CL, Wolf MT, Ambrosio F, Boninger M, Turner NJ, Weber DJ, Simpson TW, Wyse A, Brown EH, Dziki JL, Fisher LE, Brown S, Badylak SF (2014) An acellular biologic scaffold promotes skeletal muscle formation in mice and humans with volumetric muscle loss. Sci Transl Med 6(234):234ra258

Sinha M, Jang YC, Oh J, Khong D, Wu EY, Manohar R, Miller C, Regalado SG, Loffredo FS, Pancoast JR, Hirshman MF, Lebowitz J, Shadrach JL, Cerletti M, Kim MJ, Serwold T, Goodyear LJ, Rosner B, Lee RT, Wagers AJ (2014) Restoring systemic GDF11 levels reverses age-related dysfunction in mouse skeletal muscle. Science 344(6184):649–652

Sinha-Hikim I, Taylor WE, Gonzalez-Cadavid NF, Zheng W, Bhasin S (2004) Androgen receptor in human skeletal muscle and cultured muscle satellite cells: up-regulation by androgen treatment. J Clin Endocrinol Metab 89(10):5245–5255

Snijders T, Verdijk LB, van Loon LJ (2009) The impact of sarcopenia and exercise training on skeletal muscle satellite cells. Ageing Res Rev 8(4):328–338

Sonnet C, Lafuste P, Arnold L, Brigitte M, Poron F, Authier FJ, Chretien F, Gherardi RK, Chazaud B (2006) Human macrophages rescue myoblasts and myotubes from apoptosis through a set of adhesion molecular systems. J Cell Sci 119(Pt 12):2497–2507

Sousa-Victor P, Gutarra S, Garcia-Prat L, Rodriguez-Ubreva J, Ortet L, Ruiz-Bonilla V, Jardi M, Ballestar E, Gonzalez S, Serrano AL, Perdiguero E, Munoz-Canoves P (2014) Geriatric muscle stem cells switch reversible quiescence into senescence. Nature 506(7488):316–321

Summan M, Warren GL, Mercer RR, Chapman R, Hulderman T, Van Rooijen N, Simeonova PP (2006) Macrophages and skeletal muscle regeneration: a clodronate-containing liposome depletion study. Am J Physiol Regul Integr Comp Physiol 290(6):R1488–R1495

Tang AH, Rando TA (2014) Induction of autophagy supports the bioenergetic demands of quiescent muscle stem cell activation. EMBO J 33(23):2782–2797

Tatsumi R, Liu X, Pulido A, Morales M, Sakata T, Dial S, Hattori A, Ikeuchi Y, Allen RE, Mo M (2006) Satellite cell activation in stretched skeletal muscle and the role of nitric oxide and hepatocyte growth factor. Am J Physiol Cell Physiol 290(6):C1487–C1494

Tatsumi R, Sankoda Y, Anderson JE, Sato Y, Mizunoya W, Shimizu N, Suzuki T, Yamada M, Rhoads RP Jr, Ikeuchi Y, Allen RE (2009) Possible implication of satellite cells in regenerative motoneuritogenesis: HGF upregulates neural chemorepellent Sema3A during myogenic differentiation. Am J Physiol Cell Physiol 297(2):C238–C252

Thomas K, Engler AJ, Meyer GA (2015) Extracellular matrix regulation in the muscle satellite cell niche. Connect Tissue Res 56(1):1–8

Tierney MT, Aydogdu T, Sala D, Malecova B, Gatto S, Puri PL, Latella L, Sacco A (2014) STAT3 signaling controls satellite cell expansion and skeletal muscle repair. Nat Med 20(10):1182–1186

Troy A, Cadwallader AB, Fedorov Y, Tyner K, Tanaka KK, Olwin BB (2012) Coordination of satellite cell activation and self-renewal by Par-complex-dependent asymmetric activation of p38alpha/beta MAPK. Cell Stem Cell 11(4):541–553

Uezumi A, Fukada S, Yamamoto N, Takeda S, Tsuchida K (2010) Mesenchymal progenitors distinct from satellite cells contribute to ectopic fat cell formation in skeletal muscle. Nat Cell Biol 12(2):143–152

Urciuolo A, Quarta M, Morbidoni V, Gattazzo F, Molon S, Grumati P, Montemurro F, Tedesco FS, Blaauw B, Cossu G, Vozzi G, Ta R, Bonaldo P (2013) Collagen VI regulates satellite cell self-renewal and muscle regeneration. Nat Commun 4(May):1964–1964

Wagatsuma A, Sakuma K (2013) Mitochondria as a potential regulator of myogenesis. ScientificWorldJournal 2013:593267

Wagner KR, Lechtzin N, Judge DP (2007) Current treatment of adult Duchenne muscular dystrophy. Biochim Biophys Acta 1772(2):229–237

Walker RG, Poggioli T, Katsimpardi L, Buchanan SM, Oh J, Wattrus S, Heidecker B, Fong YW, Rubin LL, Ganz P, Thompson TB, Wagers AJ, Lee RT (2016) Biochemistry and biology of GDF11 and myostatin: similarities, differences, and questions for future investigation. Circ Res 118(7):1125–1141; discussion 1142

Wang Y, Thorlacius H (2005) Mast cell-derived tumour necrosis factor-alpha mediates macrophage inflammatory protein-2-induced recruitment of neutrophils in mice. Br J Pharmacol 145(8):1062–1068

Warren GL, O'Farrell L, Summan M, Hulderman T, Mishra D, Luster MI, Kuziel WA, Simeonova PP (2004) Role of CC chemokines in skeletal muscle functional restoration after injury. Am J Physiol Cell Physiol 286(5):C1031–C1036

Wilson JM, Loenneke JP, Jo E, Wilson GJ, Zourdos MC, Kim JS (2012) The effects of endurance, strength, and power training on muscle fiber type shifting. J Strength Cond Res 26(6):1724–1729

Yamaguchi M, Watanabe Y, Ohtani T, Uezumi A, Mikami N, Nakamura M, Sato T, Ikawa M, Hoshino M, Tsuchida K, Miyagoe-Suzuki Y, Tsujikawa K, Takeda S, Yamamoto H, Fukada S (2015) Calcitonin receptor signaling inhibits muscle stem cells from escaping the quiescent state and the niche. Cell Rep 13(2):302–314

Yin H, Price F, Ma R (2013) Satellite cells and the muscle stem cell niche. Physiol Rev 93(1):23–67

Young CS, Hicks MR, Ermolova NV, Nakano H, Jan M, Younesi S, Karumbayaram S, Kumagai-Cresse C, Wang D, Zack JA, Kohn DB, Nakano A, Nelson SF, Miceli MC, Spencer MJ, Pyle AD (2016) A single CRISPR-Cas9 deletion strategy that targets the majority of DMD patients restores dystrophin function in hiPSC-derived muscle cells. Cell Stem Cell 18(4):533–540

Chapter 9
The Macula Flava of the Human Vocal Fold as a Stem Cell Microenvironment

Kiminori Sato

Abstract

1. There is growing evidence to suggest that the cells in the maculae flavae are tissue stem cells of the human vocal fold and maculae flavae are a candidate for a stem cell niche.
2. The latest research shows that the cells in the human maculae flavae are involved in the metabolism of extracellular matrices that are essential for the viscoelasticity in the human vocal fold mucosa as a vibrating tissue, and considered to be important cells in the growth, development, and aging of the human vocal fold mucosa.
3. The cells in the human maculae flavae possess proteins of all three germ layers, indicating they are undifferentiated and have the ability of multipotency.
4. The cell division in the human adult maculae flavae is reflective of asymmetric self-renewal and cultured cells form a colony-forming unit. Therefore, the phenomenon gives rise to the strong possibility that the cells in the human maculae flavae are tissue stem cells.
5. Recent research suggests that the cells in the human maculae flavae arise from the differentiation of bone marrow cells via peripheral circulation.
6. The hyaluronan concentration in the maculae flavae is high and contains cells which possess hyaluronan receptors, indicating that the maculae flavae are hyaluronan-rich matrix, which is required for a stem cell niche.
7. A proper microenvironment in the maculae flavae of the human vocal fold mucosa is necessary to be effective as a stem cell niche maintaining the stemness of the contained tissue stem cells.

K. Sato (✉)
Department of Otolaryngology—Head and Neck Surgery, Kurume University School of Medicine, Kurume, Japan
e-mail: kimisato@oct-net.ne.jp

© Springer International Publishing AG 2017
A. Birbrair (ed.), *Stem Cell Microenvironments and Beyond,*
Advances in Experimental Medicine and Biology 1041,
DOI 10.1007/978-3-319-69194-7_9

Keywords Macula flava • Tissue stem cells • Stem cell niche • Vocal fold stellate cells • Vocal fold

9.1 Introduction

Among mammals, only humans can speak and sing songs throughout their lifetime. And only the human adult vocal fold has a vocal ligament, Reinke's space, and a layered structure (Kurita et al. 1986; Sato et al. 2000; Hirano 1975; Hirano and Sato 1993). Why do only human adults have such a characteristic vocal fold structure? Why and how does the newborn vocal fold mucosa grow, develop and mature? What are the factors for initiating and continuing the growth of the human vocal fold mucosa? Why does the adult vocal fold maintain its characteristic layered structure for many decades?

After birth, adult stem cells, including both germ-line stem cells and tissue stem cells, reside in a specific microenvironment termed a "niche", which varies in nature and location depending on the tissue type (Li and Xie 2005). These adult stem cells are an essential component of tissue homeostasis; they support ongoing tissue regeneration replacing cells lost due to natural cell death (apoptosis) or injury (Li and Xie 2005).

Adult tissue-specific stem cells (tissue stem cells) have the capacity to self-renew and generate functionally differentiated cells that replenish lost cells throughout an organism's lifetime. Tissue-specific stem cells reside in a niche, whereby a complex microenvironment maintains their multipotency.

Viscoelastic properties of the lamina propria of the human vocal fold mucosa determine its vibratory behavior and depend on extracellular matrices, such as collagen fibers, reticular fibers, elastic fibers, proteoglycan, glycosaminoglycan and glycoproteins. The three-dimensional structures of these extracellular matrices are indispensable to the viscoelastic properties of the human vocal fold mucosa. Fine structures of the human vocal fold mucosa influence vibrating behavior and voice quality.

Human adult maculae flavae are dense masses of cells and extracellular matrices located at the anterior and posterior ends of the membranous portion of the bilateral vocal folds. The histological structure of the maculae flavae in the human adult vocal fold mucosa is unique and not suitable for vibration. Therefore, their roles in the human vocal fold as a vibrating tissue are very interesting. However, their roles in the human vocal fold have not been clarified until recently (Lanz and Wachsmuth 1955; Subotic et al. 1984; Vecerina-Volic et al. 1988; Campos Banales et al. 1995; Fayoux et al. 2004).

The latest researches show that the human maculae flavae are involved in the metabolism of extracellular matrices that are essential for the viscoelasticity of the human vocal fold mucosa, and are considered to be an important structure in the growth, development and ageing of the human vocal fold mucosa. In addition, there is growing evidence to suggest that the cells in the maculae flavae are tissue stem cells of the human vocal fold mucosa and maculae flavae are a candidate for a stem cell niche.

In this chapter, the latest research regarding the maculae flavae of the human vocal fold as a stem cell microenvironment are summarized.

9.2   Maculae Flavae in the Human Adult Vocal Fold

The vibratory portion (membranous portion) of the human vocal fold is connected to the thyroid cartilage anteriorly via the intervening anterior macula flava and anterior commissure tendon. Posteriorly, it is joined to the vocal process of the arytenoid cartilage via the intervening posterior macula flava (Figs. 9.1 and 9.2). The vocal ligament runs between the anterior and posterior maculae flavae.

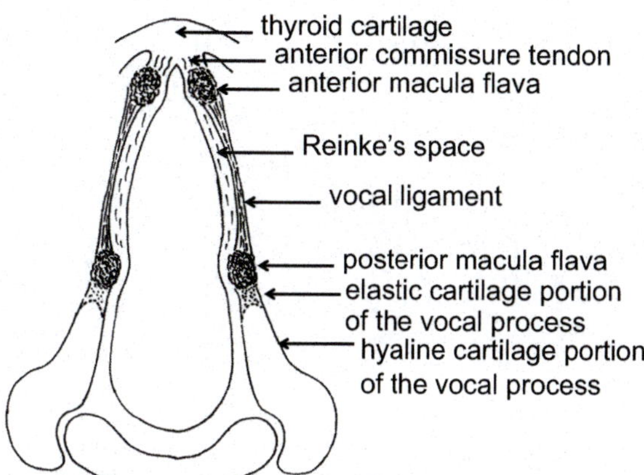

Fig. 9.1 Human adult vocal fold and maculae flavae

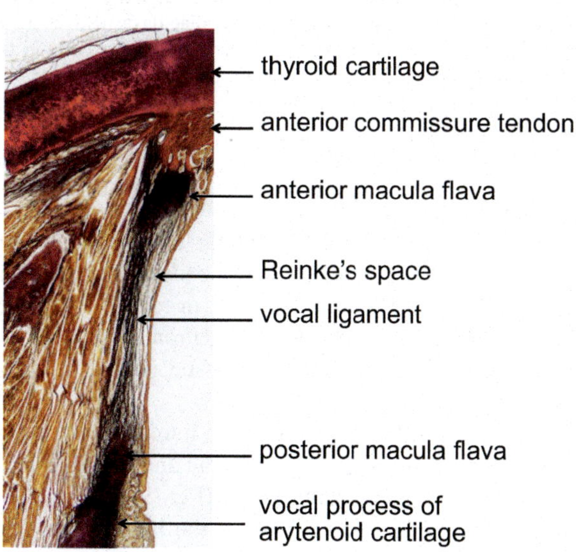

Fig. 9.2 Transverse section of human adult vocal fold (Elastica van Gieson stain)

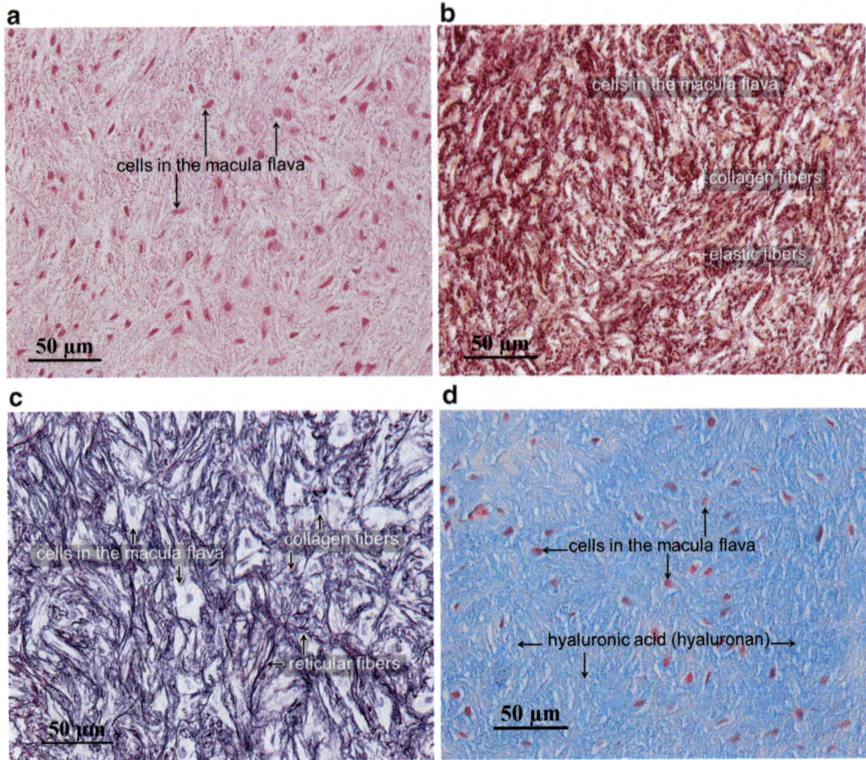

Fig. 9.3 Macula flava of the human adult vocal fold. (**a**) Human adult maculae flavae are dense masses of cells (Hematoxylin and eosin stain). (**b**) There are many collagen fibers stained red and elastic fibers stained black around the cells in the human adult maculae flavae (Elastic van Gieson stain). (**c**) There are many collagen fibers stained red and reticular fibers (type III collagen) stained black around cells in the human adult maculae flavae (Silver stain). (**d**) Much glycosaminoglycan (hyaluronan, hyaluronic acid) is situated around the cells in the human adult maculae flavae (Alcian blue stain, pH 2.5). Maculae flavae are strongly stained light blue with Alcian Blue at pH 2.5. Material in the maculae flavae that is strongly stained with Alcian Blue (pH 2.5) is digested by hyaluronidase

Human adult maculae flavae are dense masses of cells and extracellular matrices (Fig. 9.3) (Sato and Hirano 1995a; Sato et al. 2010a). The maculae flavae are located at the anterior and posterior ends of the membranous portion of the bilateral vocal folds. They are elliptical in shape and their size is approximately 1.5 mm × 1.5 mm × 1 mm (Sato and Hirano 1995a). The border between the maculae flavae and the surrounding soft tissue is relatively clearly delineated (Fig. 9.2) (Sato et al. 2003a).

The extracellular matrices of the human adult maculae flavae are composed of glycoproteins and glycosaminoglycan and fibrillar proteins such as collagen fibers, reticular fibers, and elastic fibers (Fig. 9.3). These extracellular matrices in the maculae flavae extend to those in the lamina propria (Reinke's space and vocal ligament) of the human vocal fold mucosa (Sato and Hirano 1995a).

9.3   Intermediate Filaments of the Cells in the Human Adult Maculae Flavae

The expression of proteins in the intermediate filaments of the cytoplasm is specific to cell type and differentiation (Becker et al. 2006a).

Proteins in the intermediate filaments containing cytokeratin, vimentin, glial fibrillary acidic protein (GFAP) and desmin are distributed in the cytoplasm of the cells in the adult maculae flavae (Sato et al. 2012a). Additionally, cells in the human maculae flavae express SOX 17, which is an endodermal cell marker. Consequently, the cells in the human adult maculae flavae express proteins of all three germ layers. This suggests that the cells in the maculae flavae are undifferentiated cells and have the ability of multipotency.

9.4   Radiosensitivity of the Cells in the Human Adult Maculae Flavae

The radiosensitivity of the cells in the maculae flavae is morphologically higher than that of fibroblasts in Reinke's space of the human vocal fold mucosa, indicating that the cells in the maculae flavae are not yet as fully differentiated as fibroblasts (Sato et al. 2008b).

9.5   Telomerase of the Cells in the Human Adult Maculae Flavae

In multicellular organisms, telomerase resides mainly in the germ cells that give rise to sperm and eggs, and in a few other kinds of proliferating normal cells such as stem cells (Becker et al. 2006b).

The cells in the maculae flavae express telomerase reverse transcriptase, indicating the special DNA polymerase called telomerase resides in the cells in the maculae flavae (Sato et al. 2012b). This suggests that the cells in the human maculae flavae are a tissue stem cell of the human vocal fold mucosa.

9.6   Cell Cycle of the Cells in the Human Adult Maculae Flavae

Cells express Ki-67 during proliferation (G1-, S-, G2- M-phase) in the cell cycle, but cells that are in an arrested state (G0-phase) do not express Ki-67 (Schlüter et al. 1993).

The cells in the human maculae flavae do not express Ki-67, indicating that they are resting cells (G0-phase), as are other stem cells (Sato et al. 2012b).

9.7   Vocal Fold Stellate Cells in the Human Adult Maculae Flavae

Interstitial cells with a star-like appearance in the human adult maculae flavae were discovered in our laboratory in 2001 (Fig. 9.5) (Sato et al. 2001a, 2003b, 2004, 2010b; Sato and Nakashima 2005). These cells had no nomenclature and were thus designated "vocal fold stellate cells" in the series of our study. Vocal fold stellate cells are stellate in shape and possess vitamin A-storing lipid droplets (Sato et al. 2003b). There are a number of morphological differences between vocal fold stellate cells and fibroblasts in the human vocal fold mucosa. Along the surface of the vocal fold stellate cells, a number of vesicles are present and constantly synthesize extracellular matrices which are essential for the viscoelastic properties of the human vocal fold mucosa (Sato et al. 2001a).

As a result of this heterogeneity, it is uncertain whether the vocal fold stellate cells derive from the same embryonic source as fibroblasts in the human vocal fold mucosa. The vocal fold stellate cells in the maculae flavae form an independent cell category that are considered a new category of cells in the human vocal fold mucosa.

9.8   Cell Division of Cells in the Human Maculae Flavae

In vitro culturing of the human maculae flavae yields interesting results. After a few weeks of primary culture in an MF-start primary culture medium (Toyobo, Osaka, Japan), two types of cells, fibroblast-like spindle cells (Group A) and cobblestone-like squamous cells (Group B), grow from the human macula flava fragments (Fig. 9.6) (Sato et al. 2016a). After removing the two types of cells by cell scraper, each type of cell is individually subcultured in an MF-medium (Mesenchymal Stem Cell Growth Medium) (Toyobo, Osaka, Japan) to proliferate the cells.

After a week of first subculture, subcultured Group A cells become stellate in shape and possess slender cytoplasmic processes (Fig. 9.7a). Small lipid droplets are present in the cytoplasm. The nuclei are oval in shape and their nucleus-cytoplasm ratios are low. These cells are morphologically similar to vocal fold stellate cells.

After a week of second subculture, subcultured Group B cells form a colony-forming unit (Fig. 9.7b), indicating these cells are mesenchymal stem cells or stromal stem cells in the bone marrow.

Therefore, the colony-forming phenomenon gives rise to the possibility that the cells in the human maculae flavae are tissue stem cells (Sato et al. 2016a).

As mentioned above, the cell division in the human adult maculae flavae with mesenchymal stem cell growth medium is reflective of asymmetric self-renewal (Sato et al. 2016a). Asymmetry in cell division gives rise to the possibility that the maculae flavae in the human adult vocal fold is a stem cell niche containing tissue stem cells (Sato et al. 2016a).

9.9 Transition Area Between the Human Adult Maculae Flavae and Surrouding Tissue

The transition area between the maculae flavae and their surrounding tissue is interesting.

The posterior macula flava is attached to the vocal process of the arytenoid cartilage posteriorly (Figs. 9.1 and 9.2). Elastic cartilage located at the tip of the vocal process facilitates movement of the vocal process during adduction and abduction (Sato et al. 1990). The transition of cells and extracellular matrices between the posterior macula flava and the elastic cartilage portion of the vocal process is gradual, and the border between them is not clearly delineated. The cells in the posterior macula flava appear to differentiate into chondrocytes in the tip of the vocal process (Sato et al. 2012b).

The cells in the human maculae flavae express CD44 (mesenchymal stem cell marker). Most of the fibroblasts in the tissue surrounding the maculae flavae do not express CD44. However, CD44-positive fibroblasts are observed at the periphery of the maculae flavae. The cells in the macula flava appear to differentiate into fibroblasts in the surrounding tissue (Sato et al. 2012b).

These findings raise the possibility that the cells in the maculae flavae generate functionally differentiated cells, such as chondrocytes and fibroblasts in the human vocal fold mucosa (Sato et al. 2012b). Additional investigations are needed to determine whether the cells in the maculae flavae have the capacity to self-renew and generate functionally differentiated cells (multipotency) that replenish lost cells throughout an organism's lifetime.

9.10 Hierarchy of Tissue Stem Cells in the Human Maculae Flavae

Here, the question arises whether the vocal fold stellate cells are tissue stem cells or progenitor cells (transit-amplifying cells).

Both colony-forming subcultured cells (cobblestone-like squamous cells) and non-colony-forming subcultured cells (fibroblast-like spindle cells) (Fig. 9.6) express cytoplasmic cytokeratin, vimentin, GFAP and desmin (Kurita et al. 2015). Consequently, both colony-forming cells (cobblestone-like squamous cells) and non-colony-forming cells (fibroblast-like spindle cells) express ectoderm and mesoderm germ layers. This suggests that they are undifferentiated cells and have the ability of multipotency (Kurita et al. 2015).

The vocal fold stellate cells are possibly transit-amplifying cells, that is, progenitor cells (Sato et al. 2016a). However, at the present state of our investigation, it is difficult to clarify the stem cell system and hierarchy of stem cells in the human maculae flavae and determine whether the vocal fold stellate cells are tissue stem cells or progenitor cells.

9.11 Microenvironment, Hyaluronan-Rich Matrix, of the Maculae Flavae as a Stem Cell Niche in the Human Vocal Fold

The structural and biochemical microenvironment that confers stemness upon cells in multicellular organisms is referred to as the stem cell niche. A stem cell niche is composed of a group of cells in a special tissue location for the maintenance of stem cells (Li and Xie 2005).

Hyaluronan serves as an important niche component for numerous stem cell populations (Haylock and Nilsson 2006; Preston and Sherman 2011). After the discovery of hyaluronan, it was assumed that its major functions were in the biophysical and homeostatic properties of tissues. However, current studies lead to understanding that hyaluronan also plays a crucial role in cell behavior (Toole 1991). A hyaluronan-rich matrix, which is composed of the glycosaminoglycan hyaluronan and its transmembrane receptors (cell surface hyaluronan receptors), is able to directly affect the cellular functions of stem cells in a stem cell niche (Haylock and Nilsson 2006; Preston and Sherman 2011).

The maculae flavae in the human adult vocal fold are strongly stained light blue with Alcian Blue at pH 2.5 (Fig. 9.4). The materials in the maculae flavae that are strongly stained with Alcian Blue (pH 2.5) are digested by hyaluronidase. A great deal of glycosaminoglycan (hyaluronan) is situated around the cells in the human adult maculae flavae and hyaluronan concentration is high. The border between dense masses of hyaluronan (macula flava) and the surrounding tissue is

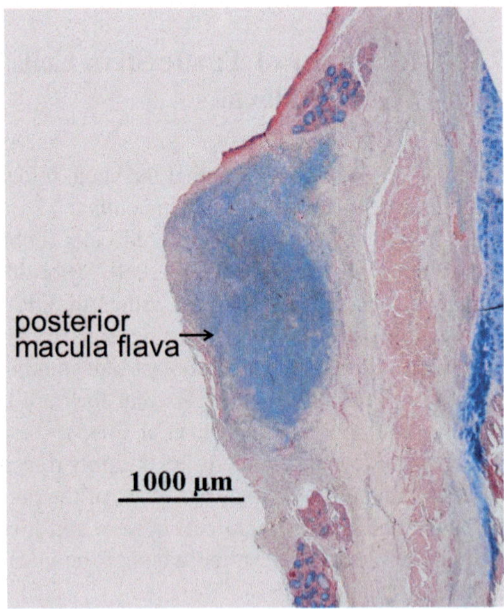

Fig. 9.4 Coronal section of the posterior macula flava (Alcian blue stain, pH 2.5)

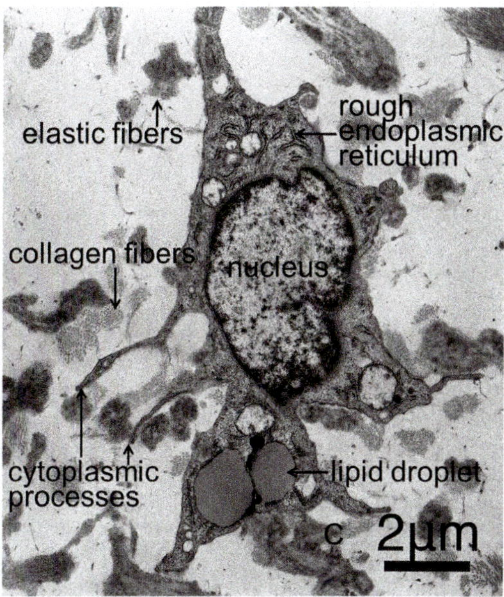

Fig. 9.5 Transmission electron micrograph of a vocal fold stellate cell in the human macula flava (uranyl acetate and lead citrate stain). Vocal fold stellate cells are stellate in shape and possess slender cytoplasmic processes. They possess vitamin A-storing lipid droplets and intracellular organelles such as rough endoplasmic reticulum and Golgi apparatus

Fig. 9.6 Primary culture of macula flava with MF-start primary culture medium (Toyobo, Osaka, Japan) (Phase-contrast microscopy). Two types of cells, cobblestone-like squamous cells and fibroblast-like spindle cells, grow from the macula flava fragments in the primary culture

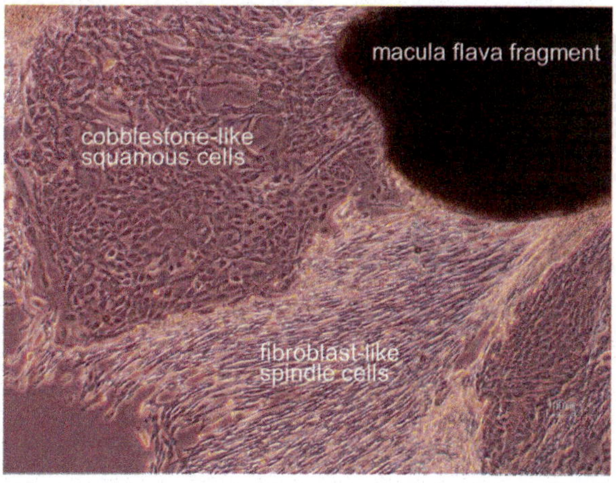

clearly delineated. Additionally, most of the cells in the maculae flavae express CD44 (cell surface hyaluronan receptors) (Fig. 9.8). This indicates that the human maculae flavae are hyaluronan-rich pericellular matrix (Sato et al. 2012b).

Since the cells in the human maculae flavae have cell surface hyaluronan receptors and are surrounded by a high concentration of hyaluronan (Sato et al. 2012b),

Fig. 9.7 Individual subculture of each type of cell in an MF-medium (Mesenchymal Stem Cell Growth Medium) (Toyobo, Osaka, Japan) to proliferate the cells (Phase-contrast microscopy). (a) Stellate cells. Fibroblast-like cells in the primary culture become stellate in shape and possess slender cytoplasmic processes and have small lipid droplets in the cytoplasm. (b) Colony-forming unit. Cobblestone-like squamous cells in an MF-medium form a colony-forming unit

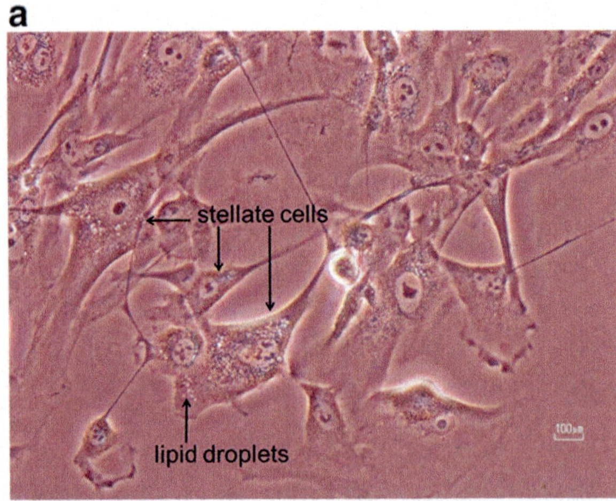

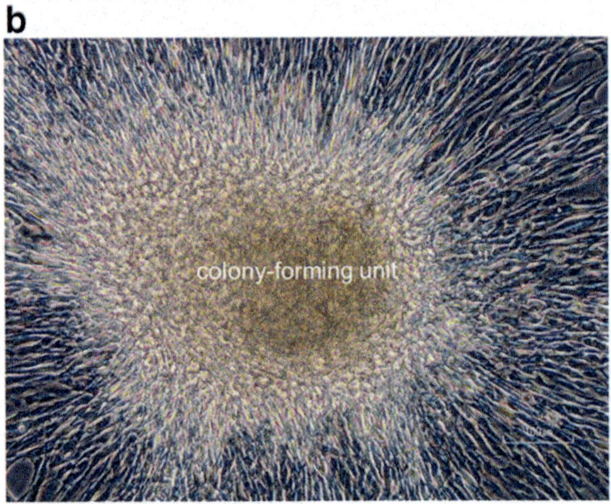

Fig. 9.8 CD44 on cytoplasm of cells in the human adult macula flava, shown by immunohistochemical staining

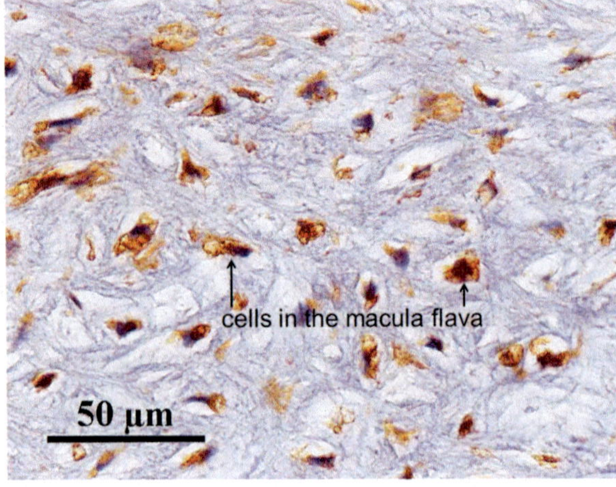

the maculae flavae are a candidate for a stem cell niche, which is a microenvironment nurturing a pool of tissue stem cells (Sato et al. 2012b).

9.12   Origin of Cells in the Human Maculae Flavae

Bone marrow-derived cells have received a great deal of attention with regard to tissue development and regeneration. Bone marrow-derived cells are considered to contain bone marrow derived mesenchymal stem cells, which are multipotent cells capable of self-renewal (Pittenger et al. 1999; Prockop 1997), and to be the origin of circulating fibrocytes, which are associated with wound healing and tissue fibrosis (Bucala et al. 1994). They circulate in the peripheral blood and are distributed to organs under normal conditions. When tissue is injured, they contribute to tissue repair by cell differentiation and migrate into injured tissue as needed (Forbes et al. 2004; Brittan et al. 2002).

The cells in the human maculae flavae express CD34 (hematopoietic stem cell marker). They also express CD45 (leukocyte common antigen), and collagen type I (Kurita et al. 2015). These proteins (CD34, CD45 and collagen type I), which are major makers of bone marrow derived circulating fibrocytes, are present in the cells in the human maculae flavae (Kurita et al. 2015).

Circulating fibrocytes were first described as blood-born fibroblast-like cells by Bucala, et al. (Bucala et al. 1994) They were found to be unique cells because they co-expressed hematopoietic markers as well as collagen type I and other mesenchymal markers. CD34, CD45 and collagen type I are major markers for circulating fibrocytes derived from bone marrow (Abedi 2012). CD34, CD45 and collagen type I are expressed in the cells in the human adult maculae flavae. Consequently, cells in the human maculae flavae quite possibly arise, not from resident interstitial cells of the vocal fold mucosa, but from the differentiation of bone marrow cells via peripheral circulation (Kurita et al. 2015).

9.13   Side Population Cells in the Vocal Fold Mucosa

Side population cells are regarded as a cell population enriched with stem cells or progenitor cells, and recognized as a candidate for tissue stem cells.

In one recent study, side population cells were identified in the epithelium and subepithelial tissue including the anterior and posterior maculae flavae (Yamashita et al. 2007). In another recent study, side population cells increased significantly in Reinke's space of an injured vocal fold starting on day 3, with a peak at day 7, followed by a decrease back to baseline values on day 14 (Gugatschka et al. 2011). These cells in the maculae flavae participated in the early stages of wound healing (Gugatschka et al. 2011). The two investigation cited here suggest that the anterior and posterior maculae flavae contain stem cells or progenitor cells and these cells have the capacity to play essential roles in tissue regeneration.

9.14 Vocal Fold Stem Cells and Their Niche in the Human Newborn Vocal Fold Mucosa

The results of our studies are consistent with the hypothesis that the cells in the maculae flavae are tissue stem cells or progenitor cells of the human newborn vocal fold mucosa (Sato et al. 2016b).

Newborns already have maculae flavae at the same sites as in adult vocal folds (Hirano and Sato 1993; Sato and Nakashima 2005; Sato and Hirano 1995b). The newborn maculae flavae are composed of relatively dense masses of cells and situated at the anterior and posterior ends of the bilateral vocal fold mucosae. The cells in the newborn maculae flavae possess some features of mesenchymeal cells (Sato and Nakashima 2005; Sato and Hirano 1995b). The cells in the human newborn maculae flavae possess proteins of all three germ layers (Sato et al. 2016b). They are undifferentiated cells which arise not from resident interstitial cells but from the differentiation of bone marrow cells (Sato et al. 2016b).

Extracellular matrices composed of collagen fibers, reticular fibers, elastic fibers and ground substances are not abundant in the newborn maculae flavae. A newborn's macula flava is in the process of acquiring a hyaluronan-rich matrix making it a candidate for a stem cell niche.

At birth, the cells have already been supplied, likely from the bone marrow, into the maculae flavae in the newborn vocal fold and are ready to start the growth and development of the human vocal fold mucosa as a vibrating tissue (Sato and Hirano 1995b).

9.15 Mechanical Regulation (Cellular Mechanotransduction) of the Cells in the Human Maculae Flavae

Current scientific findings suggest that the magnitude and frequency of tensile strain are particularly important in determining the type of mechanically induced differentiation that stem cells will undergo (Kurpinski et al. 2010). The macula flava is the microenvironment where the magnitude and frequency of tensile strain during vocal fold vibration are greatest (Titze and Hunter 2004). The function and fate of the cells in the human maculae flavae are regulated by various microenvironmental factors. In addition to chemical factors, mechanical factors also modulate the behavior of cells in the human maculae flavae.

We hypothesize that the tensions caused by phonation (vocal fold vibration) after birth stimulate cells in the anterior and posterior maculae flavae to accelerate production of extracellular matrices and form the vocal ligament, Reinke's space and the layered structure (Sato et al. 2001b, 2016b). The results of our studies (morphologic differences are detected between adult vocal fold mucosae that have been phonated and those that have remained unphonated since birth (Sato et al. 2008, 2012, 2015)) are consistent with this hypothesis.

We also hypothesize that after the layered structure of the adult vocal fold is completed, the tensions caused by phonation (vocal fold vibration) stimulate cells in the anterior and posterior maculae flavae to accelerate production of extracellular matrices and maintain the layered structure of the human adult vocal fold mucosa as a vibrating tissue. The results of our study (morphologic differences are detected between the adult vocal fold mucosae that have remained phonated and those that have been unphonated for a long period (Sato et al. 2011)) are consistent with this hypothesis.

The bending stresses on the vocal fold associated with phonation (vocal fold vibration) are greatest in the region of the maculae flavae located at both ends of the vocal fold mucosa (Titze and Hunter 2004). Tension caused by phonation seems to regulate the behavior of the cells (mechanical regulation) in the maculae flavae of the human vocal fold. It is of interest whether the mechanical forces caused by vocal fold vibration from outside the cells in the maculae flavae contacts influence intracellular signaling cascades through cell-matrix that ultimately alter many cellular behaviors.

"Mechanotransduction" is the term for the ability of living tissues to sense mechanical stress and respond by tissue remodeling. Cellular mechanotransduction is the mechanism by which cells convert mechanical stimuli into biomechanical responses. More recently, mechanotransduction has expanded to include the sensation of stress, its translation into a biochemical signal and the sequence of biological responses it produces. Mechanical stress has become increasingly recognized as one of the primary and essential factors controlling biological functions, ultimately affecting the functions of the cells, tissue, and organs (Mofrad and Kamm 2010). It is very likely that the mechanical stress caused by phonation (vocal fold vibration) is one of the primary and essential factors controlling biological functions, ultimately affecting the function of the cells in the macula flava of the human vocal fold mucosa. However, the role of mechanotransduction in the vibrating vocal fold mucosa remains unclear.

It is readily apparent that tensile and compressive strains can have direct effects on cell morphology and structure, including changes in the cell membrane, shape, and volume as well as cytoskeletal structure and organization (Kurpinski et al. 2010). These physical changes can be converted into changes in cell signaling and transcriptional activities in the nucleus to cause alterations in cellular differentiation, proliferation, and migration (Kurpinski et al. 2010).

The function and fate of stem cells are regulated by various microenvironmental factors (Kurpinski et al. 2010). In addition to chemical factors, mechanical factors can also modulate stem cell survival, organization, migration, proliferation, and differentiation (Kurpinski et al. 2010). Stem cells are potentially one of the main players in the phenotype determination of a tissue in response to mechanical loading (Kurpinski et al. 2010).

The cells in the human maculae flavae may be sensing the mechanical forces, and these tissue-specific mechanical forces (vocal fold vibration) could promote cell differentiation toward the phenotype of the cell residing within the vocal fold tissue. However, little is known about how force affects biological signaling. It is suggested that the combination of multiple mechanical and chemical factors may be involved in more complicated signaling mechanisms and assessment of the relative importance of each factor needs further investigations.

9.16   Future Prospects

As a result of the latest research, there is growing evidence to suggest that the cells in the human maculae flavae are adult multipotent stem cells, tissue stem cells or progenitor cells in the human vocal fold mucosa and that the human maculae flavae are a candidate for a stem cell niche.

Investigations concerning how to regulate these cells contained in the human maculae flavae are challenging but important in the field of regenerative medicine of the human vocal fold.

The manipulation, not only of cells, but also their microenvironment, is one of the strategies in regenerative medicine. Artificial manipulations of these cells using cutting-edge methods (e.g. via chemical biology) could lead to advanced developments in vocal fold regeneration. Understanding the mechanisms responsible for microenvironmental regulation of the cells in the human maculae flavae will provide the tools needed to manipulate cells through their microenvironment for the development of therapeutic approaches to diseases and tissue injuries of the vocal fold. Translational medicine focused on how to regulate cells and extracellular matrices (microenvironments) contained in the maculae flavae of the vocal folds will contribute to our ability to restore and regenerate human vocal fold tissue.

References

Abedi M (2012) Hematopoietic origin of fibrocytes. In: Bucala R (ed) Fibrocytes in health and disease. World Scientific Publishing, Singapore, pp 1–15

Becker WM, Kleinsmith LJ, Hardin J (2006a) Intermediate filament. In: The world of the cell, 6th edn. Benjamin Cummings, San Francisco, CA, pp 446–450

Becker WM, Kleinsmith LJ, Hardin J (2006b) The cell cycle, DNA replication, and mitosis. In: The world of the cell, 6th edn. Benjamin Cummings, San Francisco, CA, pp 554–571

Brittan M, Hunt T, Jeffery R, Poulsom R, Forbes SJ, Hodivala-Dilke K, Goldman J, Alison MR, Wright NA (2002) Bone marrow derivation of pericryptal myofibroblasts in the mouse and human small intestine and colon. Gut 50:752–257

Bucala R, Spiegel LA, Chesney J, Hogan M, Cerami A (1994) Circulating fibrocytes define a new leukocyte subpopulation that mediates tissue repair. Mol Med 1:71–81

Campos Banales ME, Perez Pinero B, Rivero J, Ruiz Casal E, Lopez Aguado D (1995) Histological structure of the vocal fold in the human larynx. Acta Otolaryngol 115:701–704

Fayoux P, Devisme L, Merrot O, Chevalier D, Gosselin B (2004) Histologic structure and development of the laryngeal macula flava. Ann Otol Rhinol Laryngol 113:498–504

Forbes SJ, Russo FP, Rey V, Burra P, Rugge M, Wright NA, Alison MR (2004) A significant proportion of myofibroblasts are of bone marrow origin in human liver fibrosis. Gastroenterol 126:955–963

Gugatschka M, Kojima T, Ohno S, Kanemaru S, Hirano S (2011) Recruitment patterns of side population cells during wound healing in rat vocal folds. Laryngoscope 121:1662–1667

Haylock DN, Nilsson SK (2006) The role of hyaluronic acid in hemopoietic stem cell biology. Regen Med 1:437–445

Hirano M (1975) Phonosurgery. Basic and clinical investigations. Otologia (Fukuoka) 21(Suppl 1):239–260

Hirano M, Sato K (1993) Histological color atlas of the human larynx. Singular Publishing Group Inc., San Diego, CA

Kurita S, Nagata K, Hirano M (1986) Comparative histology of mammalian vocal folds. In: Kirchner JA (ed) Vocal fold histopathology. College Hill Press, San Diego, pp 1–10

Kurita T, Sato K, Chitose S, Fukahori M, Sueyoshi S, Umeno H (2015) Origin of vocal fold stellate cells in the human macula flava. Ann Otol Rhinol Laryngol 124:698–705

Kurpinski K, Janairo R, Chien S, Li S (2010) Mechanical regulation of stem cells: implications in tissue remodeling. In: Mofrad M, Kamm R (eds) Cellular mechanotransduction. Diverse perspectives from molecules to tissues. Cambridge University Press, New York, NY, pp 403–416

Lanz TV, Wachsmuth W (1955) Praktische anatomie hals. Springer, Berlin, p 282

Li L, Xie T (2005) Stem cell niche: structure and function. Annu Rev Cell Dev Biol 21:605–631

Mofrad M, Kamm R (2010) Preface. In: Mofrad M, Kamm R (eds) Cellular mechanotransduction. Diverse perspectives from molecules to tissues. Cambridge University Press, New York, NY, p xi

Pittenger MF, Mackay AM, Beck SC, Jaiswal RK, Douglas R, Mosca JD, Moorman MA, Simonetti DW, Craig S, Marshak DR (1999) Multilineage potential of adult human mesenchymal stem cells. Science 284:143–147

Preston M, Sherman LS (2011) Neural stem cell niches: roles for the hyaluronan-based extracellular matrix. Front Biosci 3:1165–1179

Prockop DJ (1997) Marrow stromal cells as stem cells for nonhematopoietic tissues. Science 276:71–74

Sato K, Chitose S, Kurita T, Umeno H (2016a) Microenvironment of macula flava in the human vocal fold as a stem cell niche. J Laryngol Otol 130:656–661

Sato K, Chitose S, Kurita T, Umeno H (2016b) Cell origin in the macula flava of the human newborn vocal fold. J Laryngol Otol 130:650–655

Sato K, Hirano M (1995a) Histologic investigation of the macula flava of the human vocal fold. Ann Otol Rhinol Laryngol 104:138–143

Sato K, Hirano M (1995b) Histological investigation of the macula flava of the human newborn vocal fold. Ann Otol Rhinol Laryngol 104:556–562

Sato K, Hirano M, Nakashima T (2000) Comparative histology of the maculae flavae of the vocal folds. Ann Otol Rhinol Laryngol 109:136–140

Sato K, Hirano M, Nakashima T (2001a) Stellate cells in the human vocal fold. Ann Otol Rhinol Laryngol 110:319–325

Sato K, Hirano M, Nakashima T (2001b) Fine structure of the human newborn and infant vocal fold mucosae. Ann Otol Rhinol Laryngol 110:417–424

Sato K, Hirano M, Nakashima T (2003a) 3D structure of the macula flava in the human vocal fold. Acta Otolaryngol 123:269–273

Sato K, Hirano M, Nakashima T (2003b) Vitamin A-storing stellate cells in the human vocal fold. Acta Otolaryngol 123:106–110

Sato K, Hirano M, Nakashima T (2004) Age-related changes in vitamine A-storing stellate cells of human vocal folds. Ann Otol Rhinol Laryngol 113:108–112

Sato K, Kurita T, Chitose S, Umeno H, Nakashima T (2015) Mechanical regulation of human vocal fold stellate cells. Ann Otol Rhinol Laryngol 124:49–54

Sato K, Kurita S, Hirano M, Kiyokawa K (1990) Distribution of elastic cartilage in the arytenoids and its physiologic significance. Ann Otol Rhinol Laryngol 99:363–368

Sato K, Nakashima T (2005) Vitamine A-storing stellate cells in the human newborn vocal fold. Ann Otol Rhinol Laryngol 114:517–524

Sato K, Nakashima T, Nonaka S, Harabuchi Y (2008a) Histopathologic investigations of the unphonated human vocal fold mucosa. Acta Otolaryngol 128:694–701

Sato K, Shirouzu H, Nakashima T (2008b) Irradiated macula flava in the human vocal fold mucosa. Am J Otolaryngol 29:312–318

Sato K, Umeno H, Nakashima T (2010a) Functional histology of the macula flava in the human vocal fold. Part 1. Its role in the adult vocal fold. Folia Phoniatr Logop 62:178–184

Sato K, Umeno H, Nakashima T (2010b) Functional histology of the macula flava in the human vocal fold. Part 2: its role in the growth and development of the vocal fold. Folia Phoniatr Logop 62:263–270

Sato K, Umeno H, Nakashima T (2012a) Vocal fold stellate cells in the human macula flava and the diffuse stellate cell system. Ann Otol Rhinol Laryngol 121:51–56

Sato K, Umeno H, Nakashima T (2012b) Vocal fold stem cells and their niche in the human vocal fold. Ann Otol Rhinol Laryngol 121:798–803

Sato K, Umeno H, Nakashima T, Nonaka S, Harabuchi Y (2012) Histopathologic investigations of the unphonated human child vocal fold mucosa. J Voice 26:37–43

Sato K, Umeno H, Ono T, Nakashima T (2011) Histopathologic study of human vocal fold mucosa unphonated over a decade. Acta Otolaryngol 131:1319–1325

Schlüter C, Duchrow M, Wohlenberg C, Becker MH, Key G, Flad HD, Gerdes J (1993) The cell proliferation-associated antigen of antibody Ki-67: a very large, ubiquitous nuclear protein with numerous repeated elements, representing a new kind of cell cycle-maintaining proteins. J Cell Biol 123:513–522

Subotic R, Vecerina S, Krajina Z, Hirano M, Kurita S (1984) Histological structure of vocal fold lamina propria in foetal larynx. Acta Otolaryngol 97:403–406

Titze IR, Hunter EJ (2004) Normal vibration frequencies of the vocal ligament. J Acoust Soc Am 115:2264–2269

Toole BP (1991) Proteoglycans and hyaluronan in morphogenesis and differentiation. In: Hay E (ed) Cell biology of extracellular matrix, 2nd edn. Plenum Press, New York, NY, pp 305–341

Vecerina-Volic S, Hirano M, Karovic-Krzelj V (1988) Macula flava in the vocal fold of human fetus. Acta Otolaryngol 105:144–148

Yamashita M, Hirano S, Kanemaru S, Tsuji S, Suehiro A, Ito J (2007) Side population cells in the human vocal fold. Ann Otol Rhinol Laryngol 116:847–852

Chapter 10
Oesophageal Stem Cells and Cancer

Maria P. Alcolea

Abstract Oesophageal cancer remains one of the least explored malignancies. However, in recent years its increasing incidence and poor prognosis have stimulated interest from the cancer community to understand the pathways to the initiation and progression of the disease.

Critical understanding of the molecular processes controlling changes in stem cell fate and the cross-talk with their adjacent stromal neighbours will provide essential knowledge on the mechanisms that go awry in oesophageal carcinogenesis. Advances in lineage tracing techniques have represented a powerful tool to start understanding changes in oesophageal cell behaviour in response to mutations and mutagens that favour tumour development.

Environmental cues constitute an important factor in the aetiology of oesophageal cancer. The oesophageal epithelium is a tissue exposed to harsh conditions that not only damage the DNA of epithelial cells but also result in an active stromal reaction, promoting tumour progression. Ultimately, cancer represents a complex interplay between malignant cells and their microenvironment. Indeed, increasing evidence suggests that the accumulation of somatic mutations is not the sole cause of cancer. Instead, non-cell autonomous components, coming from the stroma, can significantly contribute from the earliest stages of tumour formation.

The realisation that stromal cells play an important role in cancer has transformed this cellular compartment into an attractive and emerging field of research. It is becoming increasingly clear that the tumour microenvironment provides unique opportunities to identify early diagnostic and prognostic markers, as well as potential therapeutic strategies that may synergise with those targeting tumour cells.

M.P. Alcolea (✉)
Wellcome Trust-Medical Research Council Cambridge Stem Cell Institute,
Tennis Court Road, CB2 1QR, Cambridge, UK

Department of Oncology, University of Cambridge, Hutchison/MRC Research Centre,
Hills Road, CB2 0XZ, Cambridge, UK

A. Birbrair (ed.), *Stem Cell Microenvironments and Beyond*,
Advances in Experimental Medicine and Biology 1041,
DOI 10.1007/978-3-319-69194-7_10

This chapter compiles recent observations on oesophageal epithelial stem cell biology, and how environmental and micro-environmental changes may lead to oesophageal disease and cancer.

Keywords Oesophageal cancer • Oesophageal stem cells • Oesophageal models • Lineage tracing • Early tumorigenesis

10.1 Outline

Oesophageal tissue maintenance, self-renewal and regenerative potential remains a largely unexplored field in epithelial stem cell biology. However, the increasing incidence and poor prognosis of oesophageal cancer have stimulated interest from the cancer and stem cell community to understand the cellular and molecular mechanisms underlying oesophageal stem cell biology, and how dysregulation of tissue homeostasis can lead to epithelial diseases such as cancer.

Evidence indicates that environmental cues represent an important factor in the aetiology of oesophageal carcinogenesis. The oesophageal epithelium is a tissue exposed to harsh environmental conditions; alcohol and tobacco consumption as well as gastric refluxate represent only a portion of the aggressions that the oesophagus has to endure. This certainly dictates the way this tissue is maintained and functions, and makes it susceptible to the accumulation of genetic mutations and the development of cancer.

In this chapter, I will revise recent observations in oesophageal epithelial stem cell biology, and how environmental changes may lead to oesophageal disease and cancer.

10.2 Oesophagus

The oesophagus is a relatively uncomplicated tube that connects our external environment with our stomach, providing means to transport food and liquids for their subsequent digestion and absorption into our bodies (Fig. 10.1). Although this organ forms part of the gastrointestinal tract, its mere function is to transport ingested substances unidirectionally, no food processing or absorption happens here (Goetsch 1910).

Given its piping function, the architecture of this organ is relatively simple compared to other gastrointestinal organs such as the stomach and the intestine. Although histological differences exist between different animals, the oesophagus is constituted by a layer of epithelial tissue or mucosa at the outer lumen side, underlying submucosa where vascular and connective tissue can be found, and the muscularis external. This muscularis muscle layer grades from skeletal to smooth muscle towards the stomach side of the oesophagus. This muscular grading allows for voluntary swallowing to become a reflex towards the end of

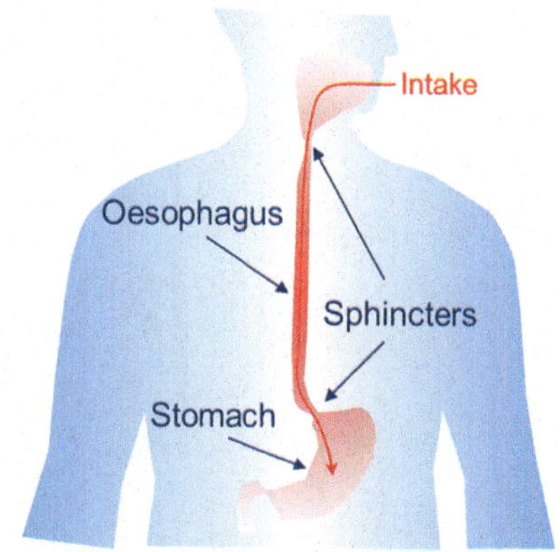

Fig. 10.1 Representation of the human oesophagus. This tissue has a simple anatomy; it represents a tube that connects our external environment with our stomach. Sphincters ensure a unidirectional transport of ingested material under normal conditions

the oesophagus, ensuring food or drink to be delivered to the stomach for digestion. At the gastroesophageal junction the sphincter prevents reflux guarantying unidirectional transport (Goetsch 1910).

10.3 Environment

The outer most side of the oesophagus, the mucosa or oesophageal epithelium, is in direct contact with the outside. Of the gastrointestinal track, this and the epithelial mucosa of the oral cavity will be the part more directly exposed to unprocessed ingested material. This ranges from relatively high temperature products like hot tea infusions or coffee, to cold drinks, environmental pollutants, including cigarette smoke in case of smokers, alcohol consumption and chemicals such as drugs but also endless food preservatives, colouring and texturizing agents (Lin et al. 2016; Tetreault 2015; Fitzgerald 2005). All this is aggravated by the constant physical abrasion of the tissue by undigested food fragments.

The constant wear and tear to which this tissue is exposed necessitates a resistant lining to ensure functionality, endurance and, ultimately, survival. This is achieved by a squamous epithelium formed by several layers of epithelial cells with high turnover frequency that stratify towards the surface forming a multi-layered highly resilient tissue (Alcolea and Jones 2015). Studies using thymidine analogue incorporation in patients have suggested a turnover of approximately 11 days for healthy human oesophageal epithelium, double that of the intestine (Pan et al. 2013). Epithelial cells proliferate at the base of the tissue, and subsequently differentiate stratifying toward the tissue surface where they terminally differentiate and eventually shed at the outer lumen side (Barbera et al. 2015). This represents an excellent way to keep renewing cells potentially damaged by exposure to environmental factors.

However, even though the oesophageal epithelial lining is able to resists and face most day-to-day aggressions, when abused the epithelium may suffer damage and

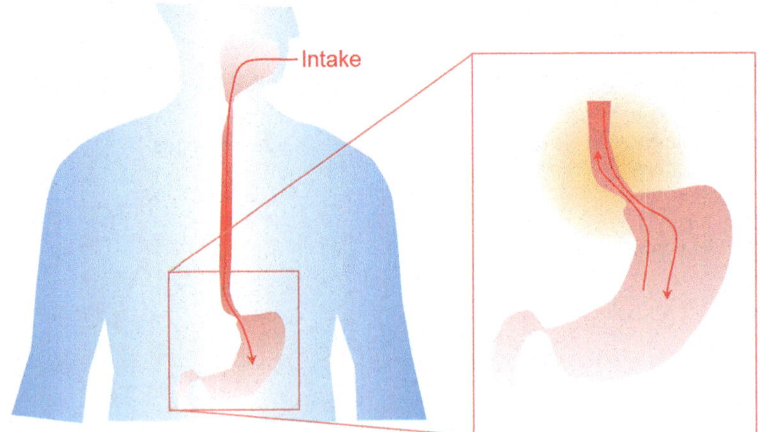

Fig. 10.2 Ingested material is normally transported unidirectionally. Under certain circumstances, gastric reflux occurs exposing the oesophageal epithelium to stomach secretions that sensitize the tissue. Continued exposure may result in epithelial metaplasia, transforming this squamous tissue in intestinal-like columnar epithelium (Barrett's oesophagus). This represents a risk factor, predisposing to adenocarcinoma transformation

result in oesophageal disease or even cancer. One clear example of this is acid reflux (Fitzgerald 2005).

Under normal conditions, all ingested substances are transported into the stomach in one direction only. The sphincter at the junction between the oesophagus and stomach relaxes to allow food down, remaining closed otherwise to protect the oesophagus from the strong acid composition of the stomach digestive secretions. Under certain conditions some of the acid is leaked back into the oesophagus, something known as gastric reflux (Fig. 10.2) (di Pietro and Fitzgerald 2013). Frequent exposure to this refluxate can lead to oesophageal inflammation, and develop into more advanced oesophageal diseases such as Barrett's oesophagus (BE), which has the potential to evolve towards oesophageal adenocarcinoma (EAC) as discussed below (Desai et al. 2012).

The continuous harsh environmental conditions to which this organ is exposed dictates the way it is maintained and functions, and makes it susceptible to abuses exceeding its tolerance that may lead to disease and cancer.

10.4 Oesophageal Cancer

Oesophageal cancer (OEC) represents the eight most common cancer and the sixth most common cause of cancer death worldwide (Rustgi and El-Serag 2014). Despite recent medical advances, this disease still presents late in the clinic and its prognosis remains poor, with a 5 year survival rate of only 10–25% of those diagnosed. There are two major histological subtypes of OEC, squamous cell carcinoma (SCC) and adenocarcinoma (AC) (Pennathur et al. 2013; Napier et al. 2014).

The incidence of the two major OEC subtypes presents clear geographic patterns that has been attributed to different environmental and nutritional factors. SCC is the major cause of OEC worldwide, representing 90% of all OEC. SCCs are predominantly high in the so-called Asian belt, encompassing Turkey, northeaster Iran, Kazakhstan, as well as northern and central China. The main risk factors for this type of cancer are tobacco and alcohol consumption; however, other factors such as diet, environmental pollutants and particularly hot beverages have been suggested to influence the distinctive geographic incidence shown by this cancer (Pennathur et al. 2013; Agrawal et al. 2012).

AC has a significantly different etiology to that of SCC. AC has been suggested to arise from abnormal glandular differentiation as a result of long-term gastric reflux (Leedham et al. 2008; Chang et al. 2007). This cancer presents one of the fastest increasing incidences in Europe and Noth America as a result of the rise in obesity, mal-dietary habits and Barrett's oesophagus, a premalignant condition resulting from gastric reflux (Pennathur et al. 2013; di Pietro et al. 2014).

10.5 Oesophageal Squamous Cell Carcinoma

SCCs have been associated with a high frequency of genetic alterations. Recent studies have shown that SCCs in the oesophagus present a greater mutational burden than breast cancer and glioblastoma multiforme (Song et al. 2014). However, the somatic mutation rate was still lower than that observed in head and neck squamous cell carcinomas (Stransky et al. 2011) and oesophageal adenocarcinoma (EAC) (Dulak et al. 2013).

Different studies in different geographical locations, including North America and China, have identified recurrent genes frequently found mutated in SCC samples. Among those, TP53, NOTCH, PIK3CA and FAT1 (FAT Atypical Cadherin), as well as copy number variations in CCND1 (Cyclin D1) and CDKN2A, seem to be common in the list of SCC mutant genes (Gao et al. 2014; Lin et al. 2014; Zhang et al. 2015; Sasaki et al. 2016).

10.6 Oesophageal Adenocarcinoma

The strongest and best-characterized risk of EAC is gastroesophageal reflux. Decades of evidence have linked EAC to Barrett's oesophagus (BE), a premalignant condition where the stratified oesophageal epithelium is replaced by a columnar intestinal epithelium in a metaplastic process in response to the strong environmental conditions of chronic gastric reflux. However, despite this knowledge, Adenocarcinoma has remained cause of concern due to its concerning rise in incidence for the last couple of decades in western and developed Countries. Efforts made to increase detection and surveillance of Barrett's oesophagus have not significantly affected this trend, given that 95% of EAC arise from patients who had not been previously diagnosed with BE (Reid et al. 2010).

Other factors that increase the risk of EAC are obesity, cigarette smoking and diet low in fruit and vegetables (Engel et al. 2003).

Sequencing studies have described the mutational signature of EAC, reflecting the high mutational burden of this disease. TP53 is the most recurrently mutated gene, other genes mutated at a lower rate inlcude CDKN2A, SMAD4, ARID1A, PIK3CA and SYNE1 (Dulak et al. 2013; Chong et al. 2013). More recently, work from Prof. Fiztgerald laboratory, has shown the highly dynamic nature of the mutational landscape of BE and EAC. This study demonstrated the polyclonal evolution of BE, with high grade dysplasia being able to arise from multiple different clones. This has significant clinical implications, as dysplasia may redevelop from residual BE left behind after treatment therapies (Ross-Innes et al. 2015).

10.7 Understanding Human Oesophagus From Mouse Models

In order to improve the poor prognosis and progressive rise in the incidence of oesophageal cancer, it is imperative to understand the etiology of this complex and heterogeneous disease. Insights as to how it originates and evolves will provide valuable information to unveil new avenues for diagnosis and therapeutics.

However, in order to do this, it is first critical to understand how this tissue is maintained under normal homeostatic conditions, how it responds to tissue perturbations such as injury or aggression, and how those rules become deregulated during oncogenesis.

Over last couple of decades there have been several studies trying to unveil the identity of a stem cell population in the oesophagus. Although, there has been some work in human tissue, the most detailed studies use mouse models (Alcolea and Jones 2014). One of the major advantages is that mice can be manipulated genetically with relative ease (van der Weyden et al. 2002). An increasing range of mouse strains covering a broad spectrum of genetic models have been instrumental in revealing changes in cell behaviour in response to oncogenes, tumour suppressor genes or just simply by allowing visualization of individual cells using fluorescent reporters. These valuable research tools are also extremely versatile, making possible the tight control over gene expression in vivo in a temporal, spatial or tissue specific manner, something that has revolutionized our knowledge in epithelial stem cell biology in the last couple of decades (Alcolea and Jones 2013).

Additionally, most of the basic principles of stem cell biology and tumour development have been established in mouse models and have been shown to be conserved from human (van der Weyden et al. 2002; Yuspa et al. 1994). Making the use

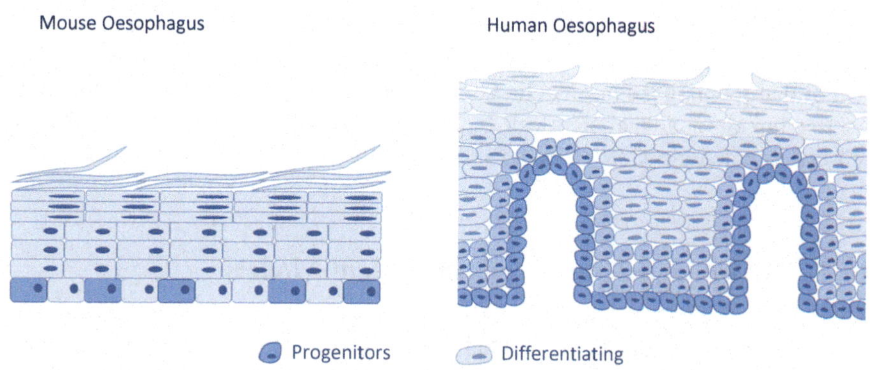

Fig. 10.3 Schematic representation depicting differences between human and mouse oesophagus

of mouse models, not only critical tools for basic research, but also for preclinical trials to target specific molecular pathways in order to tests their therapeutic potential.

Given the advantages of available mouse models, it is not surprising that researchers have made use of them to understand oesophageal stem cell dynamics. However, these rodent models also presents some caveats, as fundamental differences between mouse and human oesophagus exist.

The human oesophagus is a squamous non-keratinized epithelium organized around structures called papillae that divide the tissue into papillary and interpapillary zones (Fig. 10.3). Proliferation takes place in the first 5–6 layers from the basement membrane. On commitment to differentiation, cells exit cell cycle and stratify into the suprabasal layers, migrating to the tissue surface from which they are eventually shed. Unlike the mouse oesophagus, the human oesophagus lacks a cornified protective layer at the tissue surface, making it more vulnerable to the chemical and physical properties of the substances we ingest. This is circumvented to some extent by having additional cell layers that form a thicker epithelium, as well as by the presence of submucosal glands that release mucous and acid neutralizing agents exerting a protective role (Goetsch 1910; Barbera et al. 2015; Seery 2002; Marques-Pereira and Leblond 1965).

Generally speaking, mouse oesophagus presents a simpler structure. It is also lined by a squamous epithelium that consists of layers of keratinocytes. However, unlike humans, proliferation is confined to the basal cell layer, and no glands, papillae or other accessory structures are found (Doupe et al. 2012; Messier and Leblond 1960; Rosekrans et al. 2015).

Despite the benefit and advantages of using mouse models to understand human oesophageal biology, the significant differences between the two species makes it critical to ultimately test animal observations for their validity in human models.

The recent development of organoid cultures in different epithelial tissues, including the oesophagus, has provided an extraordinary opportunity to translate observations from mouse models into humans (Fatehullah et al. 2016; DeWard et al.

2014; Sato et al. 2011). This 3D in vitro culture method allows for the formation of organized cellular structures with different cellular subtypes and a function reminiscent of that found in the original tissue. This opens new venues to thoroughly characterise human epithelial stem cell biology in health and disease by considering cell-cell interactions while retaining spatial resolution, at least to some extent.

10.8 Oesophageal Stem Cells in Rodent Models

The oesophagus represent an epithelial barrier in contact with the exterior and, as such, it requires to be in constant turnover to sustain tissue integrity in response to the continuous damage. Proliferation, confined to the basal layer in mice and first few layers in human, is required to generate new cells to maintain the tissue in homeostasis. Under normal conditions, it is critical that upon division the same number of proliferating and differentiating cells are produced in order to maintain a balanced equilibrium. An imbalance will result in the loss of cell production, compromising tissue integrity, or in an excessive cell proliferation potentially leading to cancer (Doupe et al. 2012; Frede et al. 2014; Frede et al. 2016; Alcolea et al. 2014).

Work in the late sixties, studying tritiated thymidine incorporation in the rat oesophagus had suggested that all proliferating cells were equipotent, and that the commitment and exit from the basal layer was stochastic. By performing these experiments, Leblond and co-workers observed how all the cells incorporating the labelled thymidine isotope during division were localized to the basal layer, arguing against asymmetrical division. Over time, half of the labelled cells stratified to the suprabasal layers, suggesting that cell fate making was happening after cell division in a stochastic manner (Marques-Pereira and Leblond 1965).

With the advent of the stem cell/ transit amplifying model proposed to explain epithelial tissue maintenance (Potten and Booth 2002), more recent studies attempted to unveil the identity of a discrete stem cell population in the oesophagus. These hypothesised that the oesophageal epithelium is maintained by a slow-cycling self-renewing stem cell population, generating short lived transit-amplifying cells, that terminally differentiate after a few rounds of division (Croagh et al. 2007). Based on previous studies reporting alpha 6 integrin and CD71 maker combination as a mean to identify epidermal stem cells (Li et al. 1998; Tani et al. 2000), in vivo studies looked into these in mouse oesophagus and concluded that alpha 6 integrin positive basal cells could be separated in two distinct populations CD71 dim and CD71 bight. Label retaining and in vivo reconstitution assays indicated that the CD71 dim population fulfilled the criteria of a stem cell compartment (Croagh et al. 2007). However, this population failed to manifest an enhanced colony forming potential in in vitro clonogenic assays.

A subsequent study used a Hoechst exclusion assay to identify a label retaining population in the mouse oesophagus that was enriched for CD34 expression; a known stem cell marker (Trempus et al. 2003). This population presented increased clonogenic and regenerative potential both in vitro and in vivo, showing the typical features of a potential stem cell population. Interestingly, further analysis of this putative stem cell population did not correlate with the integrin alpha 6 high/CD71 dim expression profile previously reported in mouse oesophageal stem cells (Kalabis et al. 2008).

A more recent report from DeWard et al. used a combination of basal cell surface markers to separated oesophageal cells into distinct populations with different in vitro organoid forming efficiency. This study shows that SOX2 is oesophageal basal cell maker that plays an important role in organoid formation and self-renewal. And suggests basal cells expressing the highest levels of basal markers integrin alpha 6, beta 1 and p75 represent a putative stem cell population based on their increased organoid formation efficiency. However, no differences were observed in their self-renewal potential (DeWard et al. 2014). Based on this observation, the study concludes that a non-quiescent stem cell population resides in the basal epithelium of the mouse oesophagus.

The development of new genetically engineered mouse strains expressing multicolour fluorescent reporters which expression may be controlled temporally and/or spatially by specific promoters and/or drug treatment, has revolutionized our knowledge of cell behaviour in epithelial tissues in health and disease (Alcolea and Jones 2013). By exploiting the available reporter mouse strains scientists can now label individual cells throughout the tissue with an inheritable fluorescent reporter, and track their fate over the course of time either by performing end point experiments, or by in situ live imaging in the living organism (Alcolea and Jones 2014; Park et al. 2016).

Using quantitative methods of lineage tracing, we performed a comprehensive study to reconcile previous observations on mouse oesophageal stem cell behaviour. Individual basal cells were fluorescently labelled, and their fate tracked over the course of 1 year. Large scale clone size analysis using methods of mathematical statistics revealed that a single progenitor population that divides stochastically, balancing the production of proliferating and differentiating cells, is responsible for the maintenance of the mouse oesophageal epithelium (Doupe et al. 2012). Additional, transgenic label-retaining assays based on calculating the dilution of doxycycline induced Histone-2B-GFP fusion protein (Tumbar et al. 2004) indicated that no slow-cycling epithelial cells were present in the oesophageal epithelium. Further, quantification of the Histone-2B-GFP levels in individual cells led to the conclusion that all basal cells divide at a similar rate, in agreement with the original observations by Leblond and co-workers (Marques-Pereira and Leblond 1965; Doupe et al. 2012).

In order to unveil whether tissue injury could reveal populations with distinctive regenerative potential, a refined endoscopic method was used to create a discrete incision in the mouse oesophagus. Similar genetic lineage tracing and label retaining assays were performed. Remarkably, the uniformity of the basal cell population was once more revealed in response to wounding. The widespread activation of progenitor cells around the wound rapidly produced an excess of proliferating cells in order to close the defect in the epithelium (Fig. 10.4), leading to a very efficient and rapid healing response (Doupe et al. 2012).

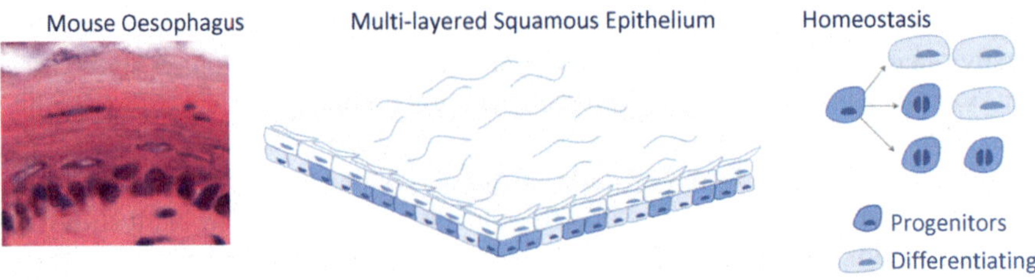

Fig. 10.4 Stochastic model of oesophageal tissue maintenance in mouse. Quantitative cell fate analysis in the mouse oesophagus has revealed that a single functionally equivalent progenitor population maintains the tissue by dividing stochastically, balancing the production of proliferating and differentiating cells. Each division can produce one of three outcomes. Symmetric fate results in two proliferating or differentiating cells, while asymmetric divisions generate one of each

10.9   Oesophageal Stem Cells in Human

As discussed above mouse and human oesophagus show certain histological differences. In essence, both tissues are formed by layers of squamous epithelial cells divided in two main compartments; the basal zone with one (mouse) or several (human) layers of small basophilic cells, and the differentiated zone where cells become progressively flatter as they approach the lumen surface where they shed from the tissue. One of the major differences in human oesophagus is the existence of a structured architecture around papillae. These arise from the invagination of the lamina propia at regular intervals and results in the tissue being divided in papillary and interpapillary epithelium (Fig. 10.3) (Goetsch 1910; Seery 2002). These defined structures have been proposed to be a potential niche for stem cells in the human oesophagus (Barbera et al. 2015; Seery 2002; Seery and Watt 2000). However, such compartmentalization is not found in mice (Doupe et al. 2012).

The number of studies available on human oesophagus have been limited by the inaccessibility of the sample, as well as the technical challenges to study stem cell behaviour and regeneration in this tissue.

Initial studies, based on PCNA staining, a proliferation maker, suggested the existence of a putative stem cell population located at the tip of the papillae (Jankowski et al. 1992). Later studies looked into cell division symmetry and found that cells in the interpapillary zone divided rarely and asymmetrically; giving rise to one basal daughter and one suprabasal differentiating cell (Seery and Watt 2000). They concluded that interpapillary basal cells attained to the expected stem cell definition at the time; stem cell fate in squamous tissue was believed to be maintained largely through division asymmetry (Watt and Hogan 2000).

More recently label retaining assays using the thymidine analogue 5-iodo-2′deoxyuridine (IdU) in patients undergoing oesophagectomy showed a higher proportion of IdU retaining cells in the papillary basal layer of healthy oesophagus. The conclusion was that a putative slow-cycling self-renewing stem cell population resides in the defined niche of the oesophageal papillae (Pan et al. 2013).

The most recent report studying human oesophageal tissue maintenance, uses a comprehensive wholemount staining technique to asses for proliferation and stem cell markers such as CD34. Data shows that proliferation and mitotic activity was highest in the interpapillary basal layer and decreased linearly towards the tip of the papilla, where a CD34 positive population resides. Additional 2D and 3D organotypic in vitro assays looked into the regenerative potential of different cell populations sorted based on CD34 and epithelial cadherin. Interestingly, no differences in self-renewal were observed when performing either single cell or population assays (Barbera et al. 2015). These observations are in agreement with earlier studies suggesting a slow cycling population resides in the papillary zone, and seem to resolve conflictive reports (Pan et al. 2013; Jankowski et al. 1992). Interestingly, this study also presents data in line with recent findings in mouse oesophagus. Progenitor cells, which can respond to injury and regenerate tissue, were found to be widespread and are not restricted to the basal layer, including cells that have already committed to epithelial differentiation (Barbera et al. 2015; Doupe et al. 2012).

10.10 Oesophageal Cell Behaviour in Tumourigenesis

The advent of in vivo linage tracing techniques has represented a powerful technique to start understanding changes in oesophageal cell behaviour in response to mutations and mutagens that favour tumour development.

Sequencing studies previously suggested that loss of function Notch mutations and loss of heterozygosity were frequently found in squamous cell carcinomas, including oesophageal SCCs (Agrawal et al. 2011, 2012; Song et al. 2014; Stransky et al. 2011; Gao et al. 2014; Lin et al. 2014). Using a lineage tracing approach similar to that previously used to study mouse oesophageal tissue maintenance (Doupe et al. 2012), we challenged mouse oesophageal homeostasis by inhibiting Notch signalling in vivo. An engineered mouse model expressing an inducible dominant negative form of mastermind like-1 tagged to a fluorescent GFP reporter (DNM1-GFP) was used in this study (Tu et al. 2005). Quantitative clonal data revealed that Notch inhibition confers a strong competitive advantage to mutant progenitor cells, generating clones that expand rapidly over the weeks following induction. Further analysis on clonal growth and progenitor differentiation suggested that mutants present a blockage in terminal division, where dividing cells produce two differentiating cells (Fig. 10.4). As a result, mutant cells divide 3 fold faster than wild type cells, and, on average, each cell division produces an excess of progenitors over differentiating cells (Fig. 10.5) (Alcolea et al. 2014). Interestingly, the clonal advantage of these clones does not only rely on cell autonomous mechanisms but also exerts a 'bystander effect', actively eliminating wild type cells, similar to those observed in super competitor mutants in Drosophila (de la Cova et al. 2004; Moreno and Basler 2004). Additional treatment with carcinogens illustrates the potential role of Notch inhibiting mutations in tumour formation; mutant clones were seen to provide means for other less advantageous mutations to colonize the tissue when

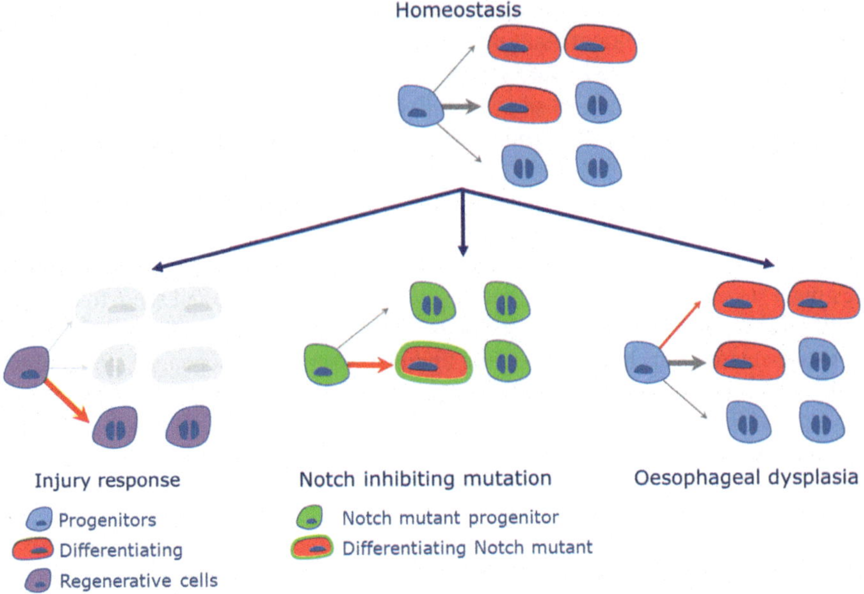

Fig. 10.5 Oesophageal progenitor cells are able to change their cell fate programme in response to tissue perturbations such as injury, neoplastic mutations and tumorigenesis (*red arrows*). Notch inhibiting mutations in progenitors showed an increased proliferation rate, favouring asymmetric cell division. Surprisingly, benign tumours developed upon cigarette smoke derived nitrosamines did not show a significant change in the rate of cell division. The perturbation seemed to be the result of a discrete bias towards proliferation

co-existing (Alcolea et al. 2014). This exemplifies how different mutations could potentially synergise during the early stages of tumour formation.

Further studies have used a combination of diethylnitrosamine (DEN) and the multikinase inhibitor Sorafenib, as a two-stage carcinogenic protocol to investigate cell dynamics during oesophageal tumorigenesis. DEN is a nitrosamine found in cigarette smoke and traditionally used to induce tumours in the oesophageal epithelium (Hoffmann et al. 1982; Rubio 1983; Rubio et al. 1987). These nicotine-derived compounds are activated in the body to form alkylating agents that cause DNA damage (Goodsell 2004). Inclusion of Sorafenib was drawn from previous observations showing the cancer promoting effect of this drug. Sorafenib was shown to lead to SCC formation in skin, and head and neck in patients treated for liver, kidney and thyroid cancers (Schneider et al. 2016; Arnault et al. 2009). DEN and Sorafenib drug combination generates early tumours forming high grade dysplasias (HGD) in the mouse oesophagus. Interestingly, lineage tracing data in the epithelial compartment points to the polyclonal origin of these tumoral lesions. Against all predictions, cells in dysplasias shared a common dynamics, with progenitor cells showing a moderate bias towards the production of dividing over non-dividing daughter cells (Fig. 10.5). Also, despite the tumour outgrowth no significant change in the rate of cell division was observed (Frede et al. 2016).

The remarkably uniform behaviour described in dysplasia contrasts with observations in squamous cell carcinomas produced in a Kras G12D mutant background. These advanced cancers were characterised by the existence of a subset of clones with a significant bias towards proliferation, reflecting the onset of cancer heterogeneity. It remains to be elucidated whether this subpopulation has an increased tumour initiating potential when compared to the bulk tumour cell population (Frede et al. 2016).

10.11 Oesophageal Cancer and Microenvironment

Cancer is a complex disease that develops in response to a concert of genetic alterations and environmental factors. The constant exposure of the oesophagus to damaging agents through ingestion, as well as gastric refluxate may result in tissue injury. This can have a significant impact on the epithelium not only by promoting the mutational burden, but also indirectly, by activating the underlying stroma (Lin et al. 2016).

It has traditionally been thought that the sole cause of cancer lays on the accumulation of genetic alterations that promote disease progression. However, increasing evidence suggest that there is an entire new dimension to it, i.e. the tumour microenvironment. Non-cell autonomous components, coming from the stroma, can significantly contribute not only to cancer progression but also to cancer initiation (Hu et al. 2012; Whiteside 2008; Tlsty and Coussens 2006).

The primary function of the stroma is to offer structural support to organs and epithelial tissues lining them. However, it also serves as a sensor orchestrating the signals required to modulate cell behaviour in response to environmental changes. Communication between epithelial and stromal cells is essential for tissue damage repair. However, stromal activation can be aberrantly triggered by the abnormal behaviour of mutant epithelial cells, misleadingly understood as an injury, promoting tumorigenesis (Arwert et al. 2012).

The tumour stroma, which consists of immune cells, fibroblasts, endothelial cells, perivascular cells, adipocytes and extracellular matrix, constitute the microenvieronment in which the tumour must develop (Arwert et al. 2012). Given that tumours have been proposed to function as an injury that is not able to heal, suggested by Dvorak (Dvorak 1986), the interplay between tumour cells and the different stromal compartments will have a significant role in tumour development and progression. The same way this interplay is central for adequate wound repair. The main difference resides in the fact that wound healing is a controlled mechanism, while tumour formation is a disorganized process (Arwert et al. 2012; Gurtner et al. 2008).

Among the risk factors promoting oesophageal cancer discussed above, cigarette smoke, alcohol, gastric reflux, obesity and dietary habits, all of them share a common feature. They all have a significant impact on the tumour stroma, mainly by promoting tissue damage. This has the inevitable consequence of fibroblast activation, increased immune response, changes in extracellular matrix and vascular reorganization, among others (Lin et al. 2016).

10.11.1 *Mesenchymal Compartment*

The main cellular component of the tumour stroma in most tumour types are fibroblasts. Tumour associated fibroblasts (TAFs) have been shown to be a heterogeneous cell population that plays an active role from the earliest stages of tumour formation. TAFs contribute to disease progression by providing the suitable environment for carcinogenesis, proliferation, angiogenesis and invasion. Growth factors, cytokines and extracellular matrix are released to promote tumour cell transformation (Joyce and Fearon 2015; Malanchi et al. 2012; Zhang and Wang 2015; Kalluri 2016). More recently, it has been shown how TAFs can also have an impact on drug resistance by signalling to tumour cells (Hirata et al. 2015; Au Yeung et al. 2016; Kaur et al. 2016).

Cancer associated fibroblasts (CAFs) have been proposed to have a critical role in the development of oesophageal cancer. Reports suggests that oesophageal CAFs can derive from different cellular populations, including normal fibroblasts and bone marrow-derived cells among others (Nouraee et al. 2013; Hutchinson et al. 2011). Transforming growth factor $\beta1$ (TGF$\beta1$) and microRNAs have been implicated in the conversion of fibroblasts to CAFs (Noma et al. 2008; Tanaka et al. 2015).

In the oesophagus, fibroblasts are localized in the submucosa layer laying directly underneath the epithelial mucosa (Goetsch 1910). Increased transforming growth factor $\beta1$ (TGF$\beta1$) and hepatocyte growth factor (HGF) have been linked to the progression from dysplasia to ESSC (Xu et al. 2013). In human ESCC, TGFβ receptor II (TβRII) was found to be downregulated in CAFs. This was associated with increased proliferation and reduced apoptosis in adjacent epithelial cells (Achyut et al. 2013). Incresed cyclooxygenase (COX)-2, the enzyme of prostaglandin E2 (PGE2), has been linked to both ESCC and EAC development via its pro-inflammatory function (Achyut et al. 2013; Taddei et al. 2014). Indeed, one of the means by which CAFs have been proposed to contribute to carcinogenesis is by producing pro-inflammatory factors.

10.11.2 *Immune Compartment*

One microenvironmental component that has become increasingly relevant in recent years due to mounting evidence probing its significant contribution to tumorigenesis and its therapeutic potential is the immune compartment (Chen and Mellman 2017).

Injury by gastric refluxate in the oesophagus has been shown to result in chronic inflammation with upregulation of cytokines, such as IL1b, IL6, and IL8 (Fitzgerald et al. 2002). Increased IL1b/IL6 signalling contributes to the metaplastic and dysplastic conversion of BE, as well as its evolution towards oesophageal adenocarcinoma (Quante et al. 2012). A mouse model overexpressing interleukin-1b developed human Barrett-like metaplasia and EAC in an interleukin 6 dependent manner. This phenotype was accelerated by exposure to bile acids, one component of gastroduodenal reflux, or nitrosamines, generated at the oesophageal junction when salivary nitrite is reduced in response to gastric secretions (Winter et al. 2007). Lineage tracing data suggested that Lgr5 positive cells of gastric origin were the origin of the Barrett's lesion in this IL1b-IL6 immune permissive environment. The results also indicated that the oesophageal to columnar transition happens under the control of Notch Delta1-dependent signalling (Quante et al. 2012).

The role of inflammation is also important for ESCC (Sadanaga et al. 1994). It has been shown that the main risk factors for this type of cancer, smoking and alcohol, favour an inflammatory response via direct chemical irritation of the oesophageal epithelium, as well as production of reactive oxygen species (Sadanaga et al. 1994; Kubo et al. 2014). A conditional mouse model where p120catenin was lost in the oesophagus revealed the role of the tumour microenvironment as a tumour driver. ESSC development in this model was associated to inflammation, immune cell infiltration, and increased NFkB/Stat-3 cross-talk in tumours (Stairs et al. 2011). A later study reinforced the important role of the immune response in ESCC development. Conditional SOX2 overexpression in the oesophagus was insufficient to drive SCC formation. Transformation of oesophageal progenitor cells required cooperation of increased Sox2 and microenvironment-activated Stat3, leading to tumorigenesis (Liu et al. 2013).

Several immune cell types have been involved in tumour development. Although the main function of our immune system is to protect our organism from invasion, the same must have mechanisms that protect us against persistent or dysregulated immune reactions. This is a critical function for our survival. Tumour cells have been proposed to hijack some of these mechanisms in order to persist and evolve. For instance, regulatory T cells that under normal conditions maintain tolerance to self-antigens, preventing autoimmune disease, if aberrantly activated in response to oesophageal cancer, promote tumour immune suppression favouring disease progression (Nabeki et al. 2015). Myeloid-derived suppressor cells (MDSCs) are immature cells that also suppress the immune reaction by induction of regulatory T cells, and inhibition of protective cell types such as T cells and natural killer cells. This cell population was found to be increased in ESCC mouse models (Stairs et al. 2011; Chen et al. 2014). Similarly, macrophages switching from M1 to M2 start producing cytokines and growth factors that favour oesophageal tumour development (Miyashita et al. 2014). Another immune suppressive mechanism hijacked by cancer cells is the modulation of immune checkpoints. Programmed cell death protein ligand (PD-L1) is a protein expressed on the surface of several tumour cells, and it is thought to play a role in immune escape by inhibiting T cell function. PD-L1 has shown a significant potential as melanoma target treatment and also presents good prospects for oesophageal cancer (Raufi and Klempner 2015).

Ultimately, cancer represents a complex interplay between malignant cells and their neighbouring stromal compartment. The realisation of the increased genetic stability of stromal cells compared to cancer cells has made them an attractive cellular compartment, formerly disregarded. Mounting line of evidence indicate the largely unexplored potential of tumour microenvironment, not only as a source of plausible therapeutic targets, but also of diagnostic and prognostic markers (Lin et al. 2016).

All in all, the oesophagus has proven to be an excellent model to understand basic epithelial stem cell biology. Its multi-layered stratified architecture, constant turnover, interaction with the environment and cross-talk with the microenvironment render this an ideal tissue where to explore stem cell behaviour in health and disease. Despite good progress in the field, further research is still needed to identify how stromal changes govern epithelial cell behaviour, and how those contribute to cancer development. The new research tools now widely accessible, such as a broad spectrum of genetically engineered mouse models, organoid cultures and recent developments in CRIPR technology represent an exciting prospect for oesophageal stem biology.

References

Achyut BR et al (2013) Inflammation-mediated genetic and epigenetic alterations drive cancer development in the neighboring epithelium upon stromal abrogation of TGF-beta signaling. PLoS Genet 9:e1003251

Agrawal N et al (2011) Exome sequencing of head and neck squamous cell carcinoma reveals inactivating mutations in NOTCH1. Science 333:1154–1157

Agrawal N et al (2012) Comparative genomic analysis of esophageal adenocarcinoma and squamous cell carcinoma. Cancer Discov 2:899–905

Alcolea MP, Jones PH (2013) Tracking cells in their native habitat: lineage tracing in epithelial neoplasia. Nat Rev Cancer 13:161–171

Alcolea MP, Jones PH (2014) Lineage analysis of epidermal stem cells. Cold Spring Harb Perspect Med 4:a015206

Alcolea MP, Jones PH (2015) Cell competition: winning out by losing notch. Cell Cycle 14:9–17

Alcolea MP et al (2014) Differentiation imbalance in single oesophageal progenitor cells causes clonal immortalization and field change. Nat Cell Biol 16:615–622

Arnault JP et al (2009) Keratoacanthomas and squamous cell carcinomas in patients receiving sorafenib. J Clin Oncol 27:e59–e61

Arwert EN, Hoste E, Watt FM (2012) Epithelial stem cells, wound healing and cancer. Nat Rev Cancer 12:170–180

Au Yeung CL et al (2016) Exosomal transfer of stroma-derived miR21 confers paclitaxel resistance in ovarian cancer cells through targeting APAF1. Nat Commun 7:11150

Barbera M et al (2015) The human squamous oesophagus has widespread capacity for clonal expansion from cells at diverse stages of differentiation. Gut 64:11–19

Chang CL et al (2007) Retinoic acid-induced glandular differentiation of the oesophagus. Gut 56:906–917

Chen DS, Mellman I (2017) Elements of cancer immunity and the cancer-immune set point. Nature 541:321–330

Chen MF et al (2014) IL-6-stimulated CD11b+ CD14+ HLA-DR- myeloid-derived suppressor cells, are associated with progression and poor prognosis in squamous cell carcinoma of the esophagus. Oncotarget 5:8716–8728

Chong IY et al (2013) The genomic landscape of oesophagogastric junctional adenocarcinoma. J Pathol 231:301–310

Croagh D, Phillips WA, Redvers R, Thomas RJ, Kaur P (2007) Identification of candidate murine esophageal stem cells using a combination of cell kinetic studies and cell surface markers. Stem Cells 25:313–318

de la Cova C, Abril M, Bellosta P, Gallant P, Johnston LA (2004) Drosophila myc regulates organ size by inducing cell competition. Cell 117:107–116

Desai TK et al (2012) The incidence of oesophageal adenocarcinoma in non-dysplastic Barrett's oesophagus: a meta-analysis. Gut 61:970–976

DeWard AD, Cramer J, Lagasse E (2014) Cellular heterogeneity in the mouse esophagus implicates the presence of a nonquiescent epithelial stem cell population. Cell Rep 9:701–711

di Pietro M, Alzoubaidi D, Fitzgerald RC (2014) Barrett's esophagus and cancer risk: how research advances can impact clinical practice. Gut Liver 8:356–370

di Pietro M, Fitzgerald RC (2013) Research advances in esophageal diseases: bench to bedside. F1000Prime Rep 5:44

Doupe DP et al (2012) A single progenitor population switches behavior to maintain and repair esophageal epithelium. Science 337:1091–1093

Dulak AM et al (2013) Exome and whole-genome sequencing of esophageal adenocarcinoma identifies recurrent driver events and mutational complexity. Nat Genet 45:478–486

Dvorak HF (1986) Tumors: wounds that do not heal. Similarities between tumor stroma generation and wound healing. N Engl J Med 315:1650–1659

Engel LS et al (2003) Population attributable risks of esophageal and gastric cancers. J Natl Cancer Inst 95:1404–1413

Fatehullah A, Tan SH, Barker N (2016) Organoids as an in vitro model of human development and disease. Nat Cell Biol 18:246–254

Fitzgerald RC (2005) Barrett's oesophagus and oesophageal adenocarcinoma: how does acid interfere with cell proliferation and differentiation? Gut 54(Suppl 1):i21–i26

Fitzgerald RC et al (2002) Inflammatory gradient in Barrett's oesophagus: implications for disease complications. Gut 51:316–322

Frede J, Adams DJ, Jones PH (2014) Mutation, clonal fitness and field change in epithelial carcinogenesis. J Pathol 234:296–301

Frede J, Greulich P, Nagy T, Simons BD, Jones PH (2016) A single dividing cell population with imbalanced fate drives oesophageal tumour growth. Nat Cell Biol 18(9):967–978

Gao YB et al (2014) Genetic landscape of esophageal squamous cell carcinoma. Nat Genet 46:1097–1102

Goetsch E (1910) The structure of the mammalian esophagus. Am J Anat 10:1–39

Goodsell DS (2004) The molecular perspective: nicotine and nitrosamines. Stem Cells 22:645–646

Gurtner GC, Werner S, Barrandon Y, Longaker MT (2008) Wound repair and regeneration. Nature 453:314–321

Hirata E et al (2015) Intravital imaging reveals how BRAF inhibition generates drug-tolerant microenvironments with high integrin beta1/FAK signaling. Cancer Cell 27:574–588

Hoffmann D, Adams JD, Brunnemann KD, Rivenson A, Hecht SS (1982) Tobacco specific N-nitrosamines: occurrence and bioassays. IARC Sci Publ 41:309–318

Hu B et al (2012) Multifocal epithelial tumors and field cancerization from loss of mesenchymal CSL signaling. Cell 149:1207–1220

Hutchinson L et al (2011) Human Barrett's adenocarcinoma of the esophagus, associated myofibroblasts, and endothelium can arise from bone marrow-derived cells after allogeneic stem cell transplant. Stem Cells Dev 20:11–17

Jankowski J, McMenemin R, Yu C, Hopwood D, Wormsley KG (1992) Proliferating cell nuclear antigen in oesophageal diseases; correlation with transforming growth factor alpha expression. Gut 33:587–591

Joyce JA, Fearon DT (2015) T cell exclusion, immune privilege, and the tumor microenvironment. Science 348:74–80

Kalabis J et al (2008) A subpopulation of mouse esophageal basal cells has properties of stem cells with the capacity for self-renewal and lineage specification. J Clin Invest 118:3860–3869

Kalluri R (2016) The biology and function of fibroblasts in cancer. Nat Rev Cancer 16:582–598

Kaur A et al (2016) sFRP2 in the aged microenvironment drives melanoma metastasis and therapy resistance. Nature 532(7598):250–254

Kubo N et al (2014) Oxidative DNA damage in human esophageal cancer: clinicopathological analysis of 8-hydroxydeoxyguanosine and its repair enzyme. Dis Esophagus 27:285–293

Leedham SJ et al (2008) Individual crypt genetic heterogeneity and the origin of metaplastic glandular epithelium in human Barrett's oesophagus. Gut 57:1041–1048

Li A, Simmons PJ, Kaur P (1998) Identification and isolation of candidate human keratinocyte stem cells based on cell surface phenotype. Proc Natl Acad Sci U S A 95:3902–3907

Lin EW, Karakasheva TA, Hicks PD, Bass AJ, Rustgi AK (2016) The tumor microenvironment in esophageal cancer. Oncogene 35:5337–5349

Lin DC et al (2014) Genomic and molecular characterization of esophageal squamous cell carcinoma. Nat Genet 46:467–473

Liu K et al (2013) Sox2 cooperates with inflammation-mediated Stat3 activation in the malignant transformation of foregut basal progenitor cells. Cell Stem Cell 12:304–315

Malanchi I et al (2012) Interactions between cancer stem cells and their niche govern metastatic colonization. Nature 481:85–89

Marques-Pereira JP, Leblond CP (1965) Mitosis and differentiation in the stratified squamous epithelium of the rat esophagus. Am J Anat 117:73–87

Messier B, Leblond CP (1960) Cell proliferation and migration as revealed by radioautography after injection of thymidine-H3 into male rats and mice. Am J Anat 106:247–285

Miyashita T et al (2014) Impact of inflammation-metaplasia-adenocarcinoma sequence and inflammatory microenvironment in esophageal carcinogenesis using surgical rat models. Ann Surg Oncol 21:2012–2019

Moreno E, Basler K (2004) dMyc transforms cells into super-competitors. Cell 117:117–129

Nabeki B et al (2015) Interleukin-32 expression and Treg infiltration in esophageal squamous cell carcinoma. Anticancer Res 35:2941–2947

Napier KJ, Scheerer M, Misra S (2014) Esophageal cancer: a review of epidemiology, pathogenesis, staging workup and treatment modalities. World J Gastrointest Oncol 6:112–120

Noma K et al (2008) The essential role of fibroblasts in esophageal squamous cell carcinoma-induced angiogenesis. Gastroenterology 134:1981–1993

Nouraee N et al (2013) Expression, tissue distribution and function of miR-21 in esophageal squamous cell carcinoma. PLoS One 8:e73009

Pan Q et al (2013) Identification of lineage-uncommitted, long-lived, label-retaining cells in healthy human esophagus and stomach, and in metaplastic esophagus. Gastroenterology 144:761–770

Park S, Greco V, Cockburn K (2016) Live imaging of stem cells: answering old questions and raising new ones. Curr Opin Cell Biol 43:30–37

Pennathur A, Gibson MK, Jobe BA, Luketich JD (2013) Oesophageal carcinoma. Lancet 381:400–412

Potten CS, Booth C (2002) Keratinocyte stem cells: a commentary. J Invest Dermatol 119:888–899

Quante M et al (2012) Bile acid and inflammation activate gastric cardia stem cells in a mouse model of Barrett-like metaplasia. Cancer Cell 21:36–51

Raufi AG, Klempner SJ (2015) Immunotherapy for advanced gastric and esophageal cancer: preclinical rationale and ongoing clinical investigations. J Gastrointest Oncol 6:561–569

Reid BJ, Li X, Galipeau PC, Vaughan TL (2010) Barrett's oesophagus and oesophageal adenocarcinoma: time for a new synthesis. Nat Rev Cancer 10:87–101

Rosekrans SL, Baan B, Muncan V, van den Brink GR (2015) Esophageal development and epithelial homeostasis. Am J Physiol Gastrointest Liver Physiol 309:G216–G228

Ross-Innes CS et al (2015) Whole-genome sequencing provides new insights into the clonal architecture of Barrett's esophagus and esophageal adenocarcinoma. Nat Genet 47:1038–1046

Rubio CA (1983) Epithelial lesions antedating oesophageal carcinoma. I Histologic study in mice. Pathol Res Pract 176:269–275

Rubio CA, Liu FS, Chejfec G, Sveander M (1987) The induction of esophageal tumors in mice: dose and time dependency. In Vivo 1:35–38

Rustgi AK, El-Serag HB (2014) Esophageal carcinoma. N Engl J Med 371:2499–2509

Sadanaga N et al (1994) Local immune response to tumor invasion in esophageal squamous cell carcinoma. The expression of human leukocyte antigen-DR and lymphocyte infiltration. Cancer 74:586–591

Sasaki Y et al (2016) Genomic characterization of esophageal squamous cell carcinoma: insights from next-generation sequencing. World J Gastroenterol WJG 22:2284–2293

Sato T et al (2011) Long-term expansion of epithelial organoids from human colon, adenoma, adenocarcinoma, and Barrett's epithelium. Gastroenterology 141:1762–1772

Schneider TC et al (2016) (Secondary) solid tumors in thyroid cancer patients treated with the multi-kinase inhibitor sorafenib may present diagnostic challenges. BMC Cancer 16:31

Seery JP (2002) Stem cells of the oesophageal epithelium. J Cell Sci 115:1783–1789

Seery JP, Watt FM (2000) Asymmetric stem-cell divisions define the architecture of human oesophageal epithelium. Curr Biol 10:1447–1450

Song Y et al (2014) Identification of genomic alterations in oesophageal squamous cell cancer. Nature 509:91–95

Stairs DB et al (2011) Deletion of p120-catenin results in a tumor microenvironment with inflammation and cancer that establishes it as a tumor suppressor gene. Cancer Cell 19:470–483

Stransky N et al (2011) The mutational landscape of head and neck squamous cell carcinoma. Science 333:1157–1160

Taddei A et al (2014) Cyclooxygenase-2 and inflammation mediators have a crucial role in reflux-related esophageal histological changes and Barrett's esophagus. Dig Dis Sci 59:949–957

Tanaka K et al (2015) miR-27 is associated with chemoresistance in esophageal cancer through transformation of normal fibroblasts to cancer-associated fibroblasts. Carcinogenesis 36:894–903

Tani H, Morris RJ, Kaur P (2000) Enrichment for murine keratinocyte stem cells based on cell surface phenotype. Proc Natl Acad Sci U S A 97:10960–10965

Tetreault MP (2015) Esophageal cancer: insights from mouse models. Cancer Growth Metastasis 8:37–46

Tlsty TD, Coussens LM (2006) Tumor stroma and regulation of cancer development. Annu Rev Pathol 1:119–150

Trempus CS et al (2003) Enrichment for living murine keratinocytes from the hair follicle bulge with the cell surface marker CD34. J Invest Dermatol 120:501–511

Tu L et al (2005) Notch signaling is an important regulator of type 2 immunity. J Exp Med 202:1037–1042

Tumbar T et al (2004) Defining the epithelial stem cell niche in skin. Science 303:359–363

van der Weyden L, Adams DJ, Bradley A (2002) Tools for targeted manipulation of the mouse genome. Physiol Genomics 11:133–164

Watt FM, Hogan BL (2000) Out of Eden: stem cells and their niches. Science 287:1427–1430

Whiteside TL (2008) The tumor microenvironment and its role in promoting tumor growth. Oncogene 27:5904–5912

Winter JW et al (2007) N-nitrosamine generation from ingested nitrate via nitric oxide in subjects with and without gastroesophageal reflux. Gastroenterology 133:164–174

Xu Z et al (2013) TGFbeta1 and HGF protein secretion by esophageal squamous epithelial cells and stromal fibroblasts in oesophageal carcinogenesis. Oncol Lett 6:401–406

Yuspa SH et al (1994) Role of oncogenes and tumor suppressor genes in multistage carcinogenesis. J Invest Dermatol 103:90S–95S

Zhang Y, Wang XF (2015) A niche role for cancer exosomes in metastasis. Nat Cell Biol 17:709–711

Zhang L et al (2015) Genomic analyses reveal mutational signatures and frequently altered genes in esophageal squamous cell carcinoma. Am J Hum Genet 96:597–611

Chapter 11
Oral Cancer Stem Cells Microenvironment

Prajna Paramita Naik, Prashanta Kumar Panda, and Sujit K. Bhutia

Abstract Cancer stem cells (CSCs) play important role in tumor growth and metastasis coupled with increased recurrences and acquired therapeutic resistance in oral cancer. The tumor microenvironment imposes intense pressure in cancer evolution in response to adverse growth conditions, resource limitation and immune predation. Here, we discussed the dynamic interplay between cancer stem cells and tumor microenvironment in the formation of intratumoral heterogeneity to modulate tumor progression. The CSCs niche provide a special microhabitat for survival, maintenance of stemness and tumor re-propagation. Moreover, adaptive cellular behavior might be driven by tough tumor microenvironmental selective forces which highly regulate alterations in the gene expression leading to the reprogramming of signaling pathways generating stem-like characteristics, adaptive metabolic plasticity and energy fueling with autophagy to permit the CSCs to sustain in the ever changing microenvironments during tumor progression. On the other hand, CSCs also direct the tumor microenvironment modulation and remodeling in its favour. The cytokines, chemokines and growth factors released from CSCs regulates neoangiogensis, differentiation, degradation of matrix protein and immune suppression favoring tumor-promoting conditions and initiates multiple signaling cascades augmenting the tumor progression.

Keywords Oral cancer • Tumor heterogeneity • Cancer stem cell • Tumor microenvironment

11.1 Introduction

Oral cancer or oral squamous cell carcinoma (OSCC) accounts for more than 90% of malignant oral lesions with about 300,000 new cases registered each year worldwide. According to GLOBOCAN database, the anatomic subsites where the cancer of oral cavity occurs; includes the base of tongue, palate, nasopharynx,

P.P. Naik • P.K. Panda • S.K. Bhutia (✉)
Department of Life Science, National Institute of Technology, Rourkela, Odisha, India
e-mail: sujitb@nitrkl.ac.in; bhutiask@gmail.com

© Springer International Publishing AG 2017
A. Birbrair (ed.), *Stem Cell Microenvironments and Beyond*,
Advances in Experimental Medicine and Biology 1041,
DOI 10.1007/978-3-319-69194-7_11

oropharynx, hypopharynx, tonsils, pyriform sinus and other and ill-defined sites of the lip, oral cavity and pharynx (Ferlay et al. 2010). In the year 2012, there were approximately 300,373 new cases of lip/oral cavity cancer and 142,387 new cases of other pharyngeal (i.e. excluding the nasopharynx) cancer worldwide as per the most recent GLOBOCAN estimates (Ferlay et al. 2012). In India, the global epicenter of oral cancer, the disease occurrence trends vary by region and investigators estimate that the total number of new mouth cancer cases will increase from 45,859 in 2010 to 64,525 in 2020 (Takiar et al. 2010). Betel quid chewing, excessive alcohol consumption, tobacco smoke, HPV infection and radiation exposure are the chief risk factors for on setting the disease (Brandwein-Gensler et al. 2005; Marron et al. 2010; Licitra et al. 2006; Chen et al. 2008). Along with the conventional anticancer approaches such as surgery, radiotherapy and cytotoxic chemotherapy, several selective treatment modalities are also available which is based on the increased understanding of tumor biology and specific tumor subtypes (Fan et al. 2011). Though there is availability of high end multidimensional treatment regimen, the oral cancer treatment is quite unpromising with significant functional and aesthetic deficits such as facial disfigurement aside from functional deformity like speech impairment and difficulty in swallowing (Naik et al. 2016). Near about 30% of oral cancer cases are often coupled with an exorbitant rate of post-treatment loco-regional recurrence, ipsilateral and bilateral lymph node metastasis due to the dissemination of neoplastic cells via abundant lymphatic submucosal plexus present in the oral cavity, ultimately leading to death (Fan et al. 2011). Further, the 5 year survival rate of oral cancer patients is restricted to only 50–60% (Leemans et al. 1994).

The ever increasing failure rate of contemporary treatment motilities, rising intrinsic and acquired therapeutic resistance, persistent recurrences and relapse of oral cancer are mostly due to the random cytoreduction strategies that are designed to target only the bulk tumor cells setting aside a small subpopulation of therapy tolerant cells. Such rare and therapy tolerant subpopulation of cells with exclusive ability of self-renewal, progeny differentiation and tumorigenicity is termed as "Cancer Stem Cells" (CSCs). The CSCs are also reported to have the unique properties of enhanced DNA damage responses, apoptotic evasion, active drug efflux potential and epithelial to mesenchymal transition (EMT) which offer CSCs the supremacy to tumorigenesis, sustained growth and therapeutic resistance (Fig. 11.1) (Costea et al. 2006). The tumor microenvironment (TME) imposes intense pressure in cancer evolution in response to adverse growth conditions, resource limitation and immune predation. Moreover, the ability of the tumor cells to organize its surrounding environment in its favor along with the ability of microenvironment to shelter tumor cells from adverse exogenous constraints determines fate of the disease progression (Sottoriva et al. 2013; McGranahan and Swanton 2017). TME varies in terms of nutrients, oxygen, growth factors, cytokines, pH, extra-cellular matrix (ECM), vascularization and stromal components including fibroblasts and immune cells (Hjelmeland et al. 2011). This distinction within the tumor landscape creates various functional niches that govern the sensitivity of the genetically similar cancer cells to same treatment module (Almendro et al. 2013; Holohan et al. 2013).

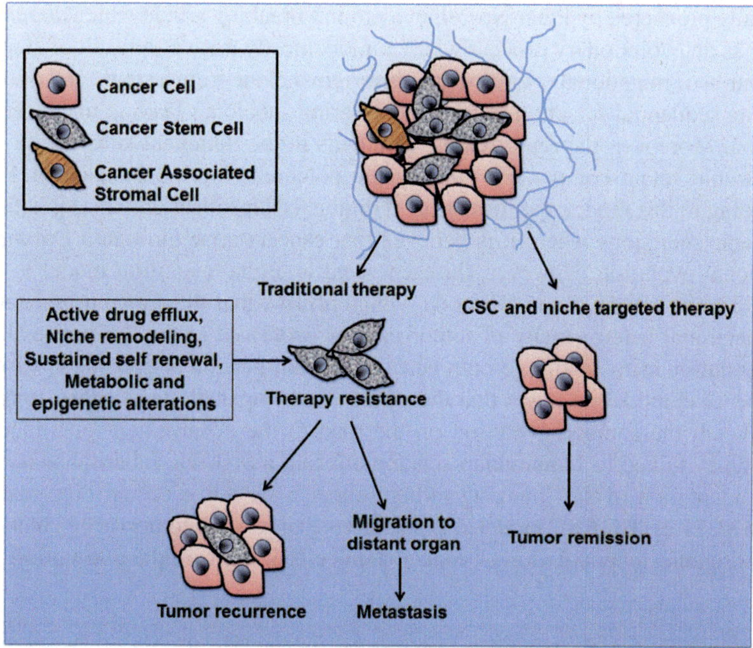

Fig. 11.1 Therapeutic relevance of oral cancer stem cells. Conventional or traditional therapy targets only bulk cells of oral cancer sparing the cancer stem cell population by virtue of abrupt developmental signaling, altered drug metabolism, niche remodeling and epigenetic reprogramming which subsequent in survival and maintenance of oral CSCs post-therapy followed to therapeutic resistance and recurrence of oral cavity cancers. Combinatorial therapy of targeting both CSCs and their microenvironmental link ups will be more effective in tumor remission

11.2 Tumor Heterogeneity and Tumor Niche

The tumor "onco-genotype" evolves gradually over time due to inherent genomic instability that makes nearly every tumor cell population unique and increases the clinical challenges for effective treatment as it also differs among patients. Many reports claim that tumor mass often show significant intratumor heterogeneity encompassing various discernable phenotypic features including cellular morphology, genetic heterogeneity, metabotypes, proliferative, angiogenic, immunogenic, and metastatic potential. Such phenotypic and functional heterogeneity among the cells within the same tumor occur as a consequence of integration of both genetic and non-genetic influences including genetic alteration, environmental variations and reversible changes in cell properties (Marusyk et al. 2012; Meacham and Morrison 2013). The pedigrees of intratumoral heterogeneity are extremely disputed and various cellular mechanisms are hypothesized to rationalize the diversity within a tumor. The two major frameworks that explain the intratumor phenotypic heterogeneity are "Clonal Evolution Model" or "Stochastic Model" and "Cancer Stem Cell (CSC) Model" or "Hierarchical Model". The clonal evolution model was originally pioneered by Peter Nowell in a ground-breaking article which documents cancer as an evolutionary process where tumors mostly arise as a result of stepwise acquisition of mutational events within the original clone from a single cell of origin allowing sequential selection of more persevering subclones leading to cancer progression. Moreover, the report claims that cells in the dominant subclone populations would retain comparable tumorigenic potential (Fig. 11.2) (Nowell 1976). According to this model, the formation of tumors is dependent on the acquisition of oncogenic mutations where it is believed that cancers arise through a Darwinian-like clonal evolution. However, this gene centric clonal evolution model is challenged by the cancer stem cell model which affirms that the apparent phenotypic and functional heterogeneity of tumor may be professed to the differences in the differentiation status owing to both genetic and non-genetic or epigenetic variability. The CSC model proposes that the genetic and epigenetic landscapes considerably switch the somatic evolution on the road to the achievement of a lucrative phenotype attuned to Lamarckian scheme offering a first-rate inheritable state for better adaptation to the changing milieu; either a stem-like or/and drug-resistant state (Fig. 11.2). The CSC model also postulates that cancers comprise of a hierarchy of tumorigenic subpopulation of cancer stem cells along with the non-tumorigenic

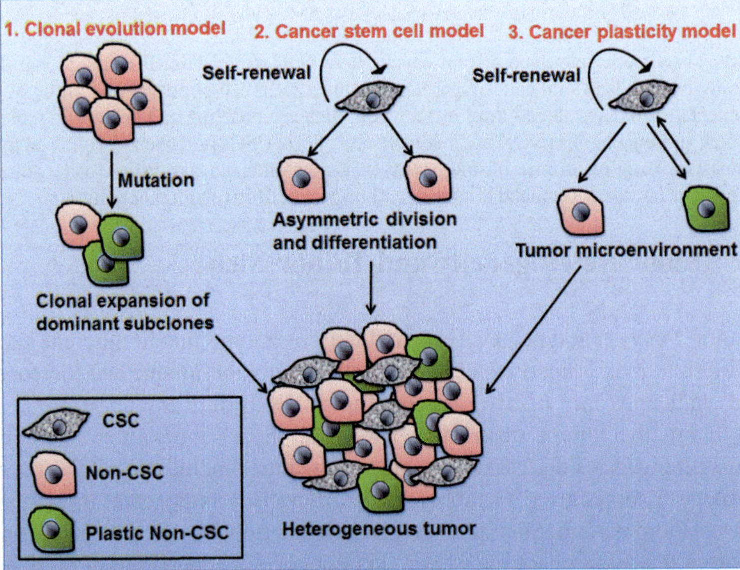

Fig. 11.2 Tumor heterogeneity model. The first model or clonal evolution theory to explains that various cancer cell populations evolve progressively by multistep acquisition of mutation finally generating heterogeneous tumor with clonal expansion of dominant subclones. The second model or cancer stem cell model describes that tumor heterogeneity arises due to the stem properties of a rare population of cancer cells which may get differentiated into any cell type within the tumor. Importantly, this model proposes that CSC-to-non-CSC conversion is a unidirectional process. The third model or cancer cell plasticity model posits the bidirectional conversions between non-CSCs and CSCs in response to the changing microenvironment

progeny populations where the CSCs rest at the apex of positional hierarchy and drive tumor growth and disease progression. It is highly essential to note that the clonal evolution model and CSC models are not mutually exclusive and the progression of intratumoral heterogeneity is a highly complex process (Shackleton et al. 2009). Intriguingly, recent reports emphasize that the concept of intratumor heterogeneity not only relevant to the cancer cells but also to the interaction of cancer cells with the diverse microenvironmental components. The intratumoral heterogeneity and evolutionary processes happening therein are influenced by topological niches and corresponded by functional heterogeneity (Marusyk et al. 2012). This alternative cancer cell plasticity model emphasizes that microenvironmental cues encourage self-renewal mechanisms to acquire CSC characteristics-a reversible process that is inherently transitory allowing the interconversion of CSCs to non-CSCs (Fig. 11.2). Cancer cell plasticity model is bidirectional involving the interconversion of tumogenic and non-tumorigenic cells within tumor adding extra complexity to the CSCs and clonal model to explain the intratumoral heterogeinity. This model suggests that both CSCs and non-CSCs are highly adaptable populations that readily switch between tumorigenic and non-tumorigenic cell states owing to appropriate microenvironmental stimuli which are capable of inducing transient evolution and plasticity (Cabrera et al. 2015). Moreover, a comprehensible role of microenvironment in tumorigenicity was recently demonstrated in melanoma. Quintana et al. experimentally showed that same melanoma cells have different tumorigenic capacity depending on the transplant conditions i.e. type of mouse strain, use of matrigel and duration of experiment suggesting microenvironmental regulation of tumor heterogeneity and tumorigenicity (Quintana et al. 2008).

11.3 Concept of Cancer Stem Cells in Oral Cancer

The exorbitant inefficacies of contemporary treatment modalities coupled with increased recurrences and metastasis in oral cancer are alleged to the CSCs which challenges the traditional concept and supports the existence of a small subpopulation of intratumoral cells called as Cancer Stem Cells (CSCs) or Cancer Initiating Cells (CICs) or Tumor Initiating Cells (TICs) with exclusive self-renewal capacity, tumorigenesis and metastatic potential. Though the concept of cancer stem cells was proposed in the late 1970s by a German physiologist Rodulf Virchow, who found similarities between the embryonic tissues and cancer tissues (Visvader and Lindeman 2008; Huntly and Gilliland 2005); Bonnet et al. (1997) were the first to isolate CSCs from acute myeloid leukaemia samples (Dick 1997). The CSCs from solid tumors were first isolated and identified in breast cancer by Al-Hajj et al. (2003). They isolated a subpopulation of CD44$^+$/ CD24$^-$ cells within the breast cancer tissues having the high tumorigenic capacity (Al-Hajj et al. 2003). The existence of CSCs is also described in other solid tumors including prostate, melanoma, lung, colon, brain, liver, HNSCC, gastric and pancreatic cancer (Singh et al. 2003; Kim et al. 2005; Li et al. 2007; Prince et al. 2007; Lessard and Sauvageau 2003).

Prince et al. for the first time accomplished the isolation of CSCs in HNSCC in 2007. They segregated a pool of cells in HNSCC with a high CD44 expression that exhibited stem cell-like characteristics like self-renewal, generation of differentiated progeny, lack of differentiation markers, and expression of immature cell markers. These CD44[+] cells were shown to have the exclusive tumorigenic capacity when introduced in immunosuppressed mice (Prince et al. 2007). Unlike to the bulk tumor cells, CSCs are akin to normal stem cells having the unique ability of unlimited self-renewal and hierarchical differentiation. Like normal stem cells, CSCs are also prophesied to have the properties of enhanced resistance to DNA damage and evasion of apoptosis. In addition to this, CSCs show some exceptional features; like epithelial to mesenchymal transition (EMT) enhanced invasive capacity and metastatic proficiency that helps in tumorigenesis, sustained growth and therapeutic resistance (Visvader and Lindeman 2008). Most of the contemporay therapy modules target only the bulk tumor population escaping the assassin CSCs that serve as a reservoir for post treatment tumor repopulation.

Oral CSCs may arise from normal adult epithelial stem cells which maintain the self-renewal machinery of a pre-existing normal stem cell rather than developing new self- renewal pathways like Notch, Hedgehog and Wnt signaling (Reya and Clevers 2005). Additional oncogenic mutations that drive the de-differentiation process followed by self-renewal in progenitors or partly differentiated cells would also allow the re-acquisition of the stem-like properties (Zhou et al. 2009). Moreover, de-differenciation of terminally differentiated adult oral epithelial cells via the acquisition of stem-like mutations can also result in the origin and development of CSCs (Zhang et al. 2013). Several factors are reported to be involved in the generation of CSC phenotypes in oral cancer. Nutrient starvation, reduced blood supply, hypoxia, mild therapeutic stress and challenged microenvironment are documented to reprogram the genetic and epigenetic landscapes that induce the acquisition of an adapted inheritable state such as drug-resistant state or/and stem-like state (Pisco and Huang 2015). One of the studies showed that long-term exposure with nicotine elevates the ALDH1 population and enhanced the stemness gene expression, upregulated EMT mediators and increased the self-renewal and sphere-forming primary oral cancer epithelial cells (Sinha et al. 2013). Moreover, smoking induced drug resistance was reversed by inhibiting nicotinic acetylcholine receptors suggesting nicotine as a potential inducer in oral CSC generation (An et al. 2012). In addition, recent report indicates that chronic arecoline exposure to oral epithelial cells enhances stem population by overexpressing, stemness-related transcription factors Oct4, Nanog and Sox2 (Wang et al. 2016). Human Papilloma Virus (HPV) proteins E6 and E7 activate Wnt signaling pathway in HPV16-positive oropharyngeal SCC that may cause the de-differentiation of oral cancer cells to CSCs (Rampias et al. 2010). The isolation and characterization of CSCs in oral cancer has been achieved successfully via the use of different techniques which mostly involve cell lines, primary tumor specimens and xenograft models. CSC isolation and characterization can be performed using flow cytometry and FACS based on the expression of specific cell surface markers, such as CD133, CD44 and ALDH1. Moreover, sorting the side populations (SP) of tumor cells via intracellular Hoechst 33,342 exclusion has also been used for the identification and characterization of CSCs. The orosphere culture system is also efficient in separating CSCs from oral cancer cell lines or tumors. Finally, the gold standard quantitative xenotransplantation assay that assess the *in vivo* tumorigenicity and self-renewing potential of putative CSCs is finally performed for functional characterization of isolated CSCs (Lin et al. 2011).

11.4 Tumor Microenvironment

Now-a-days tumors are considered as an organ with distinct tumor vasculature in which the cancer cells are accompanied by a protumor microenvironment (Egeblad et al. 2010). Rudolf Virchow was the first person to demonstrate the presence of leucocytes in tumor tissues and proposed that non-cancerous tissue elements came from the field of inflammation might affect tumorigenesis (Balkwill and Mantovani 2001). TME comprises (1) aberrantly proliferating cancer cells, (2) cancer stem cells (3) extracellular matrix (ECM) (4) infiltrating immune cells (neurophill, eosinophil, basophil, B and T lymphocytes, mast cells, natural killer cells), antigen presenting cells (APC) (macrophages, dendritic cells) and Tumor associated macrophages (TAMs) (5) angiogenic endothelial cells and their precursors including pericytes and (6) Stromal cells like fibroblast cells, myofibroblasts and cancer associated fibroblasts (CAFs) (Hanahan and Coussens 2012; Friedl and Alexander 2011). TME shelters cancer cells and provides protection against various genetic and epigenetic insults (Hovinga et al. 2010; Folkins et al. 2007; Tlsty and Coussens 2006; Yun 2008). Tumor niches are discrete and dynamic domains ensuring precise functional characteristics that helps in forming a suitable habitat for certain cells with specific fates. Likely, CSC niche provides a special microhabitat for survival, maintenance of stemness and post-therapy tumor re-propagation. Moreover, specific stimuli from the microenvironment prop up the CSCs to maintain its exclusive properties (Fuchs et al. 2004; Xie and Li 2007; Morrison and Spradling 2008). Though century ago, it has been established that TME plays a crucial role in the tumor progression and metastasis, till date it is not clear whether CSCs direct the TME modulation, generate their own "micro-niches", or exploit the pre-existing environment in its favor (Paget 1889). However, the existence of a supportive perivascular niche is evidenced in HNSCC and majority of the CSCs are located within a 100 μm-radius of blood vessels in primary tumors (Krishnamurthy et al. 2010).

11.4.1 Extracellular Matrix

Extracellular matrix is a highly complex and dynamic array of interacting proteins that are being regularly synthesized, processed and assembled during cell homeostasis and adhesion, migration, wound repair and tumor development and progression. Regardless of once viewed only as an architectural support, the ECM is now documented as a key element in controlling cellular processes. Any variation in the composition of the ECM by altered processing, secretion, or expression contributes greatly to tumorigenesis (Ziober et al. 2006). In oral carcinoma, alterations in cell–cell and cell–extracellular matrix interactions regulate invasion of malignant cells into the underlying connective tissue and migration of malignant cells to form metastases at distant sites (Lyons and Jones 2007). The major ECM molecules involved in oral cancer development and progression include fibronectin, laminin and collagens and their ECM receptors and integrins ($\alpha2\beta1$, $\alpha3\beta1$, $\alpha5\beta1$, and $\alpha6\beta4$) (Ziober et al. 2006). An immunohistochemical study of extracellular matrix, decorin and vitronectin in OSCC proposes that in primary tumors of metastatic cases, the expression of laminin, type IV collagen, heparansulphate proteoglycan, decorin and vitronectin obviously decreased, while the expression of fibronectin and tenascin increased when compared with those of the non-metastatic cases (Harada et al. 1994). The neo-expressed Tenascin-C and one of its integrin receptors $\alpha v \beta6$ in oral SCC and tumor stromal environment influences the oral SCC behavior (Ramos et al. 1997).

11.4.2 Infiltrating Immune Cells

The host-tumor immune response is extremely complex, multifactorial and dynamic. Though the primary function of immune system is to destroy the foreign cells; tumor cells within the tumor microenvironment escape the immune surveillance not due to the absence of immune cells, but due their aberrant activity. Tumors exploit the host response by creating a favorable microenvironment during cancer progression by scheming chronic inflammation to establish a habitat that favor tumor survival and growth (Whiteside 2006a). Moreover, it is reported that established tumors have the ability to lessen the immune response and also they are relatively poor in initiating an immune response (Whiteside 2006b). Therefore, understanding the immune response within tumor microenvironment holds great importance in tumor biology. The adaptive immune response contributes in a variety of ways to tumorigenesis through the immune interactions in the TME. According to reports a heterogeneous distribution of tumor infiltrating leucocytes were seen in solid tumor microenvironment which includes granulocytes, mast cells, macrophages, myeloid-suppressor cells, natural killer (NK) cells, CD8$^+$ T-cells (CTL), T-memory cells, T-regulatory (T-reg) cells and dendritic cells (DC) (Fridman et al. 2012; Senovilla et al. 2012). The creation of an immune suppressed microenvironment and evasion of the adaptive immune response may be executed through decreased expression of major histocompatibility complexes (MHC I) or induction of T cell apoptosis (Young 2006; Ogino et al. 2006; Grandis et al. 2000). In head and neck cancer, a reduced expression of MHC-I is reported which is mediated by the overexpression of gangliosides in tumor cells (Tourkova et al. 2005). Moreover, it is reported that oral carcinoma cells contain membranous FasL-positive vesicles which triggers T-cell apoptosis induction that helps in evading the cytotoxic response (Young

2006). Furthermore, an elevated incidence of intratumoral Tregs and IL-17[+] non-Th17 cells are documented to be associated with poor outcome in oropharyngeal carcinoma patients (Punt et al. 2016). Cytokines like IL-10 and TGF-β were reported to let local naive T cells to convert into suppressor T cells and take advantage of the suppressive functions of existing Treg cells (Ferris et al. 2006). The antigen presenting cells, DCs when exposed IL-10 encourage immunetolerance and CD4[+] T cells differentiation into suppressive Treg cells (Jonuleit et al. 2000). Another study illustrates that higher density of immature DCs and Treg cells and a lower density of mature DCs and activated CTLs in metastatic head and neck cancer indicates an immunosuppressive microenvironment which could be involved in the spread of neoplastic cells to cervical lymph nodes (Gonçalves et al. 2013). Moreover, Quan et al., reported that an adaptive immune response is driven by both mature, antigen-experienced T and B cells within the microenvironment of oral carcinoma (Quan et al. 2016). In HNSCC, the intratumoral cytotoxic CD8[+] T cells were shown to have increased expression of programmed death-1(PD-1), a surface protein that blocks function of T lymphocytes. Moreover, programmed death ligand -1 (PD-L1) expression is associated with augmented CD8[+] T cell apoptosis (Cho et al. 2011). Recent studies have emphasized the interaction between CSCs and immune system in tumorigenesis of oral cancer. It is reported that CD44[+] HNSCC cells showed selective PD-L1 expression compared to CD44[−] cells leading to its decreased immunogenicity which can be partially restored by inhibiting its expression (Lee et al. 2016). Moreover, according to reports CD44[+] HNSCC cells displayed decreased HLA-A2 and TAP2 expression, the latter of which is indispensable for assembling the MHC Class I-tumor antigen peptide complex suggesting that CSCs in HNSCC may be less immunogenic than the bulk of the tumor cells (Chikamatsu et al. 2011).

Among the microenvironment components, tumor-associated macrophages (TAMs or tumor infiltrating macrophages) are the major inflammatory component of the tumor which encourages tumor progression by prompting tumor invasion, migration, and angiogenesis (Shieh et al. 2009). Usually, macrophages can be categorized into two distinct polarized states with opposite responses. The first one is the classically activated pro-inflammatory (M1) state which possesses antitumor activity whereas the second one is the alternatively activated suppressive (M2) state which promotes tumor invasion and metastasis (Lúcio et al. 2016; Hu et al. 2016). Accordingly, M1 TAMs participate in anti-tumor immune response via the production of proinflammatory cytokines like INF- γ, IL-12, IL-23 (Sica et al. 2006) whereas M2 TAMs contribute to protumor immune response via the production of various suppressive cytokines such as IL-10 and TGF-β and its accumulation near blood vessels promotes angiogenesis (Martinez and Gordon 2014; Li et al. 2002; El-Rouby 2010). The CD68 recognizes both tumoricidal M1 TAMs and anti-inflammatory M2 TAMs while CD163 recognizes only M2 TAMs (He et al. 2014). The substantial reports suggested a significant correlation between TAMs and poor prognosis in patients with oral cancer (Lúcio et al. 2016). According to reports, TAMs in oral SCC mostly possess a M2-like phenotype. Moreover, M2 TAMs are documented to stimulate tissue remodeling and hinder anti-tumor cytotoxic effects of M1 TAMs (Sica et al. 2006; Zamarron and Chen 2011). Higher level of TAM infiltration is described to associate with higher tumor state and lymph node metastasis (Li et al. 2002; Marcus et al. 2004; Liu et al. 2008). Furthermore, TAMs play a protumor role in mucoepidermoid carcinoma and it is highly correlated with angiogenesis, invasion and migration (He et al. 2014). Moreover, it is reported that Axl signaling in oral cancer promotes polarizing of TAMs toward a M2 phenotype via Axl/PI3/Akt/NF-kB pathway which subsequent in poor prognosis (Chiu et al. 2015). Recently, it has been proposed that infiltrating CD11b[+] myeloid cells in the vascularization showed characteristics of M2 macrophages and promoted neovascularization and tumor progression in recurrence after irradiation compared to non-irradiated tumors in oral carcinoma. Moreover, CD11b[+] myeloid cells and CD206[+] M2 macrophages get intensified during recurrence after radiotherapy in human oral cancer specimens (Okubo et al. 2016). In HNSCC, TAMs are reported to produce an elevated level of inflammatory cytokine called macrophage migration inhibitory factor (MIF) that stimulates neutrophils. The neutrophils are recruited to HNSCC tumor by MIF via a CXCR2 mechanism which further leads to tumor invasiveness (Dumitru et al. 2011). The activated neutrophil induces ROS-mediated genetic instability, increases invasion via HGF and triggers angiogenesis via MMP9 and VEGF (Galdiero et al. 2013). Another report indicates that VEGF and IL-18 released by neutrophils promote neoangiogenesis and encourage the benign tumour cells to acquire metastatic phenotype in the early stage of oral cavity cancer (Karin and Greten 2005). Expression of signal regulatory protein α (SIRPα), a surface protein significantly correlates with the expression of CD68 and CD163 on macrophages. Again, the inhibition of SIRPα was found to reduce the phagocytosis ability and IL-6 and TNF-α production of macrophages (Ye et al. 2016). Furthermore, it is reported that TME and the peripheral blood in HNSCC contains an increased loads of TAMs with increased levels TGF-β which trigger immunosuppression and tumor growth in oral cancer (Costa et al. 2013). Moreover, TAMs produces ROS and prostaglandins which support inflammation and tumorigenesis (Karin and Greten 2005; Coussens and Werb 2002).

11.4.3 Tumor-Associated Fibroblasts Cells

Fibroblasts shape the structural framework, the tumor stroma of tissues by synthesizing ECM component (Brouty-Boyé 2005). Fibroblasts in tumor stroma have been designated peritumoral fibroblasts, reactive stroma, cancer-associated fibroblasts (CAFs) and myofibroblasts. Activated fibroblasts are prominent contributors in carcinogenesis of oral cancer. It reported that fibroblasts inhibit early stages of tumor progression however at later stages in cancer CAFs promote both tumor growth and progression. Generally, CAFs evolve from circulating fibroblasts and co-evolve along with tumor giving rise to a distinct phenotype. A common marker indicating CAF phenotype is the α-smooth muscle actin (α-SMA) and a greater proportion of α-SMA-positive peritumoral fibroblasts have been demonstrated to associate with poor prognosis in many cancers (Bhowmick et al. 2004). Other markers includes tenascin-C, periostin, NG-2, PDGF receptor-a/b, FSP (S100A4), FAP vimentin, type I collagen, prolyl 4-hydroxylase and fibroblast surface protein. For the first time, in 2004, CAFs in stroma of oral SCC were reported (Barth et al. 2004) and since then, massive studies have emphasized their importance in disease progression. High frequency of CAFs in OSCCs is significantly correlated with invasion, disease progression, tumor recurrence and poor patient prognosis (Kawashiri et al. 2009; Vered et al. 2010). The growth factors, matrix proteins and proteases secreted by fibroblasts and tumor cells create a tumor microenvironment that facilitates tumor maintenance, invasion and metastasis via the paracrine interaction of epithelial tumor cells and stromal cells (Mueller and Fusenig 2002). Moreover, in HNSCC CAFs are also characterized by expression of integrin $\alpha6$, an important molecule in cell adhesion and surface signaling. Integrin $\alpha6$ is reported to bind with ECM component laminins and interacts with CDKN1A, which alters the cell cycle progression. Also, upregulation of α-SMA and integrin-$\alpha6$ was demonstrated to be highly correlated with poor clinical outcome in oral cancer (Lim et al. 2011). The CAFs secrete various cytokines like TGF-β, CXCL12, paracrine motility factor and hepatocyte growth factor (HGF) (Leef and Thomas 2013). The TGF-β promotes immune suppression whereas upon binding with CXCR4, CXCL12 upregulates MMP9 and HIF-1α expression that promotes EMT (Ishikawa et al. 2009; De Wever et al. 2008). Recently, it is proposed that fibroblast-derived HGF and SDF-1 seem to portray a crucial role in the mutual interactions between oral SCC cells and underlying stromal fibroblasts that consequents in the local invasion (Daly et al. 2008). Moreover, report elucidates that the paracrine interaction between oral carcinoma cells SCC-25 and CAFs provides a mechanistic background for the gene regulation of MMPs which causes to poor clinical outcome in head and neck cancer (Fullár et al. 2012). Furthermore, the presence of MMP-1 (Collagenase 1) expression in the stromal compartment of invasive head and neck cancer suggests that its expression in peritumoral fibroblasts is induced in paracrine manner by tumor cells and tumor-infiltrating inflammatory cells (Johansson et al. 1997; Westermarck et al. 2000). Comparison of the secretome profiles between CAFs and normal oral fibroblasts (NOF) using mass spectrometry-based proteomics and biological network analysis reveals that proteins engaged in ECM organization and disassembly and collagen metabolism are highly upregulated. Moreover, the expression of type I collagen N-terminal propeptide (PINP) was found to associate *in vivo* with CAFs in the tumor front and promoted shortened survival of oral cancer patients (Bagordakis et al. 2016).

Myofibroblasts are cells that display a phenotype between fibroblasts and smooth muscle cells α-SMA expression (Chaponnier et al. 2006). Myofibroblasts secrete a large repertoire of chemokines, cytokines, inflammatory mediators, growth factors, neurotransmitters, hormones, adhesion proteins and most abundantly ECM proteins (Powell et al. 2005). Barth et al. in 2004 for the first time produced evidence in favor of a role for myofibroblasts in OSCC and later of this year, Lewis et al. demonstrated that myofibroblasts when induced TGF-$\beta1$ by release HGF to promotes OSCC invasion *in vitro* (Barth et al. 2004; Lewis et al. 2004). Myofibroblasts activation in the vicinity of tumor cells in stroma encourages the setting up of reciprocal paracrine interactions between the two cell compartments. In oral SCC, αvβ6-dependent activation of TGFβ1 mediated fibroblast–myofibroblast transdifferentiation occurs during tumor invasion and factors released from myofibroblasts favors tumor growth (Kellermann et al. 2008; Marsh et al. 2011). Myofibroblast-released factors like activin A, induces invasion and triggers the release of matrix MMP by OSCC tumor cells. Moreover, presence of myofibroblasts in OSCCs correlates with augmented production of MMP-2 and MMP-9 (Sobral et al. 2011).

11.4.4 *Tumor-Associated Endothelial Cells*

Tumor angiogenesis is a multifaceted process where formation of new blood vessels occurs in response to interactions between tumor cells and endothelial cells (ECs), growth factors, and ECM components. Tumor vessels are reported to stimulate the progression of many human solid tumors, including HNSCC (Zeng et al. 2005). Tumor-associated ECs (TAEs) and tumor vessels vary in many respects from their normal counterparts. Unlike normal vessels, tumor vessels possess different structural features, such as leakiness, uneven thickness of the basement membrane and fewer pericytes (Akino et al. 2009). Moreover, TAEs expressed typical endothelial cell markers like CD31. Unexpectedly, TAEs are found to be cytogenetically abnormal and relatively large and heterogeneous nuclei (Hida et al. 2004). Substantial reports have documented that tumor cells secrete angiogenic growth factors that stimulate EC proliferation to induce angiogenesis (Folkman 2002; Sparmann and Bar-Sagi 2004). New tumor vessels penetrate into neoplastic growths and enhance nutrients and oxygen supplying and oxygen and removes waste materials (Folkman 2002). According to reports, the angiogenic factors like VEGF and IL-8 released by tumor or stromal cells are documented to directly bind to their receptors on ECs to trigger angiogenesis by encouraging endothelial sprouting, branching, differentiation and survival (Folkman 2002; Sparmann and Bar-Sagi 2004). Notch ligand Jagged1 play a critical role in angiogenesis in HNSCC. Report indicates that tumor cells induced by growth factors via MAPK triggered Notch activation in neighboring endothelial cells which promoted capillary-like sprout formation suggesting the direct interplay between tumor cells and ECs that promotes angiogenesis. Jagged1 enhances neovascularization and HNSCC growth *in vivo* and it expression is significantly correlated with tumor blood vessel content (Zeng et al. 2005).

11.5 Crosstalk Between Cancer Stem Cells and Tumor Microenvironment

Recently, two interesting model of tumor evolution was proposed by Castaño et al. keeping an analogy with bed and bug to connect the association between the TME and CSCs. One of models proposes that "The bed (TME) determines the bug (CSC) fate" in which microenvironmental factors are believed to play critical role in niche formation and intratumoral heterogeneity (i. e. 'niche drives clone'). On the other hand, another model proposes "The bug makes their own bed" in which tumor sub-clones with certain mutations are thought to confer selective advantage to populate the tumor (i.e. 'clone drives niche') (Castaño et al. 2012). Both the components, TME and CSCs work synergistically in cancer progression (Fig. 11.3).

11.5.1 Tumor Microenvironment Determines Fates of Cancer Stem Cells

According to literature, any non-stem cancer cells can spontaneously give rise to a stem-like state in response to specific microenvironmental stimuli. Growth of cancer cells in a confined microenvironment leads to the alterations in metabolic and physicochemical milieu and the reciprocal interaction between tumor cells and TME promotes tumor progression (Chaffer et al. 2011). TME continuously gets reshaped during tumor progression and prompt adaptive cellular behaviors including dormancy, invasion and metastasis and therapy resistance. This adaptive cellular behavior might be driven by tough tumor microenvironmental selective forces that highly regulated alterations in the gene expression leading to the reprogramming of signaling pathways generating stem-like characteristics (Marjanovic et al. 2013; Anderson et al. 2006). The typical triad of tumor microenvironment that plays a crucial role in driving malignant tumor cell behaviors consists of hypoxia, nutrient depletion and low pH (Heddleston et al. 2010; Keith and Simon 2007). The interdependence of tumor cells with their respective microenvironment and the intratumoral heterogeinity of cancer postulate the idea that CSCs are "bugs" that cannot live without the "bed" i.e. the TME (Castaño et al. 2012).

11.5.1.1 Hypoxia

Hypoxia is one of the most common features of malignant head and neck tumor which is also registered as a key contributor of tumor progression, metastasis and chemo-radio therapy resistance in HNSCC. Improper vascularization, poor oxygen transport in the intratumaoral region and necrotic areas may lead to the acute or chronic hypoxia in the tumor microenvironment where the oxygen demand cannot meet the oxygen supply (Vaupel and Mayer 2007; Jiang et al. 2011). As an adaptive response to the reduced oxygenation, cancer cells may produce hypoxia inducible factors (HIFs). HIFs comprises hypoxia-inducible factor 1α (HIF-1α), hypoxia-inducible factor 1 (HIF-1β), hypoxia-inducible factor 2α (HIF-2α) and hypoxia-inducible factor 3α (HIF-3α) (Brennan et al. 2005). Dimerization HIF-1α and HIF-1β leads to the transcriptional activation of genes responsible for the adaption to hypoxia which includes the genes involved in angiogenesis, metastasis and therapy resistance (Li et al. 2013). Reports suggest that hypoxic microenvironment generates selection pressure for the development of therapy tolerant aggressive tumor consisting of CSC phenotypes (Lin and Yun 2010). Moreover, hypoxia plays a crucial role in maintaining CSCs in their undifferentiated state and permits the accumulation of genetic and epigenetic insults over an extended period of time and aids in self-renewal. HIFs are reported to target genes like Wnt, cMyc, Notch, Oct4 which are involved in maintaining stem properties (Fig. 11.3) (Keith and Simon 2007; Winquist et al. 2009; Qing and Simon 2009). Recent reports advocate that hypoxic microenvironment may promote the stem-like biological properties of laryngeal cancer cell lines by the intensification of the CD133+ stem cell

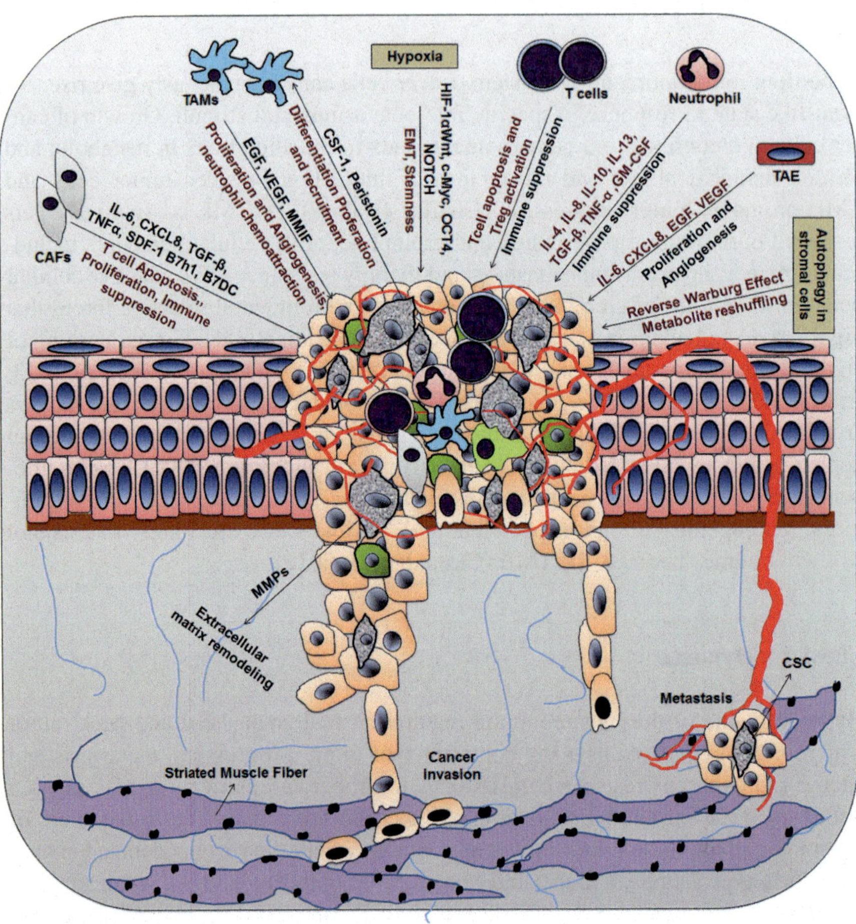

Fig. 11.3 Crosstalk between cancer stem cells and microenvironment. Tumor associated cells including fibroblast, macrophages, T cells, B cells, neutrophills and endothelial cells secretes of growth factors, interleukin and chemokines that helps in proliferation, angiogenesis and immune evasion. CSCs encourage the proliferation of tumor associated cells by secreting secrete growth factors and chemokines. Hypoxia in the tumor microenvironment induces the production of HIF-1α, Oct4, Wnt and Notch and endorses stem like-features in cancer cells

subpopulation. The laryngeal carcinaoma cells Hep-2 and AMC-HN-8 when grown under hypoxic condition stimulated HIF-1α production along with the acceleration of stemness signature gene OCT4, SOX2 and NANOG. Hypoxia also augmented the laryngeal CSC marker CD133 expression and increased the proliferation, invasion, colony formation and sphere formation capacity (Wu et al. 2014). Reports also explore that hypoxia can induce stemness in laryngeal carcinoma by enriching the percentage of CD133$^+$ cells post radiation treatment. After 10 Gy of irradiation for 24 h the hypoxic Hep2 cells introduced G1 cell cycle arrest and acquired stem like characteristics by overexpressing the stemness marker CD133 in laryngeal carcinoma (Wang et al. 2011). Moreover, hypoxia is also reported enhance the ALDH1^{high} CSC population in a syngeneic mouse model of HNSCC (Duarte et al. 2012). In orthotopic immunocompetent murine models of HNSCC, hypoxia induced autophagy is described to promote the evolution of aggressive phenotypes (Vigneswaran et al. 2011).

11.5.1.2 Autophagy

Autophagy is an evolutionarily conserved catabolic pathway involving degradation of cytoplasmic content which recycles ATP and metabolites in response to nutrient deprivation or metabolic stress, hypoxia, chemo/radiotherapy and activated oncogenes. Autophagy offers survival advantage to tumors by virtue of its nutrient recycling capacity as tumors cells are frequently exposed to metabolic stress owing to hypoxia and nutrient deprivation and promotes tumorigenesis (Bhutia et al. 2013). The autophagic stroma model of cancer emphasizes the induction of oxidative stress, mitochondrial dysfunction and autophagy/mitophagy in tumor invasion and metastasis. Intriguingly, autophagy/mitophagy induction in the tumor stromal compartment helps the cancer cells to directly "feed off" of stromal-derived energy-rich metabolites (glutamine, pyruvate, and ketones/BHB) and chemical building blocks (amino acids, nucleotides) (Fig. 11.3). It is important to note that in the tumor microenvironment, the aggressive cancer cells are "eating" the CAFs via autophagy/mitophagy (Pavlides et al. 2010). The solid tumor core which is a hypoxic and nutritionally challenged environment constructs a compensatory environment around them by turning the CAFs into their "metabolic slaves" (Roy and Bera 2016). Interestingly, it reported that loss of caveolin-1 (Cav-1) in stromal cells drives the activation of the metabolic reprogramming of CAFs and upregulates the expression of pyruvate kinase M2 (PKM2), a glycolytic enzyme resulting in the induction of autophagy and glycolysis. Enhanced glycolysis fuels the mitochondrial metabolism of nearby cancer cells leads to high ATP generation and cell survival (Capparelli et al. 2012). RAS-dependent and NF-κB–dependent HNSCC cell line were each able to induce metabolic reprogramming of CAFs via oxidative stress resulting in a lactate shuttling process that feeds the cancer cells fueling anabolic growth via and MCT1/MCT4 metabolic couple between the tumor and the stroma (Curry et al. 2014).

11.5.1.3 Tumor Metabolism

The "Warburg effect", as proposed by Otto Warburg in the 1920s, postulates that even in presence of abundant oxygen, cancer cells are more reliant on aerobic glycolysis than the oxidative phosphorylation (OXPHOS). Though glycolysis is less efficient in terms of ATP production, during the proliferation of cancer cells, it provides metabolites for the synthesis of macromolecules. However, contradictorily "Reverse Warburg Effect" posits that tumor cells exploit normal stroma through H_2O_2 paracrine signaling resulting in oxidative stress in stromal fibroblasts thereby trigger mitochondrial dysfunction, mitophagy and glycolytic metabolism. This helps in the release of metabolic intermediate metabolites like lactate, glutamine and ketone bodies to be used for oxidative phosphorylation in cancer cells. According to report, the non-physiological high glucose and oxygen concentration favor a glycolytic phenotype. However, when patient-derived, low-passage CSCs is investigated OXPHOS was found to be preferred energy metabolism of CSCs. Moreover, it is reported that when OXPHOS is blocked CSCs are able to switch to a glycolytic phenotype. This observed adaptive metabolic plasticity might permit the CSCs to sustain in the microenvironments during tumor progression (Neiva et al. 2009; Dong et al. 2013).

There exists a multicompartment model of energy metabolism in oral cancer. It is also reported that it may be a three metabolic compartments in OSCC, where the peripheral tumors cells relies on OXPHOS and cells in the deeper layer tumor are more glycolytic (aerobic or anaerobic) whereas the third metabolic compartment represented as cells in tumor stroma undergoing aerobic glycolysis. This three compartment metabolism was demonstrated through higher level of expression of MCT4 in tumor stroma and deeper tumor, whereas MCT1 level was more in the leading tumor edge. Energy metabolism through OXPHOS in the leading tumor edge was confirmed by functional mitochondrial metabolism markers TOMM20 and LDHb (Curry et al. 2014). In differentiated cancer cells, the glycolytic phenotypes predominate over OXPHOS phenotype. CSCs instead might rely more on oxidative metabolism for their energy production. The CSCs also appear to be metabolically plastic and when OXPHOS is blocked they can eventually develop resistance by acquiring an intermediate glycolytic/oxidative phenotype. For the first time Curry et al. reported the connection between cancer stemness with lactate and ketone uptake and mitochondrial metabolism in HNSCC (Fig. 11.4). "Three compartment tumor metabolism" involving (1) proliferative and mitochondrial-rich cancer cells (Ki-67+/TOMM20+/COX+/MCT1+); (2) non-proliferative and mitochondrial-poor cancer cells (Ki-67/TOMM20/COX/MCT1); and (3) non-proliferative and mitochondrial-poor stromal cells (Ki-67/TOMM20/COX/MCT1) in HNSCC displayed metabolic symbiosis where the non-proliferative stromal cells provide metabolites for OXPHOS in highly proliferating cancer cells (Bagordakis et al. 2016). Again, metabolic stress in the confined TME is recently reported to assist in emergence and sustenance of CSC-like phenotypes. Chronic metabolic stress (CMS) due to long-term nutrient deprivation in the TME persuades a Wnt-dependent phenoconversion of non-CSCs toward CSCs through stochastic state transition (Lee et al. 2015).

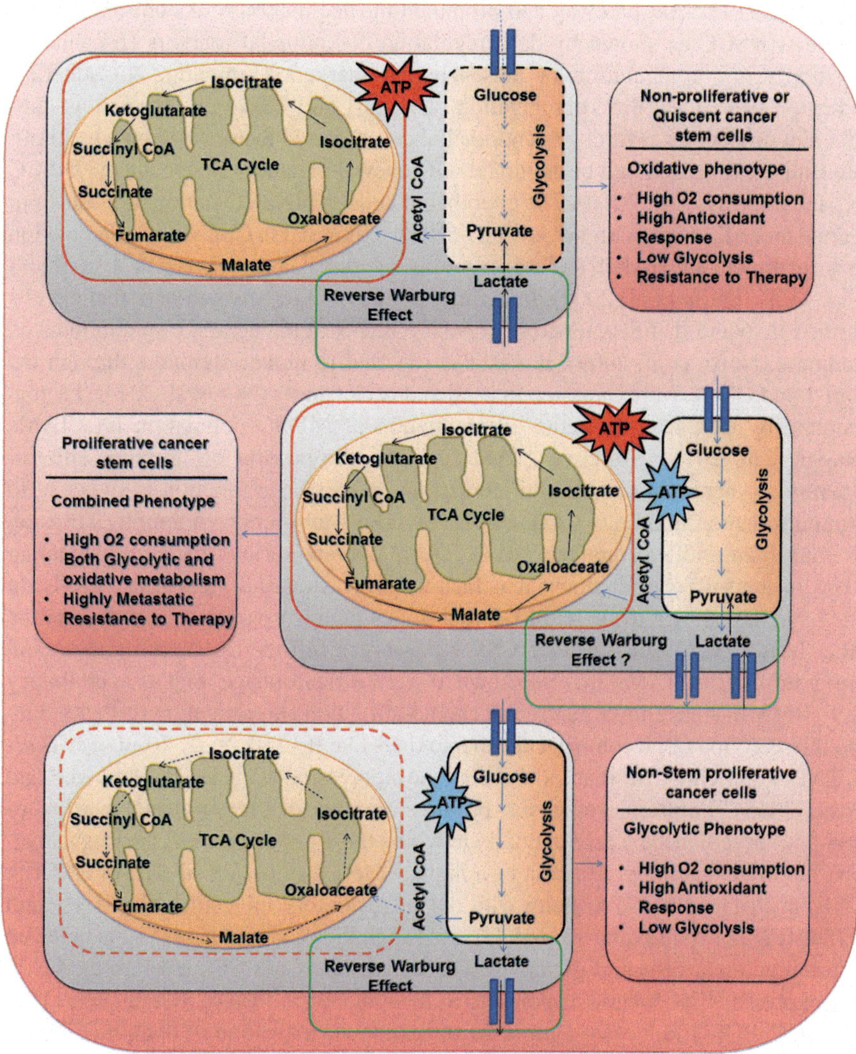

Fig. 11.4 Three compartment metabolism in tumor stroma. The non-proliferating cancer stem cells rely on the mostly on OXPHOS instead of glycolysis. However, proliferating cancer stem cells rely on both OXPHOS and glycolysis. The differentiated and proliferating non-stem cancer cell relies only on glycolysis. Autophagy mediated generation of metabolites in the tumor stromal and lactate shuttling process via the reverse Warburg effect offers metabolic coupling between the tumor and the stromal cells

11.5.1.4   Interaction Between Stromal Cells and Cancer Stem Cells

Endothelial cell-secreted factors are reported to enhance the invasive migration and resistance to anoikis in oral CSCs (Fig. 11.3) (Neiva et al. 2009). According to recent study, endothelial cell-secreted EGF induced EMT via Snail induction through the PI3k-Akt pathway and promoted the acquisition of a stem-like pheno-type in HNSCC as shown by downregulation of epithelial markers (E-cadherin, Desmoplakin), upregulation of mesenchymal markers (Vimentin, N-Cadherin), induction of cell motility, upregulation of ALDH and CD44 and growth as non-adherent orospheres. Moreover, endothelial cell-secreted factors is shown to stimu-late Bmi-1 expression and promote the self-renewal of cancer stem cells in HNSCC (Krishnamurthy et al. 2010). The endothelial cells in HNSCC tumors are shown to secrete more IL-6 and a higher level IL-6R are expressed in CSCs. Their interaction in turn directs the JAK/STAT to encourage survival and self-renewal of CSCs (Fig. 11.3) (Duffy et al. 2008). Impeding the interaction between endothelial cells derived IL-6 and IL-6R is found to prevent STAT-3 signaling and cisplatin induced stemness (Neiva et al. 2009). It was also reported to induce stemness through the Bmi-1 and CD44 overexpression in head and neck cancer (Nör et al. 2014). Factors secreted by CSCs promote the differentiation of MDSC population into TAMs (Raggi et al. 2016). Moreover, TAMs undergo the process of 'immune edition' where the potentially danger antitumorigenic macrophage function is switched to protumorigenic immune suppression favoring antitumor immunity evasion (Ramanathan and Jagannathan 2014). TAMs operate various paracrine signaling loops in the microenvironment that facilitates invasion and metastasis. TAMs, in one hand, are shown to secrete EGF that drives tumor progression whereas on the other hand OSCC cells produce CSF-1 that urges further TAM proliferation and tumor infiltration. TAMs are also shown to secrete macrophage migration inhibitory factor (MMIF) that allows neutrophil chemo-attraction and activation further induc-ing the secretion of protumorigenic chemokines like IL-4, IL-8, IL-10, IL-13, TGF-β, TNF-α and GM-CSF that induce immunosuppression (Fig. 11.3) (Markwell and Weed 2015). Furthermore, CAFs promote immunosuppression and evasion of immune surveillance by preferentially inducing the T cell apoptosis and Treg activa-tion in OSCC via the production of a higher level of cytokines like IL-6, CXCL8, TNF, TGFβ1 and VEGFA along with the co-regulatory molecules like B7H1 and B7DC (Fig. 11.3) (Takahashi et al. 2015). The SDF-1/CXCR4 signaling is believed to promote metastasis and therapy resistance attributes of CSCs wherein the CSCs overexpresses CXCR4 and responds to a chemotactic gradient of SDF-1 (Fig. 11.3). In OSCC, SDF-1 induces lymph node metastasis via activation of both the ERK1/2 and Akt/PKB cascades (Uchida et al. 2013).

11.5.1.5 Matrix Remodeling Enzymes and Cancer Stem Cells

Matrix metalloproteinases (MMPs) are important class of zinc-dependent endopep-tidases that are involved in degradation of matrix proteins that facilitates invasion and metastasis. The over expression of MMPs is also reported to induce acquisition of CSC-like properties in oral cancer (Fig. 11.3). Up regulation of membrane type 1 matrix metalloproteinase (MT1-MMP), a cell surface matrix degrading proteinase decreased the expression of epithelial markers (E-cadherin, cytokeratin18 and β-catenin) and increased the expression of mesenchymal markers (vimentin and fibronectin) to promte EMT in SCC-9 cells. Moreover, it increased the level of Twist and ZEB and promoted expression of CSCs surface markers, self-renewal ability, resistance to therapies and apoptosis (Yang et al. 2013). A disintegrin and metalloproteinase domain-containing protein 17 (ADAM17) is recently reported to be associated with metastasis in oral cancer. Increased levels of ADAM17 expression and concomitant CD44 cleavage is shown to be regulating CD44 cleavage which is critical for orasphere formation or stemness and HNSCC tumorigenesis (Fig. 11.3) (Kamarajan et al. 2013).

11.5.2   Cancer Stem Cells Remodel Tumor Microenvironment

Dynamic interplay between CSCs and TAMs modulates the molecular, functional and phenotypic identities of both types of cell. The bidirectional interaction of polarized macrophages with stem and progenitor cells plays a significant role in tissue repair and remodeling by tumor-promoting conditions in the TME (Fig. 11.3). The tumor cell produces CSF-1, IL-10, chemokines (CCL2, CCL18, CCL17 and CXCL4) and ECM components and induces M2-like polarization of macrophages promoting cancer progression (Fig. 11.3) (Raggi et al. 2016). In HNSCC, CSCs overexpresses IL-6 R which upon interaction with IL-6 promotes tumorigenesis (Nör et al. 2014). CSCs releases Peristonin that act on $\alpha 3\beta v$ integrin to recruit TAMs. Moreover, it is also reported that M-CSF secreted by CSCs promotes the differentiation of MDSCs to TAMs (Fig. 11.3) (Raggi et al. 2016). It also shows that CSCs secrete VEGF to promote neoangiogensis and support a local vascular environment (Gilbertson and Rich 2007). The CSCs are reported to remodel the extracellular matrix by promoting the degradation of matrix protein via various proteinases which facilitates the migration of CSCs to distant sites. Hyaluronic acid (HA), an ECM component upon interaction with stem cell surface marker CD44 initiates multiple signaling cascades augmenting the oral tumor progression. The transmembrane protein CD44 is also a co-receptor for several receptor tyrosine kinases (RTKs) including c-MET and EGFR. In oral cancer, ERK1/2 interaction with CD44 induces tumor aggressiveness (Judd et al. 2012). Moreover, HA synthesizing enzyme hyaluronan synthase 2 (HAS2) are highly synthesized in oral cancer and HAS2 down regulation leads to CD44 dependent decrease of tumor cell migration (Wang et al. 2013). The CSCs in HNSCC are characterized as high CD44 expressing phenotypes which is a docking receptor necessary for MMP-9 (Zhang et al. 2012). Over expression of transcription factor Snail promotes EMT in cancer cells to endorse stem-like properties and in this context snail mediated overexpression of MMPs is shown to be associated with the high invasion capacity in UMSCC1 cell lines (Lin et al. 2011). Moreover, EMMPRIN or CD147 (Extracellular matrix metallo-protease inducer) is a cell surface protein and oral stem cell marker which mediates ECM remodeling during invasion and metastasis via MMP induction (Huang et al. 2013). In HNSCC, EMMPRIN-2 overexpression induces secretion of MMP-2, uPA, Cathepsin which in turn promotes ECM remodelling and angiogenesis thereby paving the way towards invasion and metastasis (Huang et al. 2014).

11.6 Conclusion and Prospective

The molecular crosstalk between cancer stem cells and tumor microenvironment play an important role in formation of intratumoral heterogeneity to modulate tumor progression in oral cancer. TME accommodates cancer cells including CSCs and creates an immunosuppressive atmosphere that safeguards CSCs from various genetic and epigenetic offenses. Tumor niches are discrete domains that assure specific functional characteristics and encourage the formation of suitable microhabitat for CSCs to maintain its stemness, invasiveness, metastatic proficiency, tumorigenicity and therapy resistance properties. Moreover, specific stimuli including hypoxia, autophagy, tumor metabolism and tumor associated stromal cells in the microenvironment stimulate reprogramming of signaling pathways in the way to acquire stem-like characteristics. In HNSCC, CSCs express specific receptors, cytokines and chemokines to recruit TAMs, promotes angiogenesis, differentiation of MDSCs to TAMs to support immune evasion and tumor growth. In addition, CSCs triggers extensive remodeling of ECM by inducing the degradation of matrix protein via various proteinases which facilitates invasion and metastasis. Although the molecular interplay between CSCs and TME is well established, many cellular events in this microhabitat remain unidentified. How CSCs especially circulating tumor cells influence the stromal cells at new sites during metastasis and what is molecular signaling for polarizing tumor associated cells? Does high autophagy flux in CSCs associate intracellular antigen presentation to reprogram in generating tumor promoting macrophages and other immune cells? It is also not known what is the detail molecular circuit in establishing connection between CSCs and epidermal keratinocytes during tumor progression in oral cancer. In conclusion, understanding the CSCs and its tumor microenvironment with identification of key molecular pathway might facilitate the development strategy to improve therapeutic outcomes through precise intervention to treat oral cancer.

Acknowledgement Research support was partly provided by Department of Biotechnology [Grant Number: BT/PR7791/BRB/10/1187/2013]; Science and Engineering Research Board (SERB), Department of Science and Technology [Grant Number: SR/SO/BB-0101/2012]; Council of Scientific and Industrial Research (CSIR) [Grant Number: 37(1608)/13/EMR-II] Human Resource Development Group, Government of India; Science and Technology Department, Government of Odisha. PPN is obliged to DST SERB, New Delhi, India for providing fellowship.

Conflict of Interest
The authors declare no conflicts of interest.

References

Akino T, Hida K, Hida Y, Tsuchiya K, Freedman D, Muraki C et al (2009) Cytogenetic abnormalities of tumor-associated endothelial cells in human malignant tumors. Am J Pathol 175:2657–2667

Al-Hajj M, Wicha MS, Benito-Hernandez A, Morrison SJ, Clarke MF (2003) Prospective identification of tumorigenic breast cancer cells. Proc Natl Acad Sci U S A 100:3983–3988

Almendro V, Marusyk A, Polyak K (2013) Cellular heterogeneity and molecular evolution in cancer. Annu Rev Pathol 8:277–302

An Y, Kiang A, Lopez JP, Kuo SZ, Yu MA, Abhold EL et al (2012) Cigarette smoke promotes drug resistance and expansion of cancer stem cell-like side population. PLoS One 7:e47919

Anderson AR, Weaver AM, Cummings PT, Quaranta V (2006) Tumor morphology and phenotypic evolution driven by selective pressure from the microenvironment. Cell 127:905–915

Bagordakis E, Sawazaki-Calone I, Macedo CCS, Carnielli CM, de Oliveira CE, Rodrigues PC et al (2016) Secretome profiling of oral squamous cell carcinoma-associated fibroblasts reveals organization and disassembly of extracellular matrix and collagen metabolic process signatures. Tumor Biol 37:9045–9057

Balkwill F, Mantovani A (2001) Inflammation and cancer: back to Virchow? Lancet 357:539–545

Barth PJ, Zu Schweinsberg TS, Ramaswamy A, Moll R (2004) CD34+ fibrocytes, α-smooth muscle antigen-positive myofibroblasts, and CD117 expression in the stroma of invasive squamous cell carcinomas of the oral cavity, pharynx, and larynx. Virchows Arch 444:231–234

Bhowmick NA, Neilson EG, Moses HL (2004) Stromal fibroblasts in cancer initiation and progression. Nature 432:332–337

Bhutia SK, Mukhopadhyay S, Sinha N, Das DN, Panda PK, Patra SK et al (2013) Autophagy: cancer's friend or foe? Adv Cancer Res 118:61

Brandwein-Gensler M, Teixeira MS, Lewis CM, Lee B, Rolnitzky L, Hille JJ et al (2005) Oral squamous cell carcinoma: histologic risk assessment, but not margin status, is strongly predictive of local disease-free and overall survival. Am J Surg Pathol 29:167–178

Brennan P, Mackenzie N, Quintero M (2005) Hypoxia-inducible factor 1α in oral cancer. J Oral Pathol Med 34:385–389

Brouty-Boyé D (2005) Developmental biology of fibroblasts and neoplastic disease. In: Developmental biology of neoplastic growth. Springer, Berlin, pp 55–77

Cabrera MC, Hollingsworth RE, Hurt EM (2015) Cancer stem cell plasticity and tumor hierarchy. World J Stem Cells 7:27–36

Capparelli C, Whitaker-Menezes D, Guido C, Balliet R, Pestell TG, Howell A et al (2012) CTGF drives autophagy, glycolysis and senescence in cancer-associated fibroblasts via HIF1 activation, metabolically promoting tumor growth. Cell Cycle 11:2272–2284

Castaño Z, Fillmore CM, Kim CF, McAllister SS (2012) The bed and the bugs: interactions between the tumor microenvironment and cancer stem cells. Semin Cancer Biol 22:462–470

Chaffer CL, Brueckmann I, Scheel C, Kaestli AJ, Wiggins PA, Rodrigues LO et al (2011) Normal and neoplastic nonstem cells can spontaneously convert to a stem-like state. Proc Natl Acad Sci U S A 108:7950–7955

Chaponnier C, Desmoulière A, Gabbiani G (2006) Tissue repair, contraction and the myofibroblast. Springer, US

Chen YJ, Chang JTC, Liao CT, Wang HM, Yen TC, Chiu CC et al (2008) Head and neck cancer in the betel quid chewing area: recent advances in molecular carcinogenesis. Cancer Sci 99:1507–1514

Chikamatsu K, Takahashi G, Sakakura K, Ferrone S, Masuyama K (2011) Immunoregulatory properties of CD44+ cancer stem-like cells in squamous cell carcinoma of the head and neck. Head Neck 33:208–215

Chiu KC, Lee CH, Liu SY, Chou YT, Huang RY, Huang SM et al (2015) Polarization of tumor-associated macrophages and Gas6/Axl signaling in oral squamous cell carcinoma. Oral Oncol 51:683–689

Cho YA, Yoon HJ, Lee JI, Hong SP, Hong SD (2011) Relationship between the expressions of PD-L1 and tumor-infiltrating lymphocytes in oral squamous cell carcinoma. Oral Oncol 47:1148–1153

Costa NL, Valadares MC, Souza PPC, Mendonça EF, Oliveira JC, Silva TA et al (2013) Tumor-associated macrophages and the profile of inflammatory cytokines in oral squamous cell carcinoma. Oral Oncol 49:216–223

Costea D, Tsinkalovsky O, Vintermyr O, Johannessen A, Mackenzie I (2006) Cancer stem cells–new and potentially important targets for the therapy of oral squamous cell carcinoma. Oral Dis 12:443–454

Coussens LM, Werb Z (2002) Inflammation and cancer. Nature 420:860–867

Curry JM, Sprandio J, Cognetti D, Luginbuhl A, Bar-ad V, Pribitkin E et al (2014) Tumor microenvironment in head and neck squamous cell carcinoma. Semin Oncol 41(2):217–234

Daly AJ, McIlreavey L, Irwin CR (2008) Regulation of HGF and SDF-1 expression by oral fibroblasts–implications for invasion of oral cancer. Oral Oncol 44:646–651

De Wever O, Demetter P, Mareel M, Bracke M (2008) Stromal myofibroblasts are drivers of invasive cancer growth. Int J Cancer 123:2229–2238

Dick D (1997) Human acute myeloid leukemia is organized as a hierarchy that originates from a primitive hematopoietic cell. Nat Med 3:730–737

Dong C, Yuan T, Wu Y, Wang Y, Fan TW, Miriyala S et al (2013) Loss of FBP1 by Snail-mediated repression provides metabolic advantages in basal-like breast cancer. Cancer Cell 23:316–331

Duarte S, Loubat A, Momier D, Topi M, Faneca H, Pedroso de Lima MC et al (2012) Isolation of head and neck squamous carcinoma cancer stem-like cells in a syngeneic mouse model and analysis of hypoxia effect. Oncol Rep 28:1057–1062

Duffy SA, Taylor JM, Terrell JE, Islam M, Li Y, Fowler KE et al (2008) Interleukin-6 predicts recurrence and survival among head and neck cancer patients. Cancer 113:750–757

Dumitru CA, Gholaman H, Trellakis S, Bruderek K, Dominas N, Gu X et al (2011) Tumor-derived macrophage migration inhibitory factor modulates the biology of head and neck cancer cells via neutrophil activation. Int J Cancer 129:859–869

Egeblad M, Nakasone ES, Werb Z (2010) Tumors as organs: complex tissues that interface with the entire organism. Dev Cell 18:884–901

El-Rouby DH (2010) Association of macrophages with angiogenesis in oral verrucous and squamous cell carcinomas. J Oral Pathol Med 39:559–564

Fan S, Tang QL, Lin YJ, Wl C, Li JS, Huang ZQ et al (2011) A review of clinical and histological parameters associated with contralateral neck metastases in oral squamous cell carcinoma. Int J Oral Sci 3:180–191

Ferlay J, Bray F, Forman D, Mathers C, Parkin D (2010) Cancer incidence and mortality worldwide: IARC CancerBase No. 10. International Agency for Research on Cancer, Lyon, France [cited 2012 04/04/2012]

Ferlay J, Soerjomataram I, Ervik M, Dikshit R, Eser S, Mather C (2015) Globocan 2012 version 1.0: cancer incidence and mortality worldwide—IARC CancerBase No. 11. International Agency for Research on Cancer, Lyon, France

Ferris RL, Whiteside TL, Ferrone S (2006) Immune escape associated with functional defects in antigen-processing machinery in head and neck cancer. Clin Cancer Res 12:3890–3895

Folkins C, Man S, Xu P, Shaked Y, Hicklin DJ, Kerbel RS (2007) Anticancer therapies combining antiangiogenic and tumor cell cytotoxic effects reduce the tumor stem-like cell fraction in glioma xenograft tumors. Cancer Res 67:3560–3564

Folkman J (2002) Role of angiogenesis in tumor growth and metastasis. Semin Oncol 29:15–18

Fridman WH, Pages F, Sautes-Fridman C, Galon J (2012) The immune contexture in human tumours: impact on clinical outcome. Nat Rev Cancer 12:298–306

Friedl P, Alexander S (2011) Cancer invasion and the microenvironment: plasticity and reciprocity. Cell 147:992–1009

Fuchs E, Tumbar T, Guasch G (2004) Socializing with the neighbors: stem cells and their niche. Cell 116:769–778

Fullár A, Kovalszky I, Bitsche M, Romani A, Schartinger VH, Sprinzl GM et al (2012) Tumor cell and carcinoma-associated fibroblast interaction regulates matrix metalloproteinases and their inhibitors in oral squamous cell carcinoma. Exp Cell Res 318:1517–1527

Galdiero MR, Garlanda C, Jaillon S, Marone G, Mantovani A (2013) Tumor associated macrophages and neutrophils in tumor progression. J Cell Physiol 228:1404–1412

Gilbertson RJ, Rich JN (2007) Making a tumour's bed: glioblastoma stem cells and the vascular niche. Nat Rev Cancer 7:733–736

Gonçalves AS, Costa NL, Arantes DAC, Cássia Gonçalves Alencar R, Silva TA, Batista AC (2013) Immune response in cervical lymph nodes from patients with primary oral squamous cell carcinoma. J Oral Pathol Med 42:535–540

Grandis JR, Falkner DM, Melhem MF, Gooding WE, Drenning SD, Morel PA (2000) Human leukocyte antigen class I allelic and haplotype loss in squamous cell carcinoma of the head and neck: clinical and immunogenetic consequences. Clin Cancer Res 6:2794–2802

Hanahan D, Coussens LM (2012) Accessories to the crime: functions of cells recruited to the tumor microenvironment. Cancer Cell 21:309–322

Harada T, Shinohara M, Nakamura S, Oka M (1994) An immunohistochemical study of the extracellular matrix in oral squamous cell carcinoma and its association with invasive and metastatic potential. Virchows Arch 424:257–266

He KF, Zhang L, Huang CF, Ma SR, Wang YF, Wang WM et al (2014) CD163+ tumor-associated macrophages correlated with poor prognosis and cancer stem cells in oral squamous cell carcinoma. Biomed Res Int 2014:838632

Heddleston J, Li Z, Lathia J, Bao S, Hjelmeland A, Rich J (2010) Hypoxia inducible factors in cancer stem cells. Br J Cancer 102:789–795

Hida K, Hida Y, Amin DN, Flint AF, Panigrahy D, Morton CC et al (2004) Tumor-associated endothelial cells with cytogenetic abnormalities. Cancer Res 64:8249–8255

Hjelmeland AB, Wu Q, Heddleston J, Choudhary G, MacSwords J, Lathia JD et al (2011) Acidic stress promotes a glioma stem cell phenotype. Cell Death Differ 18:829–840

Holohan C, Van Schaeybroeck S, Longley DB, Johnston PG (2013) Cancer drug resistance: an evolving paradigm. Nat Rev Cancer 13:714–726

Hovinga KE, Shimizu F, Wang R, Panagiotakos G, Van Der Heijden M, Moayedpardazi H et al (2010) Inhibition of notch signaling in glioblastoma targets cancer stem cells via an endothelial cell intermediate. Stem Cells 28:1019–1029

Hu Y, He MY, Zhu LF, Yang CC, Zhou ML, Wang Q et al (2016) Tumor-associated macrophages correlate with the clinicopathological features and poor outcomes via inducing epithelial to mesenchymal transition in oral squamous cell carcinoma. J Exp Clin Cancer Res 35:12

Huang Z, Wang L, Wang Y, Zhuo Y, Li H, Chen J et al (2013) Overexpression of CD147 contributes to the chemoresistance of head and neck squamous cell carcinoma cells. J Oral Pathol Med 42:541–546

Huang Z, Tan N, Guo W, Wang L, Li H, Zhang T et al (2014) Overexpression of EMMPRIN isoform 2 is associated with head and neck cancer metastasis. PLoS One 9:e91596

Huntly BJ, Gilliland DG (2005) Leukaemia stem cells and the evolution of cancer-stem-cell research. Nat Rev Cancer 5:311–321

Ishikawa T, Nakashiro K, Klosek SK, Goda H, Hara S, Uchida D et al (2009) Hypoxia enhances CXCR4 expression by activating HIF-1 in oral squamous cell carcinoma. Oncol Rep 21:707–712

Jiang J, Tang YL, Liang XH (2011) EMT: a new vision of hypoxia promoting cancer progression. Cancer Biol Ther 11:714–723

Johansson N, Airola K, Grenman R, Kariniemi A-L, Saarialho-Kere U, Kähäri V (1997) Expression of collagenase-3 (matrix metalloproteinase-13) in squamous cell carcinomas of the head and neck. Am J Pathol 151:499

Jonuleit H, Schmitt E, Schuler G, Knop J, Enk AH (2000) Induction of interleukin 10–producing, nonproliferating CD4+ T cells with regulatory properties by repetitive stimulation with allogeneic immature human dendritic cells. J Exp Med 192:1213–1222

Judd NP, Winkler AE, Murillo-Sauca O, Brotman JJ, Law JH, Lewis JS et al (2012) ERK1/2 regulation of CD44 modulates oral cancer aggressiveness. Cancer Res 72:365–374

Kamarajan P, Shin JM, Qian X, Matte B, Zhu JY, Kapila YL (2013) ADAM17-mediated CD44 cleavage promotes orasphere formation or stemness and tumorigenesis in HNSCC. Cancer Med 2:793–802

Karin M, Greten FR (2005) NF-kappaB: linking inflammation and immunity to cancer development and progression. Nat Rev Immunol 5:749–759

Kawashiri S, Tanaka A, Noguchi N, Hase T, Nakaya H, Ohara T et al (2009) Significance of stromal desmoplasia and myofibroblast appearance at the invasive front in squamous cell carcinoma of the oral cavity. Head Neck 31:1346–1353

Keith B, Simon MC (2007) Hypoxia-inducible factors, stem cells, and cancer. Cell 129:465–472

Kellermann MG, Sobral LM, da Silva SD, Zecchin KG, Graner E, Lopes MA et al (2008) Mutual paracrine effects of oral squamous cell carcinoma cells and normal oral fibroblasts: induction of fibroblast to myofibroblast transdifferentiation and modulation of tumor cell proliferation. Oral Oncol 44:509–517

Kim CFB, Jackson EL, Woolfenden AE, Lawrence S, Babar I, Vogel S et al (2005) Identification of bronchioalveolar stem cells in normal lung and lung cancer. Cell 121:823–835

Krishnamurthy S, Dong Z, Vodopyanov D, Imai A, Helman JI, Prince ME et al (2010) Endothelial cell-initiated signaling promotes the survival and self-renewal of cancer stem cells. Cancer Res 70:9969–9978

Lee E, Yang J, Ku M, Kim N, Park Y, Park C et al (2015) Metabolic stress induces a Wnt-dependent cancer stem cell-like state transition. Cell Death Dis 6:e1805

Lee Y, Shin JH, Longmire M, Wang H, Kohrt HE, Chang HY et al (2016) CD44+ Cells in Head and Neck Squamous Cell Carcinoma Suppress T-Cell–Mediated Immunity by Selective Constitutive and Inducible Expression of PD-L1. Clin Cancer Res 22:3571–3581

Leef G, Thomas SM (2013) Molecular communication between tumor-associated fibroblasts and head and neck squamous cell carcinoma. Oral Oncol 49:381–386

Leemans CR, Tiwari R, Nauta JJ, Waal IVD, Snow GB (1994) Recurrence at the primary site in head and neck cancer and the significance of neck lymph node metastases as a prognostic factor. Cancer 73:187–190

Lessard J, Sauvageau G (2003) Bmi-1 determines the proliferative capacity of normal and leukaemic stem cells. Nature 423:255–260

Lewis M, Lygoe K, Nystrom M, Anderson W, Speight P, Marshall J et al (2004) Tumour-derived TGF-β1 modulates myofibroblast differentiation and promotes HGF/SF-dependent invasion of squamous carcinoma cells. Br J Cancer 90:822–832

Li C, Shintani S, Terakado N, Nakashiro K, Hamakawa H (2002) Infiltration of tumor-associated macrophages in human oral squamous cell carcinoma. Oncol Rep 9:1219–1223

Li C, Heidt DG, Dalerba P, Burant CF, Zhang L, Adsay V et al (2007) Identification of pancreatic cancer stem cells. Cancer Res 67:1030–1037

Li DW, Dong P, Wang F, Chen XW, Xu CZ, Zhou L (2013) Hypoxia induced multidrug resistance of laryngeal cancer cells via hypoxia-inducible factor-1α. Asian Pac J Cancer Prev 14(8):4853

Licitra L, Perrone F, Bossi P, Suardi S, Mariani L, Artusi R et al (2006) High-risk human papillomavirus affects prognosis in patients with surgically treated oropharyngeal squamous cell carcinoma. J Clin Oncol 24:5630–5636

Lim KP, Cirillo N, Hassona Y, Wei W, Thurlow JK, Cheong SC et al (2011) Fibroblast gene expression profile reflects the stage of tumour progression in oral squamous cell carcinoma. J Pathol 223:459–469

Lin Q, Yun Z (2010) Impact of the hypoxic tumor microenvironment on the regulation of cancer stem cell characteristics. Cancer Biol Ther 9:949–956

Lin CY, Tsai PH, Kandaswami CC, Lee PP, Huang CJ, Hwang JJ et al (2011) Matrix metalloproteinase-9 cooperates with transcription factor Snail to induce epithelial–mesenchymal transition. Cancer Sci 102:815–827

Liu SY, Chang LC, Pan LF, Hung YJ, Lee CH, Shieh YS (2008) Clinicopathologic significance of tumor cell-lined vessel and microenvironment in oral squamous cell carcinoma. Oral Oncol 44:277–285

Lúcio PSC, Ribeiro DC, Aguiar MC, Alves PM, Nonaka CFW, Godoy GP (2016) Tumor-associated macrophages (TAMs): clinical-pathological parameters in squamous cell carcinomas of the lower lip. Braz Oral Res 30:e95

Lyons A, Jones J (2007) Cell adhesion molecules, the extracellular matrix and oral squamous carcinoma. Int J Oral Maxillofac Surg 36:671–679

Marcus B, Arenberg D, Lee J, Kleer C, Chepeha DB, Schmalbach CE et al (2004) Prognostic factors in oral cavity and oropharyngeal squamous cell carcinoma. Cancer 101:2779–2787

Marjanovic ND, Weinberg RA, Chaffer CL (2013) Cell plasticity and heterogeneity in cancer. Clin Chem 59:168–179

Markwell SM, Weed SA (2015) Tumor and stromal-based contributions to head and neck squamous cell carcinoma invasion. Cancers (Basel) 7:382–406

Marron M, Boffetta P, Zhang ZF, Zaridze D, Wünsch-Filho V, Winn DM et al (2010) Cessation of alcohol drinking, tobacco smoking and the reversal of head and neck cancer risk. Int J Epidemiol 39:182–196

Marsh D, Suchak K, Moutasim KA, Vallath S, Hopper C, Jerjes W et al (2011) Stromal features are predictive of disease mortality in oral cancer patients. J Pathol 223:470–481

Martinez FO, Gordon S (2014) The M1 and M2 paradigm of macrophage activation: time for reassessment. F1000Prime Rep 6:12703

Marusyk A, Almendro V, Polyak K (2012) Intra-tumour heterogeneity: a looking glass for cancer? Nat Rev Cancer 12:323–334

McGranahan N, Swanton C (2017) Clonal heterogeneity and tumor evolution: past, present, and the future. Cell 168:613–628

Meacham CE, Morrison SJ (2013) Tumour heterogeneity and cancer cell plasticity. Nature 501:328–337

Morrison SJ, Spradling AC (2008) Stem cells and niches: mechanisms that promote stem cell maintenance throughout life. Cell 132:598–611

Mueller MM, Fusenig NE (2002) Tumor-stroma interactions directing phenotype and progression of epithelial skin tumor cells. Differentiation 70:486–497

Naik PP, Das DN, Panda PK, Mukhopadhyay S, Sinha N, Praharaj PP et al (2016) Implications of cancer stem cells in developing therapeutic resistance in oral cancer. Oral Oncol 62:122–135

Neiva KG, Zhang Z, Miyazawa M, Warner KA, Karl E, Nör JE (2009) Cross talk initiated by endothelial cells enhances migration and inhibits anoikis of squamous cell carcinoma cells through STAT3/Akt/ERK signaling. Neoplasia 11:583IN12–593IN14

Nör C, Zhang Z, Warner KA, Bernardi L, Visioli F, Helman JI et al (2014) Cisplatin induces Bmi-1 and enhances the stem cell fraction in head and neck cancer. Neoplasia 16:137–146

Nowell PC (1976) The clonal evolution of tumor cell populations. Science 194:23–28

Ogino T, Shigyo H, Ishii H, Katayama A, Miyokawa N, Harabuchi Y et al (2006) HLA class I antigen down-regulation in primary laryngeal squamous cell carcinoma lesions as a poor prognostic marker. Cancer Res 66:9281–9289

Okubo M, Kioi M, Nakashima H, Sugiura K, Mitsudo K, Aoki I et al (2016) M2-polarized macrophages contribute to neovasculogenesis, leading to relapse of oral cancer following radiation. Sci Rep 6:27548

Paget S (1889) The distribution of secondary growths in cancer of the breast. Lancet 133:571–573

Pavlides S, Tsirigos A, Migneco G, Whitaker-Menezes D, Chiavarina B, Flomenberg N et al (2010) The autophagic tumor stroma model of cancer: Role of oxidative stress and ketone production in fueling tumor cell metabolism. Cell Cycle 9:3485–3505

Pisco A, Huang S (2015) Non-genetic cancer cell plasticity and therapy-induced stemness in tumour relapse:'what does not kill me strengthens me'. Br J Cancer 112:1725–1732

Powell DW, Adegboyega PA, Di Mari JF, Mifflin RC (2005) Epithelial cells and their neighbors I. Role of intestinal myofibroblasts in development, repair, and cancer. Am J Physiol Gastrointest Liver Physiol 289:G2–G7

Prince M, Sivanandan R, Kaczorowski A, Wolf G, Kaplan M, Dalerba P et al (2007) Identification of a subpopulation of cells with cancer stem cell properties in head and neck squamous cell carcinoma. Proc Natl Acad Sci U S A 104:973–978

Punt S, Dronkers EA, Welters MJ, Goedemans R, Koljenović S, Bloemena E et al (2016) A beneficial tumor microenvironment in oropharyngeal squamous cell carcinoma is characterized by a high T cell and low IL-17+ cell frequency. Cancer Immunol Immunother 65:393–403

Qing G, Simon MC (2009) Hypoxia inducible factor-2α: a critical mediator of aggressive tumor phenotypes. Curr Opin Genet Dev 19:60–66

Quan H, Fang L, Pan H, Deng Z, Gao S, Liu O et al (2016) An adaptive immune response driven by mature, antigen-experienced T and B cells within the microenvironment of oral squamous cell carcinoma. Int J Cancer 138(12):2952–2962

Quintana E, Shackleton M, Sabel MS, Fullen DR, Johnson TM, Morrison SJ (2008) Efficient tumour formation by single human melanoma cells. Nature 456:593–598

Raggi C, Mousa H, Correnti M, Sica A, Invernizzi P (2016) Cancer stem cells and tumor-associated macrophages: a roadmap for multitargeting strategies. Oncogene 35:671–682

Ramanathan S, Jagannathan N (2014) Tumor associated macrophage: a review on the phenotypes, traits and functions. Iran J Cancer Prev 7:1–8

Ramos DM, Chen BL, Boylen K, Stern M, Kramer RH, Sheppard D et al (1997) Stromal fibroblasts influence oral squamous-cell carcinoma cell interactions with tenascin-C. Int J Cancer 72:369–376

Rampias T, Boutati E, Pectasides E, Sasaki C, Kountourakis P, Weinberger P et al (2010) Activation of Wnt signaling pathway by human papillomavirus E6 and E7 oncogenes in HPV16-positive oropharyngeal squamous carcinoma cells. Mol Cancer Res 8:433–443

Reya T, Clevers H (2005) Wnt signalling in stem cells and cancer. Nature 434:843–850

Roy A, Bera S (2016) CAF cellular glycolysis: linking cancer cells with the microenvironment. Tumor Biol 37:8503–8514

Senovilla L, Vacchelli E, Galon J, Adjemian S, Eggermont A, Fridman WH et al (2012) Trial watch: Prognostic and predictive value of the immune infiltrate in cancer. Oncoimmunology 1:1323–1343

Shackleton M, Quintana E, Fearon ER, Morrison SJ (2009) Heterogeneity in cancer: cancer stem cells versus clonal evolution. Cell 138:822–829

Shieh YS, Hung YJ, Hsieh CB, Chen JS, Chou KC, Liu SY (2009) Tumor-associated macrophage correlated with angiogenesis and progression of mucoepidermoid carcinoma of salivary glands. Ann Surg Oncol 16:751–760

Sica A, Schioppa T, Mantovani A, Allavena P (2006) Tumour-associated macrophages are a distinct M2 polarised population promoting tumour progression: potential targets of anti-cancer therapy. Eur J Cancer 42:717–727

Singh SK, Clarke ID, Terasaki M, Bonn VE, Hawkins C, Squire J et al (2003) Identification of a cancer stem cell in human brain tumors. Cancer Res 63:5821–5828

Sinha N, Mukhopadhyay S, Das DN, Panda PK, Bhutia SK (2013) Relevance of cancer initiating/stem cells in carcinogenesis and therapy resistance in oral cancer. Oral Oncol 49:854–862

Sobral LM, Bufalino A, Lopes MA, Graner E, Salo T, Coletta RD (2011) Myofibroblasts in the stroma of oral cancer promote tumorigenesis via secretion of activin A. Oral Oncol 47:840–846

Sottoriva A, Spiteri I, Piccirillo SG, Touloumis A, Collins VP, Marioni JC et al (2013) Intratumor heterogeneity in human glioblastoma reflects cancer evolutionary dynamics. Proc Natl Acad Sci U S A 110:4009–4014

Sparmann A, Bar-Sagi D (2004) Ras-induced interleukin-8 expression plays a critical role in tumor growth and angiogenesis. Cancer Cell 6:447–458

Takahashi H, Sakakura K, Kawabata-Iwakawa R, Rokudai S, Toyoda M, Nishiyama M et al (2015) Immunosuppressive activity of cancer-associated fibroblasts in head and neck squamous cell carcinoma. Cancer Immunol Immunother 64(11):1–11

Takiar R, Nadayil D, Nandakumar A (2010) Projections of number of cancer cases in India (2010-2020) by cancer groups. Asian Pac J Cancer Prev 11:1045–1049

Tlsty TD, Coussens LM (2006) Tumor stroma and regulation of cancer development. Annu Rev Pathol 1:119–150

Tourkova IL, Shurin GV, Chatta GS, Perez L, Finke J, Whiteside TL et al (2005) Restoration by IL-15 of MHC class I antigen-processing machinery in human dendritic cells inhibited by tumor-derived gangliosides. J Immunol 175:3045–3052

Uchida D, Kuribayashi N, Kinouchi M, Ohe G, Tamatani T, Nagai H et al (2013) Expression and function of CXCR4 in human salivary gland cancers. Clin Exp Metastasis 30:133–142

Vaupel P, Mayer A (2007) Hypoxia in cancer: significance and impact on clinical outcome. Cancer Metastasis Rev 26:225–239

Vered M, Dayan D, Yahalom R, Dobriyan A, Barshack I, Bello IO et al (2010) Cancer-associated fibroblasts and epithelial-mesenchymal transition in metastatic oral tongue squamous cell carcinoma. Int J Cancer 127:1356–1362

Vigneswaran N, Wu J, Song A, Annapragada A, Zacharias W (2011) Hypoxia-induced autophagic response is associated with aggressive phenotype and elevated incidence of metastasis in orthotopic immunocompetent murine models of head and neck squamous cell carcinomas (HNSCC). Exp Mol Pathol 90:215–225

Visvader JE, Lindeman GJ (2008) Cancer stem cells in solid tumours: accumulating evidence and unresolved questions. Nat Rev Cancer 8:755–768

Wang M, Li X, Lu X, Qu Y, Xu O, Sun Q (2011) Cancer stem cells promotes resistance of laryngeal squamous cancer to irradiation mediated by hypoxia. Lin chuang er bi yan hou tou jing wai ke za zhi. J Clin Otorhinolaryngol Head Neck Surg 25:823–826

Wang SJ, Earle C, Wong G, Bourguignon LY (2013) Role of hyaluronan synthase 2 to promote CD44-dependent oral cavity squamous cell carcinoma progression. Head Neck 35:511–520

Wang TY, Peng CY, Lee SS, Chou MY, Yu CC, Chang YC (2016) Acquisition cancer stemness, mesenchymal transdifferentiation, and chemoresistance properties by chronic exposure of oral epithelial cells to arecoline. Oncotarget 7:84072–84081

Westermarck J, Li S, Jaakkola P, Kallunki T, Grénman R, Kähäri VM (2000) Activation of fibroblast collagenase-1 expression by tumor cells of squamous cell carcinomas is mediated by p38 mitogen-activated protein kinase and c-Jun NH2-terminal kinase-2. Cancer Res 60:7156–7162

Whiteside TL (2006a) The role of immune cells in the tumor microenvironment. In: The link between inflammation and cancer. Springer, Boston, pp 103–124

Whiteside TL (2006b) Immune suppression in cancer: effects on immune cells, mechanisms and future therapeutic intervention. Semin Cancer Biol 16:3–15

Winquist RJ, Boucher DM, Wood M, Furey BF (2009) Targeting cancer stem cells for more effective therapies: Taking out cancer's locomotive engine. Biochem Pharmacol 78:326–334

Wu CP, Du HD, Gong HL, Li DW, Tao L, Tian J et al (2014) Hypoxia promotes stem-like properties of laryngeal cancer cell lines by increasing the CD133+ stem cell fraction. Int J Oncol 44:1652–1660

Xie T, Li L (2007) Stem cells and their niche: an inseparable relationship. Development 134:2001–2006

Yang CC, Zhu LF, Xu XH, Ning TY, Ye JH, Liu LK (2013) Membrane Type 1 Matrix Metalloproteinase induces an epithelial to mesenchymal transition and cancer stem cell-like properties in SCC9 cells. BMC Cancer 13:171

Ye X, Zhang J, Lu R, Zhou G (2016) Signal regulatory protein alpha associated with the progression of oral leukoplakia and oral squamous cell carcinoma regulates phenotype switch of macrophages. Oncotarget 7:81305–81321

Young MRI (2006) Protective mechanisms of head and neck squamous cell carcinomas from immune assault. Head Neck 28:462–470

Yun CO (2008) Overcoming the extracellular matrix barrier to improve intratumoral spread and therapeutic potential of oncolytic virotherapy. Curr Opin Mol Ther 10:356–361

Zamarron BF, Chen W (2011) Dual roles of immune cells and their factors in cancer development and progression. Int J Biol Sci 7:651–658

Zeng Q, Li S, Chepeha DB, Giordano TJ, Li J, Zhang H et al (2005) Crosstalk between tumor and endothelial cells promotes tumor angiogenesis by MAPK activation of Notch signaling. Cancer Cell 8:13–23

Zhang Z, Sant'Ana Filho M, Nör JE (2012) The biology of head and neck cancer stem cells. Oral Oncol 48:1–9

Zhang H, Wu H, Zheng J, Yu P, Xu L, Jiang P et al (2013) Transforming growth factor β1 signal is crucial for dedifferentiation of cancer cells to cancer stem cells in osteosarcoma. Stem Cells 31:433–446

Zhou BB, Zhang H, Damelin M, Geles KG, Grindley JC, Dirks PB (2009) Tumour-initiating cells: challenges and opportunities for anticancer drug discovery. Nat Rev Drug Discov 8:806–823

Ziober AF, Falls EM, Ziober BL (2006) The extracellular matrix in oral squamous cell carcinoma: friend or foe? Head Neck 28:740–749

Chapter 12
Fetal Membranes-Derived Stem Cells Microenvironment

Phelipe Oliveira Favaron and Maria Angelica Miglino

Abstract Recently, the regenerative medicine has been trying to congregate different areas such as tissue engineering and cellular therapy, in order to offer effective treatments to overcome several human and veterinary medical problems. In this regard, fetal membranes have been proposed as a powerful source for obtainment of multipotent stem cells with low immunogenicity, anti-inflammatory properties and nontumorigenicity properties for the treatment of several diseases, including replacing cells lost due to tissue injuries or degenerative diseases. Morpho-physiological data have shown that fetal membranes, especially the yolk sac and amnion play different functions according to the gestational period, which are direct related to the features of the microenvironment that their cells are subject. The characteristics of the microenvironment affect or controls important cellular events involved with proliferation, division and maintenance of the undifferentiated stage or differentiation, especially acting on the extracellular matrix components. Considering the importance of the microenvironment and the diversity of embryonic and fetal membrane-derived stem cells, this chapter will addressed advances in the isolation, phenotyping, characteristics of the microenvironment, and applications of yolk sac and amniotic membrane-derived stem cells for human and veterinary regenerative medicine.

Keywords Yolk sac • Amnion • Amniotic membrane • Mesenchymal stem cells • Extracellular matrix • Cell therapy

P.O. Favaron • M.A. Miglino (✉)
Surgery Department, School of Veterinary Medicine and Animal Science,
University of Sao Paulo, Sao Paulo, SP, Brazil
e-mail: miglino@usp.br

© Springer International Publishing AG 2017
A. Birbrair (ed.), *Stem Cell Microenvironments and Beyond,*
Advances in Experimental Medicine and Biology 1041,
DOI 10.1007/978-3-319-69194-7_12

12.1 Introduction

Regenerative medicine is defined as a branch of translational research that congregates important areas related to tissue engineering and cellular therapy (Mason and Dunnill 2008). When combined they provide promising alternatives to overcome several human and veterinary medical problems related to the replacement of tissues and organs that have been damage by diseases, trauma and congenital issues, which in the present scenario especially for humans, can not be solved by tissues and organs donation and transplantation (Mahla 2016).

In the last 20 years, cellular therapy has grown a lot and several different stem cells (SC) lineages from different organs and species have been established and characterized (Fernandes et al. 2012; Avasthi et al. 2008). SC are present in all organisms during the life and are defined according to the ability of keeping the undifferentiated state along the life (self-renewal capacity) or undergoing differentiation on specialized cell types, after specific stimuli, or using specific cell culture media associated with growth factors *in vitro* (Bertin et al. 2016).

Nevertheless, the obtaining and use of SC, especially those obtained from embryonic tissues that show higher capacity of proliferation and differentiation (Zare et al. 2014) are limited by political, ethical, social, and legal regulations (Gupta 2009). In order to overcome these issues, the fetal membrane tissues, an abundant, ethically acceptable and readily accessible source of SC have been investigated in the recent years (Lobo et al. 2016; Favaron et al. 2014, 2015; Mançanares et al. 2015; Fernandes et al. 2012). Compared to SC derived from their adult counterparts, the fetal membrane SC have higher proliferation abilities, greater differentiation potential being generally defined as multipotent (Bertin et al. 2016; Fratini et al. 2016), do not cause immunological troubles (Bobis et al. 2006), and they are consider a safe cell lineage for *in vivo* applications (Vidane et al. 2014).

Recently, the SC and their effects have not been studied alone, but considering their relationship with the microenvironment or the niche associated to these cells. The niche is responsible to define the morpho-physiological and biochemical characteristics of the microenvironment where quiescent SC are located before specific signals activate the process of cell differentiation. In particular, this dynamic compartment performs mainly three functions: (1) controls process related to SC proliferation, (2) determines the fate of SC daughters and (3) protects SC from exhaustion or death (Bertin et al. 2016). In this regard, there is a strict relationship between the SC and the niche, that permits to the SC by cellular mechano-transducers sense changes in the matrix elasticity, resulting in morphological and differentiation changes of the SC (Engler et al. 2006). It is known that there are common elements constituting the SC microenvironment or their niche: (1) stromal cells that support SC interacting with each other via cell surface receptors and soluble factors, (2) the vasculature and nervous system that drag systemic and physiological inputs and (3) extracellular matrix (ECM) proteins, such as: fibronectin, laminin, elastin, and the collagen system (Arenas and Zurbarán 2002), which all together supply structural organization, mechanical signals to the niche, cell migration, proliferation and differentiation (Bertin et al. 2016; Nogami et al. 2016; Choi et al. 2013; Yamazaki et al. 2011; Brown et al. 2010; Engler et al. 2006).

The role of ECM deserves special attention since in one hand, the interactions of the cells with the ECM provide essential mechanical signals and, on the other hand, the ECM can concentrate important growth factors and cytokines by binding both local and systemic biomolecules within the microenvironment (Bertin et al. 2016).

Only recently, tissue bioengineering using decellularization process described that fetal tissues possess more coiled fibers and fibronectin than adult bioscaffolds (Silva et al. 2016), which can increase the possibilities of applications in regenerative medicine. For this, a concomitant effort have been done to in one hand study the diversity of SC-derived from embryonic and fetal membranes and in other hand describe the diversity of ECM molecules in these tissues and understand their functions for cell proliferation, migration and differentiation (Favaron et al. submitted).

Considering the importance of the microenvironment for SC proliferation and differentiation and the diversity of embryonic and fetal membrane-derived SC, this chapter will addressed advances in the isolation, phenotyping, characteristics of the microenvironment, and applications of yolk sac and amniotic membrane-derived SC for human and veterinary regenerative medicine.

12.2 Yolk Sac Membrane and Derived Stem Cells

The yolk sac displays a remarkable diversity of developmental, structural and functional characteristics, mainly because this is an unique extraembryonic membrane that occurs in all vertebrates (Mossman 1987). In mammals, the yolk sac precedes and supports the chorioallantoic placentation during early stages of pregnancy (Sheng and Foley 2012) and it is responsible for maternal-fetal exchange (Favaron et al. 2012). For this, special attention has been drawn to this fetal membrane, since it maintains essential functions during early stages of gestation, including the hematopoiesis and development of the vascular system, nutrient transfer, migration of primordial germinative cells, and organogenesis (such as the development of the primitive intestine) (Mançanares et al. 2013; Hyttel et al. 2010; Jafredo et al. 2005; Nakagawa et al. 2000; Auerbach et al. 1996).

Due to the primary hematopoiesis function, so far, much data on cell differentiation from yolk sac tissues are available for hematopoietic SC from mice and humans (Jafredo et al. 2005; Auerbach et al. 1996, 1998; Yoder et al. 1997; Huang and Auerbach 1993; Globerson et al. 1987). In contrast to what was expected the precursor cells derived from pluripotent SC isolated from yolk sac tissues are not fully totipotent, but were able to differentiate into lymphocytes, granulocytes, monocytes, erythrocytes, and megakaryocytes (Zhang et al. 2003; Palis and Yoder 2001; Ikuta and Weissman 1992). Despite this, the extraembryonic hematopoietic SC derived from the yolk sac are consider a valuable source of SC to support existing transplantation therapies in clinical regenerative medicine (Sugiyama et al. 2011; Wilpshaar et al. 2002). In addition, the understand of both biology and microenvironment of the hematopoietic SC in the yolk sac membrane in different phases of the gestation is crucial to improve their use on cell therapies.

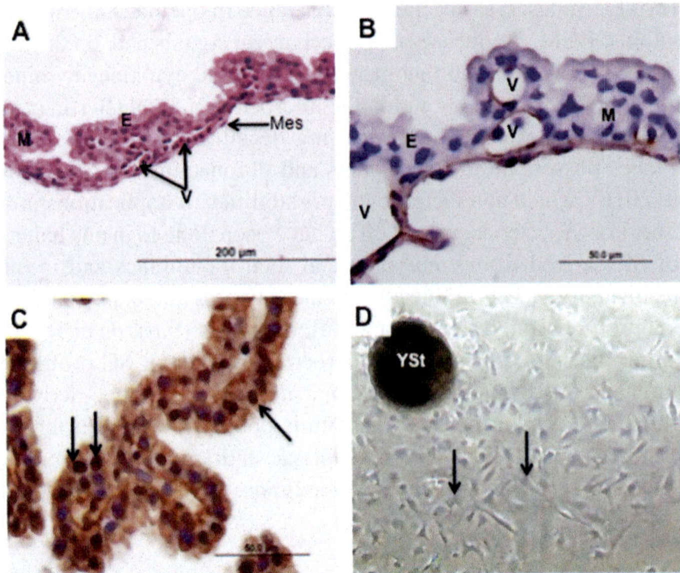

Fig. 12.1 Yolk sac morphology and cell culture. (**a**) The yolk sac membrane is composed by endodermic cells (E), mesenchyme (M) with vitelline vessels (V) and mesothium (Mes). Staining using hematoxilin and eosin. (**b, c**) Immunohistochemistry for vimentin showing the intense vascularization (V) of the yolk sac membrane and for PCNA, showing high proliferation of the endodermic cells (*arrows*). (**d**) Culture of yolk sac stem cells from rodent (Necromys lasiurus, Cricetidae). Note the yolk sac explant (Yst) with fibroblastic-like stem cells (*arrows*) with a central nucleus

However, in recent years promising studies have been done on yolk sac mesenchymal SC not only for human (Wang et al. 2008; Xiao-dong et al. 2003, 2005) and rodents (Favaron et al. 2014; Zhao et al. 2002; Zhao 2003), but also for other species such as dog (Fratini et al. 2016; Wenceslau et al. 2011) and bovine (Mançanares et al. 2015). The yolk sac membrane used for mesenchymal SC isolation usually present a villous structure for which the histological microenvironment is composes by endodermal cells, mesenchyme with vitelline vessels inside and a basal mesothelium (Fernandes et al. 2012). The vitelline vessels are abundant inside the mesenchyme and the endodermal cells high proliferative during the gestation (Fig. 12.1a–c). Using tissue explants or digestive solutions such as trypsin or collagenase, mesenchymal SC are isolated and available for culture (Fig. 12.1d). As the membrane, the mesenchymal SC-derived from the yolk sac show interesting and promising characteristics regarding to the morphology, growth and differentiation potential.

12.3 Amniotic Membrane and Derived Stem Cells

The amnion is the innermost membrane that surround the embryo/fetus during the entire pregnancy. It forms a sac that contains the amniotic fluid (Favaron et al. 2015), a novel and interesting source of SC for cell therapy, especially due to their high renewal capacity, expression of embryonic markers, multipotency to differentiate in different cell lines, low immunogenicity, anti-inflammatory properties, nontumorigenicity, and noninvasive isolation (Kim et al. 2014; Antonucci et al. 2012). Regarding to this easily sample collection, the amniotic membrane are available for sampling just after the parturition avoiding all political and ethical restrictions involved with embryonic SC.

Histologically, the amniotic membrane is composed by a very thin cubic epithelial layer, a subjacent mesoderm and dispersed small blood vessels (Fig. 12.2a, b). In contrast to the very simple histological nature (Chang et al. 2010), recently data have shown data amniotic-derived stem cells (Fig. 12.2c) collect in different gestational stages present different pattern of phenotype, methylation, immunomodulatory and stemness properties of amniotic-derived SC (Barboni et al. 2014), which are closely related to the microenvironment of this extraembryonic membrane and its functions in the different gestational phases.

The human amniotic membrane derived SC were isolated for the first time in 2004 and demonstrated the ability to differentiate into osteogenic and adipogenic cell lines (In 't Anker et al. 2004). Later, it was also demonstrated the ability of these cells for differentiation in other cell lines, such as chondrogenic in dogs (Vidane et al. 2014; Rutigliano et al. 2013) and neurogenic in sheep (Zhu et al. 2013).

In clinical practice, the potential of amniotic membrane derived SC is very wide and increasing, and their clinical applications can reach a variety of diseases, especially those associated with degenerative processes induced by inflammatory and fibrotic processes (Parolini and Caruso 2011). In this regard, the clinical applications of amniotic membrane SC show satisfactory therapeutic results, without toxicity or side effects. Studies using monkeys have demonstrated the usefulness of these cells for the treatment of injured areas of the spinal cord in the central nervous system, resulting in the regeneration of neurons (Sankar and Muthusamy 2003). Additionally, studies have shown satisfactory results for the treatment of Parkinson in rat models using amniotic membrane SC in order to produce dopamine and prevent neuronal degeneration (Kakishita et al. 2003). In vivo studies in rats also showed that after inoculation, the amniotic SC were able to restore liver function (Miki et al. 2007). In addition, transplantation in immunodeficient rats with liver problems showed evidences of synthesis and excretion of albumin after 7 days of cell transplantation (Sakuragawa et al. 2000). Amniotic membrane SC expressed pancreatic markers such as $\alpha 2B$ amylase and produced glucagon, after being induced to pancreatic differentiation (Ilancheran et al. 2007). Transplantation of amniotic membrane SC showed also satisfactory clinical results for the treatment of pulmonary disease (Insausti et al. 2010; Magatti et al. 2009) and diabetes mellitus (Uccelli et al. 2008).

For this reason, the amniotic membrane emerged as a new and important source of SC established in several species including human (In 't Anker et al. 2004; Díaz-Prado et al. 2011), horse (Cremonesi et al. 2011; Lange-Consiglio et al. 2012), sheep (Mauro et al. 2010), cat (Vidani et al. 2016; Vidane et al. 2014; Rutigliano et al. 2013), dog (Uranio et al. 2011), rat (Marcus et al. 2008) and rabbit (Borghesi et al. 2017) with different patterns of phenotype and differentiation potential, which

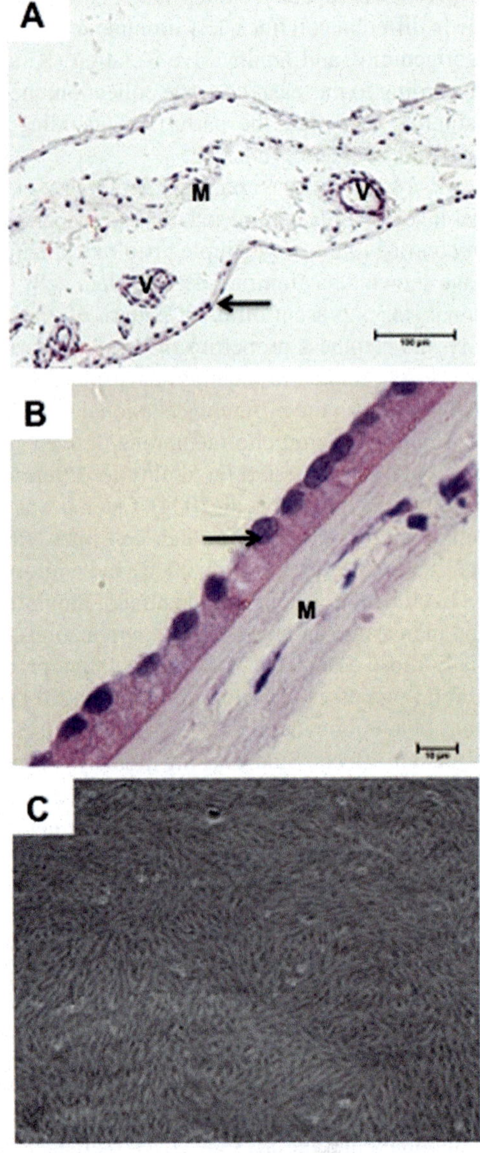

Fig. 12.2 Histology and culture of amniotic membrane. (**a**, **b**) Histology of dog and human amniotic membrane, respectively. Note the cubic epithelium (*arrow*), mesoderm (M) and the presence of small blood vessels (V). Staining using hematoxilin and eosin. (**c**) Culture of amniotic stem cells from rabbit. Note the fibroblastic-like morphology of the cells

can be associated with the histological and physiological characteristics regarding to the microenvironment associated in this membrane during the gestation.

Acknowledgments The authors thank the several members from the School of Veterinary Medicine and Animal Science, University of Sao Paulo-USP, for collaboration and technical assistance.

References

Antonucci I, Pantalone A, Tete S, Salini V, Borlongan CV, Hess D, Stuppia L (2012) Amniotic fluid stem cells: a promising therapeutic resource for cell-based regenerative therapy. Curr Pharm Des 18:1846–1863

Arenas LAS, Zurbarán CB (2002) La Matriz Extracelular: El Ecossistema de la Célula. Salud Uninorte 16:9–18

Auerbach R, Huang H, Lu L (1996) Hematopoietic stem cells in the mouse embryonic yolk sac. Stem Cells 14:269–280

Auerbach R, Wang SJ, Yu D, Gilligan B, Lu LS (1998) Role of endothelium in the control of mouse yolk sac stem cells differentiation. Dev Comp Immunol 22:333–338

Avasthi S, Srivastava RN, Singh A, Srivastava M (2008) Stem cell: past, present and future—a review article. Internet J Med Update 3:22–30

Barboni B, Russo V, Curini V, Martelli A, Berardinelli P, Mauro A, Mattioli M, Marchisio M, Signoroni PB, Parolini O, Colosimo A (2014) Gestational stage affects amniotic epithelial cells phenotype, methylation status, immunomodulatory and stemness properties. Stem Cell Rev 10:725–741

Bertin E, Piccoli M, Franzin C, Spiro G, Donà S, Dedja A, Schiavi F, Taschin E, Bonaldo P, Braghetta P, De Coppi P, Pozzobon M (2016) First steps to define murine amniotic fluid stem cell microenvironment. Sci Rep 6:37080

Bobis S, Jarocha D, Majka M (2006) Mesenchymal stem cells: characteristics and clinical applications. Folia Histochem Cytobiol 44:215–230

Borghesi J, Mario LC, Carreira AC, Miglino MA, Favaron PO (2017) Phenotype and multipotency of rabbit (*Oryctolagus cuniculus*) amniotic stem cells. Stem Cell Res Ther 8(1):27

Brown BN, Barnes CA, Kasick RT, Michel R, Gilbert TW, Beer-Stolz D, Castner DG, Ratner BD, Badylak SF (2010) Surface characterization of extracellular matrix scaffolds. Biomaterials 31(3):428–437

Chang YJ, Hwang SM, Tseng CP, Cheng FC, Huang SH, Hsu LF, Hsu LW, Tsai MS (2010) Isolation of mesenchymal stem cells with neurogenic potential from the mesoderm of the amniotic membrane. Cells Tissues Organs 192:93–105

Choi JS, Kim JD, Yoon HS, Cho YW (2013) Full-thickness skin wound healing using human placenta-derived extracellular matrix containing bioactive molecules. Tissue Eng Part A 19(3–4):329–339

Cremonesi F, Corradetti B, Lange-Consiglio A (2011) Fetal adnexa derived stem cells from domestic animals: progress and perspectives. Theriogenology 75:1400–1415

Díaz-Prado S, Muiños-López E, Hermida-Gómez T, Rendal Vázquez ME, Fuentes-Boquete I, De Toro FJ, Blanco FJ (2011) Stem cells from human amniotic membrane. Tissue Eng: Part C 17:49–59

Engler AJ, Sen S, Sweeney HL, Discher DE (2006) Matrix elasticity directs stem cell lineage specification. Cell 126:677–689

Favaron PO, Carter AM, Mess AM, Oliveira MF, Miglino MA (2012) An unusual feature of yolk sac placentation in *Necromys lasiurus* (Rodentia, Cricetidae, Sigmodontinae). Placenta 33:578–580

Favaron PO, Mess A, Will SE, Maiorka PC, de Oliveira MF, Miglino MA (2014) Yolk sac mesenchymal progenitor cells from New World mice (*Necromys lasiurus*) with multipotent differential potential. PLoS One 9:e95575

Favaron PO, Carvalho RC, Borghesi J, Anunciação AR, Miglino MA (2015) The amniotic membrane: development and potential applications—a review. Reprod Domest Anim 50(6):881–892

Favaron PO, Borghesi J, Mess AM, Castelucci P, Miglino MA (submitted) Establishment of three-dimensional scaffolds from hemochorial placentas. PLoS One

Fernandes RA, Costola-Souza C, Sarmento CAP, Gonçalves L, Favaron PO, Miglino MA (2012) Placental tissues as sources of stem cells—review. OJAS 2:166–173

Fratini P, Carreira AC, Alcântara D, de Oliveira e Silva FM, Rodrigues MN, Miglino MA (2016) Endothelial differentiation of canine yolk sac cells transduced with VEGF. Res Vet Sci 104:71–76

Globerson A, Woods V, Abel L, Morrissey L, Cairns JS, Kukulansky T, Kubai L, Auerbach R (1987) In vitro differentiation of mouse embryonic yolk sac cells. Differentiation 36:185–193

Gupta BD (2009) An introduction to stem cells and debate surrounding them. J Indian Acad Forensic Med 31:3

Huang H, Auerbach R (1993) Identification and characterization of hematopoietic stem cells from the yolk sac of the early mouse embryo. Proc Natl Acad Sci 90:10110–10114

Hyttel P, Sinowatz F, Vejlsted M (2010) Essentials of domestic animal embryology. Saunders Elsevier, Toronto, p 455

Ikuta K, Weissman IL (1992) Evidence that hematopoietic stem cells express mouse c-kit but do not depend on steel factor for their generation. Proc Natl Acad Sci 89:1452–1458

Ilancheran S, Michalska A, Peh G, Wallace EM, Pera M, Manuelpillai U (2007) Stem cells derived from human fetal membranes display multi-lineage differentiation potential. Biol Reprod 77:577–588

In 't Anker PS, Scherjon SA, Kleijburg-Van Der Keur C, De Groot-Swings GM, Claas FH, Fibbe WE, Kanhai HH (2004) Isolation of mesenchymal stem cells of fetal or maternal origin from human placenta. Stem Cells 22:1338–1345

Insausti CL, Alcaraz A, Garcia-Vizcaino EM, Mrowiec A, López-Martínez MC, Blanquer M, Piñero A, Majado MJ, Moraleda JM, Castellanos G, Nicolás FJ (2010) Amniotic membrane induces epithelialization in massive posttraumatic wounds. Wound Repair Regen 18:368–377

Jafredo T, Bollerot K, Sugiyama DG, Drevon C (2005) Tracing the hemangioblast during embryogenesis: developmental relationships between endothelial and hematopoietic cells. Int J Dev Biol 49:269–277

Kakishita K, Nakao N, Sakuragawa N, Itakura T (2003) Implantation of human amniotic epithelial cells prevents the degeneration of nigral dopamine neurons in rats with 6-hydroxydopamine lesions. Brain Res 980:48–56

Kim EY, Lee KB, Kim MK (2014) The potential of mesenchymal stem cells derived from amniotic membrane and amniotic fluid for neuronal regenerative therapy. BMB Rep 47:135–140

Lange-Consiglio A, Corradetti B, Bizzaro D, Magatti M, Ressel L, Tassan S, Parolini O, Cremonesi F (2012) Characterization and potential applications of progenitor-like cells isolated from horse amniotic membrane. J Tissue Eng Regen Med 6:622–635

Lobo SE, Leonel LC, Miranda CM, Coelho TM, Ferreira GA, Mess A, Abrão MS, Miglino MA (2016) The placenta as an organ and a source of stem cells and extracellular matrix: a review. Cells Tissues Organs 201:239–252

Magatti M, De Munari S, Vertua E, Nassauto C, Albertini A, Wengler GS, Parolini O (2009) Amniotic mesenchymal tissue cells inhibit dendritic cell differentiation of peripheral blood and amnion resident monocytes. Cell Transplant 18:899–914

Mahla RS (2016) Stem cells applications in regenerative medicine and disease therapeutics. Int J Cell Biol 2016:6940283

Mançanares CA, Leiser R, Favaron PO, Carvalho AF, Oliveira VC, Santos JM, Ambrósio CE, Miglino MA (2013) A morphological analysis of the transition between the embryonic primitive intestine and yolk sac in bovine embryos and fetuses. Microsc Res Tech 76:756–766

Mançanares CA, Oliveira VC, Oliveira LJ, Carvalho AF, Sampaio RV, Mançanares AC, Souza AF, Perecin F, Meirelles FV, Miglino MA, Ambrósio CE (2015) Isolation and characterization of mesenchymal stem cells from the yolk sacs of bovine embryos. Theriogenology 84:887–898

Marcus AJ, Coyne TM, Rauch J, Woodbury D, Black IB (2008) Isolation, characterization, and differentiation of stem cells derived from the rat amniotic membrane. Differentiation 76:130–144

Mason C, Dunnill P (2008) A brief definition of regenerative medicine. Regen Med 3:1–5

Mauro A, Turriani M, Ioannoni A, Russo V, Martelli A, Di Giacinto O, Nardinocchi D, Berardinelli P (2010) Isolation, characterization, and in vitro differentiation of ovine amniotic stem cells. Vet Res Commun 34:S25–S28

Mossman HW (1987) Vertebrate fetal membranes: comparative ontogeny and morphology: evolution; phylogenetic significance: basic functions; research opportunities. The Macmillan, London

Miki T, Mitamura K, Ross MA, Stolz DB, Strom SC (2007) Identification of stem cell marker-positive cells byimmunofluorescence in term human amnion. J Reprod Immunol 75:91–6.

Nakagawa S, Saburi S, Yamanouchi K, Tojo H, Tachi C (2000) In vitro studies on PGC or PGC-like cells in cultured yolk sac cells and embryonic stem cells of the mouse. Arch Histol Cytol 63:229–241

Nogami M, Kimura T, Seki S, Matsui Y, Yoshida T, Koike-Soko C, Okabe M, Motomura H, Gejo R, Nikaido T (2016) A human amnion-derived extracellular matrix-coated cell-free scaffold for cartilage repair: in vitro and in vivo studies. Tissue Eng Part A 22:680–688

Palis J, Yoder MC (2001) Yolk-sac hematopoiesis: the first blood cells of mouse and man. Exp Hematol 29:927–936

Parolini O, Caruso M (2011) Review: Preclinical studies on placenta-derived cells and amniotic membrane. Placenta 25:186–195

Rutigliano L, Corradetti B, Valentini L, Bizarro D, Meucci A, Crempnesi F, Lange-Consiglio A (2013) Molecular characterization and in vitro differentiation of feline progenitor-like amniotic epithelial cells. Stem Cell Res Ther 4:133

Sakuragawa N, Enosawa S, Ishii T, Thangavel R, Tashiro T, Okuyama T, Suzuki S (2000) Human amniotic epithelial cells are promising transgene carriers for allogeneic cell transplantation into liver. J Hum Genet 45:171–176

Sankar V, Muthusamy R (2003) Role of human amniotic epithelial cell transplantation in spinal cord injury repair research. Neuroscience 118:11–17

Sheng G, Foley AC (2012) Diversification and conservation of the extraembryonic tissues in mediating nutrient uptake during amniote development. Ann N Y Acad Sci 1271:97–103

Silva AC, Rodrigues SC, Caldeira J, Nunes AM, Sampaio-Pinto V, Resende TP, Oliveira MJ, Barbosa MA, Thorsteinsdóttir S, Nascimento DS, Pinto-do-Ó P (2016) Three-dimensional scaffolds of fetal decellularized hearts exhibit enhanced potential to support cardiac cells in comparison to the adult. Biomaterials 104:52–64

Sugiyama D, Inoue-Yokoo T, Fraser ST, Kulkeaw K, Mizuochi C, Horio Y (2011) Embryonic regulation of the mouse hematopoietic niche. Sci World J 11:1770–1780

Uccelli A, Moretta L, Pistoia V (2008) Mesenchymal stem cells in health and disease. Nat Rev Immunol 8:726–736

Uranio MF, Valentini L, Lange-Consiglio A, Caira M, Guaricci AC, L'abbate A, Catachio CR, Ventura M, Cremonesi F, Dell'aquila ME (2011) Isolation, proliferation, cytogenetic, and molecular characterization and in vitro differentiation potency of canine stem cells from foetal adneza: a comparative study of amniotic fluid, amnion, and umbilical cord matrix. Mol Reprod Dev 78:361–373

Vidane AS, Souza AF, Sampaio RV, Bressan FF, Pieri NC, Martins DS, Meirelles FV, Miglino MA, Ambrósio CE (2014) Cat amniotic membrane multipotent cells are nontumorigenic and are safe for use in cell transplantation. Stem Cells Cloning 7:71–78

Vidani AS, Pinheiro AO, Casals JB, Passarelli D, Hage MCFNS, Bueno RS, Martins DS, Ambrósio CE(2016) Transplantation of amniotic membrane-derived multipotent cells ameliorates and delays the progression ofchronic kidney disease in cats. Reprod Dom Anim 51:1–11.

Wang XY, Lan Y, He WY, Zhang L, Yao HY, Hou CM, Tong Y, Liu YL, Yang G, Liu XD, Yang X, Liu B, Mao N (2008) Identification of mesenchymal stem cells in aorta-ganad-mesonephros and yolk sac of human embryos. Blood 111:2436–2443

Wenceslau CV, Miglino MA, Martins DS, Ambrósio CE, Lizier NF, Pignatari GC, Kerkis I (2011) Mesenchymal progenitor cells from canine fetal tissues: yolk sac, liver, and bone marrow. Tissue Eng Part A 17:2165–2176

Wilpshaar J, Bhatia M, Kanhai HH, Breese R, Heilman DK, Johnson CS, Falkenburg JH, Srour EF (2002) Engraftment potential of human fetal hematopoietic cells in NOD/SCID mice is not restricted to mitotically quiescent cells. Blood 100:120–127

Xiao-dong N, Mei-ling Z, Zi-ping Z, Wei-hua Y, Xiu-ming Z, Peng X, Shu-nong L (2003) Purification and adipogenic differentiation of human yolk sac mesenchymal stem cells. Chin J Pathophysiol 19:1316–1319

Xiao-dong N, Wei-hua Y, Zi-ping Z, Mei-ling Z, Xiao-ying Z, Jun-xia L, Xin-min S, Chun-nong H, Xiu-ming Z, Yan L, Peng X, Shu-nong L (2005) Osteogenic and neurogenic differentiation of human yolk sac mesenchymal stem cells. Chin J Pathophysiol 21:636–641

Yamazaki S, Ema H, Karlsson G, Yamaguchi T, Miyoshi H, Shioda S, Taketo MM, Karlsson S, Iwama A, Nakauchi H (2011) Nonmyelinating Schwann cells maintain hematopoietic stem cell hibernation in the bone marrow niche. Cell 147:1146–1158

Yoder MC, Hiatt K, Dutt P, Mukherjee P, Bodine DM, Orlic D (1997) Characterization of definitive lymphohematopoietic stem cells in the day 9 murine yolk sac. Immunity 7:335–344

Zare S, Kurd S, Rostamzadeh A, Nilforoushzadeh MA (2014) Types of stem cells in regenerative medicine: a review. J Skin Stem Cell 1:e28471

Zhang J, Niu C, Ye L, Huang H, He X, Tong WG, Ross J, Haug J, Johnson T, Feng JQ, Harris S, Wiedemann LM, Mishina Y, Li L (2003) Identification of the haematopoietic stem cell niche and control of the niche size. Nature 425:836–841

Zhao ZP (2003) Investigation of murine yolk sac multipotent mesenchymal stem cells combined with collagen surface-modified CPC in vitro. PhD Thesis, Central South University, China

Zhao ZP, Na XD, Yang HF, Zhou JN (2002) Osteogenic differentiation of murine yolk sac mesenchymal stem cells in vitro. Zhongguo Yi Xue Ke Xue Yuan Xue Bao 24:41–44

Zhu X, Wang X, Cao G, Liu F, Yang Y, Li X, Zhang Y, Mi Y, Liu J, Zhang L (2013) Stem cell properties and neural differentiation of sheep amniotic epithelial cells. Neural Regen Res 8:1210–1219

Chapter 13
Current Technologies Based on the Knowledge of the Stem Cells Microenvironments

Damia Mawad, Gemma Figtree, and Carmine Gentile

Abstract The stem cell microenvironment or *niche* plays a critical role in the regulation of survival, differentiation and behavior of stem cells and their progenies. Recapitulating each aspect of the stem cell niche is therefore essential for their optimal use in in vitro studies and in vivo as future therapeutics in humans. Engineering of optimal conditions for three-dimensional stem cell culture includes multiple transient and dynamic physiological stimuli, such as blood flow and tissue stiffness. Bioprinting and microfluidics technologies, including organs-on-a-chip, are among the most recent approaches utilized to replicate the three-dimensional stem cell niche for human tissue fabrication that allow the integration of multiple levels of tissue complexity, including blood flow. This chapter focuses on the physico-chemical and genetic cues utilized to engineer the stem cell niche and provides an overview on how both bioprinting and microfluidics technologies are improving our knowledge in this field for both disease modeling and tissue regeneration, including drug discovery and toxicity high-throughput assays and stem cell-based therapies in humans.

Keywords Stem cell niche • Microenvironment • In vitro 3D models • Organoids
• Bioprinting • Tissue fabrication

D. Mawad
Faculty of Science, School of Materials Science and Engineering,
University of New South Wales, Sydney, NSW, Australia

G. Figtree
Sydney Medical School, University of Sydney, Sydney, NSW, Australia

C. Gentile (✉)
Sydney Medical School, University of Sydney, Sydney, NSW, Australia

Beth Israel Deaconess Medical Center, Harvard Medical School, Boston, MA, USA
e-mail: carmine.gentile@sydney.edu.au

© Springer International Publishing AG 2017

A. Birbrair (ed.), *Stem Cell Microenvironments and Beyond*,
Advances in Experimental Medicine and Biology 1041,
DOI 10.1007/978-3-319-69194-7_13

13.1   Introduction

The stem cell microenvironment (*niche*) plays a major role in controlling the cellular fate and therefore any application of these cells in vitro and in vivo. To date, increasing knowledge of the in vivo niche has improved the engineering of in vitro systems for stem cells by integrating critical three-dimensional features, such as specific cellular and extracellular components, tissue responsiveness to stiffness and blood flow. While some of these aspects are not fully recapitulated in in vitro models of the stem cell niche and some of these models better recapitulate some aspects of the in vivo niche versus others depending on their application, this review aims at describing briefly the latest technologies based on our knowledge of the stem cell niche and more in details current approaches for its engineering, with a particular focus on future applications for human disease modeling, including drug discovery and toxicity high-throughput assays and regenerative medicine.

13.2   Stem Cell Niche In Vivo and In Vitro

Stem cells and their applications generated a great hope for research and medicine in the past decades. However, their use in humans has been limited by several factors, including ethical conundrum on stem cell editing (Lanphier et al. 2015). From the engineering point of view, their use is very costly *per se* and they present limited survival following their injection in the host tissue, which further increases costs associated with their use for therapies in humans. To overcome these limitations, 3D cultures have been investigated based on our knowledge of a highly-defined spatial and temporal bioavailability of factors during embryogenesis and organogenesis, such as oxygen, growth factors, cytokines and extracellular matrix (ECM) molecules (please see Fig. 13.1) (Passier et al. 2016; Gunter et al. 2016; Gentile 2016; Dennis et al. 2015). For instance, survival of dissociated stem cells is less than 5% within the first days of delivery. On the contrary, 3D cultures of stem cells increased cell survival over time (Dennis et al. 2015). Similarly, incorporation of vascular cells, growth factors and cytokines improved stem cell survival into the host (Gentile 2016). Stem cells and progenitor cells also respond to the same stimuli in different ways based on their origin and differentiation state (Gentile 2016). Based on their intrinsic cell composition, the 3D in vivo niche of *pluripotent stem cells (PSCs)* typical of the blastocyst is characterized by cells of the three germ layers (*mesoderm, endoderm and ectoderm*), which is recapitulated in vitro within *embryoid bodies (EBs)* (Doetschman et al. 1985; Itskovitz-Eldor et al. 2000; Vallier and Pedersen 2005; Bratt-Leal et al. 2009; Carpenedo et al. 2009; Lanphier et al. 2015). Conversely, multipotent mesenchymal stem cells (MSCs) are characterized by a less-defined niche than PSCs, as MSCs are present within the blood stream until they reach the host tissue to differentiate. However, 3D in vitro cultures of MSCs in either *mesenspheres* or fibrin gels show improved features compared to monolayer cultures (Potapova et al. 2007; Murphy et al. 2014a). 3D cultures of neural and

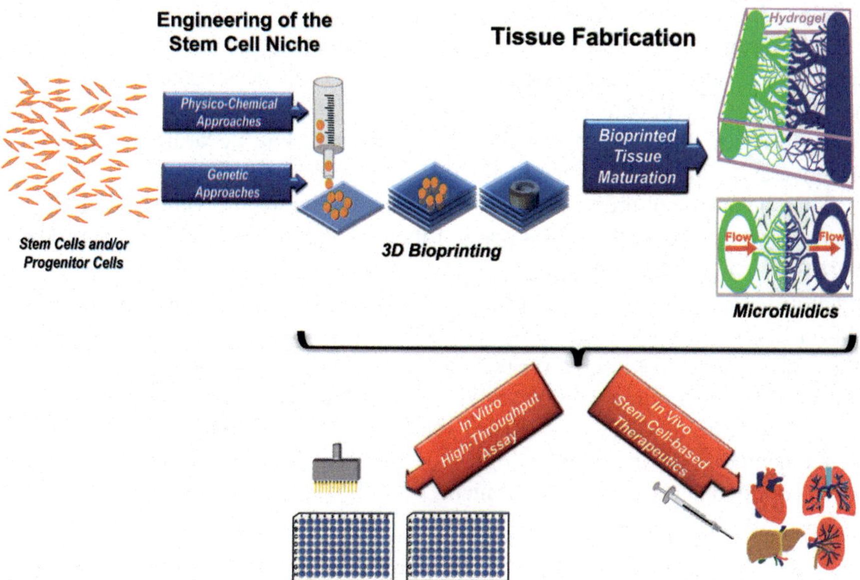

Fig. 13.1 Engineering of the stem cell niche: approaches and applications (adapted from Gentile 2016)

cardiac progenitor (*neurospheres* and *cardiospheres*, respectively) also showed improved features compared to monolayer cultures (Messina et al. 2004; Reynolds and Rietze 2005; Marban and Cingolani 2012; Baraniak et al. 2012). Due to all these reasons, engineering of the niche specific for each stem cell type becomes critical in avoiding the failure of stem cell-based therapies in humans.

13.2.1   Technologies Based on the Stem Cell Niche

Several groups have engineered the typical niche of the human body using stem cells, including the blood brain barrier and brain (Lippmann et al. 2012; Adriani et al. 2016), lung airway (Benam et al. 2016) and the hematopoietic niche (Torisawa et al. 2014). However, there are still several challenges in engineering a fully-controlled stem cell niche of these and other vital organs, such as the brain, the heart and the immune system. The most challenging features to overcome are: (i) replicating physiologically-relevant culture conditions without affecting the robustness of the cellular response to the experimental stimuli, such as blood flow and tethering effects of the extracellular matrix during development; (ii) stem cells limited survival in new culturing conditions and/or following engraftment; (iii) maintaining a specific phenotype in culture. Factors determining optimal engineering of 3D cultures of stem cells have been previously described, and include cell composition, gradient of oxygen and nutrients, growth factors, extracellular matrix and stiffness, cellular adhesion molecules, and the addition of exogenous material (see (Gentile 2016)). Aging of cells represent an additional limitation to the use of stem cells and converting them into cells with a younger phenotype has been recently achieved in haematopoietic stem cells (Guidi and Geiger 2017). Conversely, 3D cultures have been utilized to define how the niche affects cell behaviour in cardiac microtissues (Boudou et al. 2012; Polonchuk et al. 2017; Figtree et al. 2017).

Approaches to culture stem cell in vitro can be divided in: (i) _natural and synthetic scaffolds_ ((Willerth and Sakiyama-Elbert 2008, Cosson et al. 2015, He and Lu 2016)) (ii) _scaffold free-technologies_ (also described as _organoids, microtissues, spheroids, etc._) (Gentile 2016); (iii) _organs-on-a-chip_ (Van der Helm et al. 2016; Huh et al. 2013; Zhang et al. 2009; Van der Meer and Van den Berg 2012). Chemical, physical and genetic stimuli have been utilized for optimal engineering of the stem cell niche to either retain their phenotype or to differentiate them into other cells, as described in the following sections of this chapter. Bioprinting and microfluidics have emerged as the main technologies for 3D cultures of stem cells and therefore for the optimal engineering of their niche.

13.2.1.1   3D Bioprinting of the Stem Cell Niche

3D bioprinting is the layer-by-layer deposition of defined biological material (or "_bioink_") within a biopaper (or "_hydrogel_"), both engineered for optimal tissue formation and organogenesis, and it allows the inclusion of physiological features of several complexity, such as blood vessels or gradient of extracellular cues (Mironov et al. 2003; Jakab et al. 2010; Visconti et al. 2010; Murphy and Atala 2014). It can be divided in three types: (i) ink jet, (ii) laser jet; and (iii) extrusion, all fully controlled by a computer (Kamble et al. 2016). Bioprinting of 3D structures, including spheroid cultures to be used as building blocks, demonstrated higher architectural complexity and improved cell survival with liquid-like properties, like oil droplets in water and mimicking natural processes (Dennis et al. 2015; Fleming et al. 2010). During the bioprinting process utilizing spheroid cultures, critical factors to consider are: (i) spheroid diameter and composition (for the _bio-ink_); (ii) viscosity; and (iii) gelification time (for the _hydrogel_) (Gentile 2016). Main chemical, physical and genetic approaches utilized to improve the engineering of bioprinted tissues and organs are fully described in Sects. 13.2.2, 13.2.3 and 13.2.4 of this chapter.

13.2.1.2   Microfluidics and Human-on-a-Chip

Multi-compartmental bioengineered constructs have been engineered as "organs-on-a-chip" with microfluidics, representing improved complexity of physiological systems (Oleaga et al. 2016; Frey et al. 2014; Bhatia and Ingber 2014). Microfabricated structures can be used to precisely control the spatial positioning of cells and study their interactions (Hui and Bhatia 2007). These physiological systems are fabricated by combining soft lithography for cardiac bodies (Christoffersson et al. 2016), 3D bioprinting of monolayer cells at high density scaffold-free ("bio-printed heart on a chip" (Lind et al. 2016)) or in presence of scaffolds for endothe-lialization (Zhang et al. 2016). Additionally, 3D microfluidic devices can be readily monitored with several imaging modalities and closely replicate key physiological and structural features of small functional units of organs. For example, "micro-channels" can permit physiological flow rates, peristaltic contractions and the essential function of blood vessels for delivering oxygen and nutrients, while removing waste (Bhatia and Ingber 2014; Huh et al. 2010). Miniaturized models of functional biological units have already been fabricated on a chip, including models for lung, liver, kidney, intestine, heart, fat, bone marrow, cornea, skin, and the blood-brain barrier (Bhatia and Ingber 2014).

13.2.1.3 Static Versus Dynamic Stem Cell Niche

Although microfluidic-based organ-on-a-chip systems can integrate key compo-nents together and still allow precise control and measurement in a dynamic envi-ronment, certain features of static 3D spheroid cultures may still be more advantageous. For example, spheroids can generate more tissue mass, allowing sci-entists to perform analytical experiments that usually require large samples. On one hand, 3D spheroid cultures also allow the growth of macroscale architecture and highly complex and spatially heterogeneous tissues that cannot be supported at the microscale. On the other hand, microfluidic chips still offer an unprecedented flex-ibility in independently controlling and monitoring features such as flow and other mechanical cues, helping to dissect their contribution to tissue and organ function. Finally, microfluidics allow fluorescence confocal microscopy analyses of cells, trans-epithelial electrical resistance measurements, multiple electrode arrays, and other analytical systems not easily recapitulated in static 3D spheroid cultures (Bhatia and Ingber 2014, Huh et al. 2010). Therefore, there is growing interest in integrating 3D spheroid cultures with microfluidics and 3D bioprinting technology for the engineering of the stem cell niche for both static and dynamic conditions.

13.2.2 *Physico-Chemical Approaches to Engineer the Niche for an Undifferentiated Phenotype*

Decades long intensive research has established that synthetic polymers are excel-lent candidates to serve as platforms for the support of the long-term culture of stem cells. For instance, Villa-Diaz et al. (2010) showed the ability of a well-defined syn-thetic polymer, poly[2-(methacryloyloxy)ethyl dimethyl-(3-sulfopropyl)ammonium hydroxide] (PMEDSAH), to sustain long-term (~25 passages) hESC growth in sev-eral different culture media. Short term self-renewal of hESC has been also shown in a synthetic hydrogel fabricated from poly (N-isopropylacrylamide-co-acrylic acid) [p(NIPAAm-co-AAc] and crosslinked with a peptide (Li et al. 2006). These are two examples of 2D and 3D polymeric scaffolds with controlled and well-defined properties that are being developed as platforms for in vitro cell studies to investigate the interactions of cells and materials. While a 3D network mimics closely the native environment of the cell, a 2D scaffold allows more control over surface properties such as topography and roughness.

13.2.2.1 Surface Topography

Controlling the nano and micro scale topography of a polymeric substrate has a direct effect on the cell behavior (Murphy et al. 2014b). Also, generating well-defined patterns in these substrates has the potential to produce functional custom-ised tissues for applications in regenerative medicine. In a study by Kim et al. (2010) the adhesion of human adipose-derived stem cells (hASCs) is shown to be favourable on micro-patterned surfaces of poly(lactic-co-glycolic acid) (PLGA) as opposed to unpatterned surfaces. The long-term self-renewal (>3 weeks) of mouse embryonic stem cells (mESCs) cultured on 2-hydroxyethyl methacrylate-co-ethylene dimethacrylate (HEMA-EDMA) substrates is shown to be dependent on the surface roughness (Jaggy et al. 2015). Substrates with a hierarchical topography at both nano- and microscale (large agglomerates up to 9 µm in height with an average surface roughness (S_a) of 919 ± 22 nm) supported the long-term maintenance of mESCs. On the contrary, culturing of mESCs on either smooth ($S_a = 2 ± 0.4$ nm) or nano rough surfaces ($S_a = 68 ± 30$ nm) led to their fast differentiation. McMurray et al. (2011) identified a nanostructured polycaprolactone surface that retains a stem cell phenotype and maintains stem cell growth over 8 weeks. In a 3D configuration, the adhesion and expansion of hESCs was tested on electrospun nanofibers mats fabricated from three synthetic FDA approved polymers: (i) poly-ε-caprolactone (PCL), (ii) poly-L-lactic acid (PLLA) and (iii) poly lactic-co-glycolic acid (PLGA) (Kumar et al. 2015). The study reported three important factors that can modulate the colony size of stem cells: chemical nature of the polymer back-bone, the diameter size of the fibre and the fibre orientation. PCL polymer pre-sented as the most supportive substrate for hESCs self-renewal (see Fig. 13.2). Smaller diameter nanofibrous substrates (280 ± 122 nm) supported a significantly greater number of hESC colonies, relative to their larger diameter counterparts (521 ± 195 nm). Aligned nanofibrous substrates were found to be more suited than their random counterparts. With the advent of technology, various fabrication tools are available for researchers to produce controlled micro or nanoscale features on polymeric surfaces facilitating adhesion and proliferation of cells. Examples include *photolithography, printing techniques, self-assembly of block copolymers* and *instability-induced patterning* (Nie and Kumacheva 2008). These techniques allow patterning of the polymeric surface at different length scales and hence more control over directing the stem cell fate.

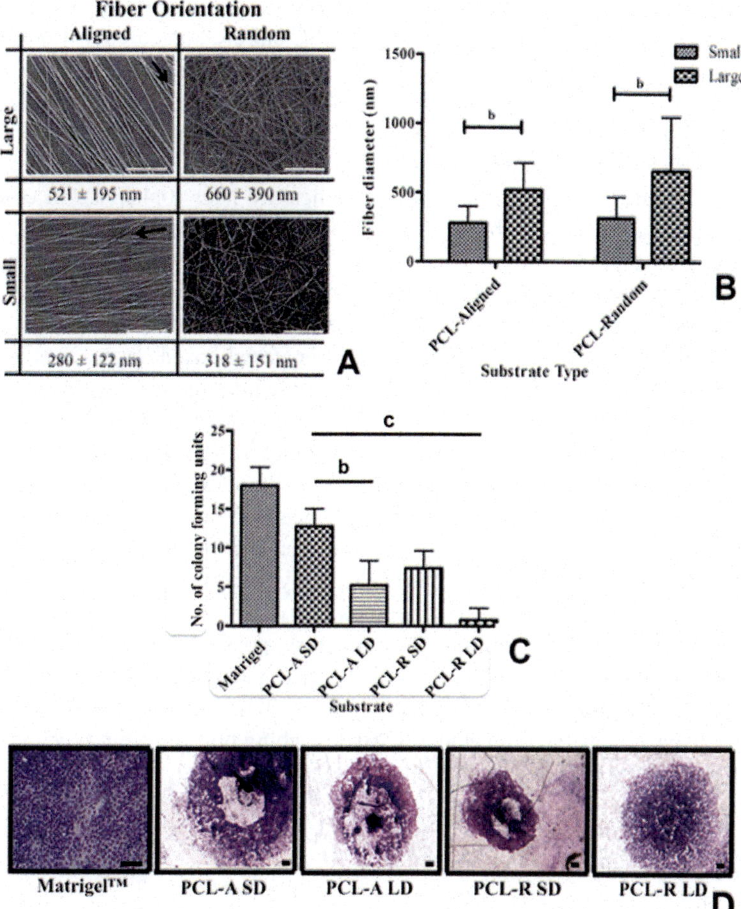

Fig. 13.2 Reduced PCL nanofibrous substrate fiber diameters enhance hESC clonogenicity. (**a**) PCL nanofibrous substrates (aligned and random) with large and small fiber diameters. Nanofiber diameters are indicated under each image. An *arrow* indicates the predominant direction of fiber orientation. Scale bar = 3 μm. (**b**) Average fiber diameters (mm) for large and small PCL fibers in aligned and random conformations. (**c**) Quantification of the number of hESC colonies recovered on PCL nanofibrous substrates with various fiber diameters, in both aligned and random conformations; in physiological normoxia (2% O_2) for 21 days. (**d**) Characterization of hESC CFU morphology on: PCL smaller fiber diameter (aligned, 280 nm; random, 318 nm), PCL larger fiber diameter (aligned, 521 nm; random 660 nm), and Matrigel™ (positive control) (adapted from Kumar et al. 2015)

13.2.2.2   Surface Stiffness

It is well established that the stiffness of the polymeric construct has a direct and strong influence on stem cell behavior (Engler et al. 2006). However, the correlation between the substrate stiffness and the effect on the stem cell fate depends on both the stiffness of the material and the type of stem cell under investigation. A study by Chowdhury et al. (2010) demonstrated that mESC cultured on soft substrates fabricated from polyacrylamide (PA) gels (0.6 kPa) maintained their pluripotent state in contrast to when they were cultured on rigid substrates such as polystyrene (>4 MPa). On the other hand, Cozzolino et al. (2016) reported that resident liver stem cells (RLSCs) differentiated within 24 h after being cultured on PA gels with an elastic modulus matching the stiffness of healthy liver (0.4 kPa). In contrast, when cultured on PA gels with a stiffness of 80 kPa (corresponding to the stiffness of fibro cirrhotic parenchyma), RLSCs maintained their phenotype, delaying the onset of hepatocyte differentiation process. In the natural biological environment, the behaviour of cells is dictated by the structure and mechanical properties of the tissue. As such, one main design criteria of biomaterial-based approaches is tuneable mechanical properties, which led to intense research interests in polymers. The diversity of existing synthetic and natural polymers, coupled with the ability to design new types of polymers can readily produce biomaterials with controlled properties. Hydrogels, which are crosslinked 3D polymeric networks, are predominantly used when the stiffness of the substrate needs to be adjusted; this is mainly because the mechanical properties of the hydrogel can be tuned by adjusting the crosslinking density (Mawad et al. 2007, 2012). Hydrogels could be fabricated covering a large range of mechanical properties (very soft, ~ < 1 kPa to very stiff, ~ 500 kPa). They can also be fabricated from natural polymers such as alginate, chitosan, gelatin or synthetic polymers such as polyacrylamide (PA), poly(vinyl alcohol) (PVA) and poly(ethylene glycol) (PEG). The reader is referred to a comprehensive review by Tsou et al. (2016) describing how different type of hydrogels interact with different type of stem cells.

13.2.2.3 Summary

Surface topography, stiffness and type of polymer all play a crucial role in determining the fate of stem cells. Although the actual mechanisms are still not well defined, protein adsorption on the poymeric surface is a key factor. The conformation, orientation and quantity of adsorbed protein allow cell attachment via integrin receptor (Dee et al. 2003; Dalby et al. 2014). Furthermore, cell surface integrins play an important role in the interaction of stem cells with the surrounding matrix and are described as vital for their self-renewal (Lee et al. 2010; Kohen et al. 2009). As such the physicochemical properties of polymers play a significant role in modulating cell-matrix interactions. For example, anionic polymers such as poly[(methyl vinyl ether)-alt-(maleic acid)] (PMVE-alt-MA) that bears carboxylic and sulfonic groups can bind to growth factors enabling cell attachment (Brafman et al. 2010). hPSCs cultured on PMVE-alt-MA exhibited higher expression levels of integrins. Consequently, these polymers are demonstrated to support *ex vivo* expansion of hPSCs while maintaining their undifferentiated state.

13.2.3 Physico-Chemical Approaches to Engineer the Niche for Stem Cell Differentiation

One of the main objectives of using polymeric biomaterials in stem cell technology is to direct the cell differentiation into a specific cell lineage. However, the complexity and dynamic nature of the stem cell niche requires that a number of physico-chemical stimuli should be incorporated in the material design to achieve a positive outcome. For instance, a material designed for neural tissue regeneration would require a polymer of low stiffness (100–500 Pa), whereas bone applications benefit from composite materials (polymer and inorganic bioactive ceramics) mimicking the native composite nature of bone itself.

13.2.3.1 Neurogenic Lineage

Over the past decade significant developments have been made in combining neural stem cells (NSCs) with natural or synthetic polymeric biomaterials. In particular, nanofiber scaffolds or hydrogels combined with stem cells and growth factors are being developed to tackle neurological diseases. Nanofiber scaffolds are attractive in neural regeneration because they create aligned neural tissue similar to the native one. Using these nanofiber scaffolds, stem cell differentiation can be controlled by the orientation and diameter of the fibres (Sperling et al. 2017, see Fig. 13.3), 2D versus 3D configurations (Jakobsson et al. 2017), and polymer chemistry (Saha et al. 2008). Synthetic polymers such as PLA (Soleimani et al. 2010), PLGA (Kramer et al. 2011) and their co-polymers (Bini et al. 2006) are extensively used as the base materials for fabricating nanofiber mats for neural tissue engineering. These polymers are biodegradable, can be functionalized and spun into ultrafine continuous fibres that closely resemble the extra cellular matrices.

On the other hand, hydrogels play a significant role in neural transplantation because stem cells can be mixed with the biomaterial in a liquid form and induced to gel following targeted injection in vivo. Hydrogels are also porous structures that facilitate nutrient and oxygen transport and their mechanical properties can be tuned to suit the application requirements. In the presence of the appropriate growth factors, hydrogels can be employed to direct the differentiation of neural stem cells. These growth factors can be either chemically grafted on the polymeric backbone or simply physically entrapped in the network. Examples of hydrogels used for differentiation of neural stem cells include those based on natural polymers such as collagen (Huang et al. 2013; Yuan et al. 2014), hyaluronic acid (HA) (Liang et al. 2013; Preston and Sherman 2011) and hyaluronan derivative (Moshayedi and Carmichael 2013), or synthetic polymers such as PEG (Mckinnon et al. 2013), and polyurethane (Hsieh et al. 2015). In recent years, a new class of "smart" synthetic polymers, conducting polymers (CPs), is being explored for scaffold fabrication for nerve regeneration. CPs are conjugated polymers capable of conducting electrons. Coupled with their organic nature that matches the mechanical properties of tissue,

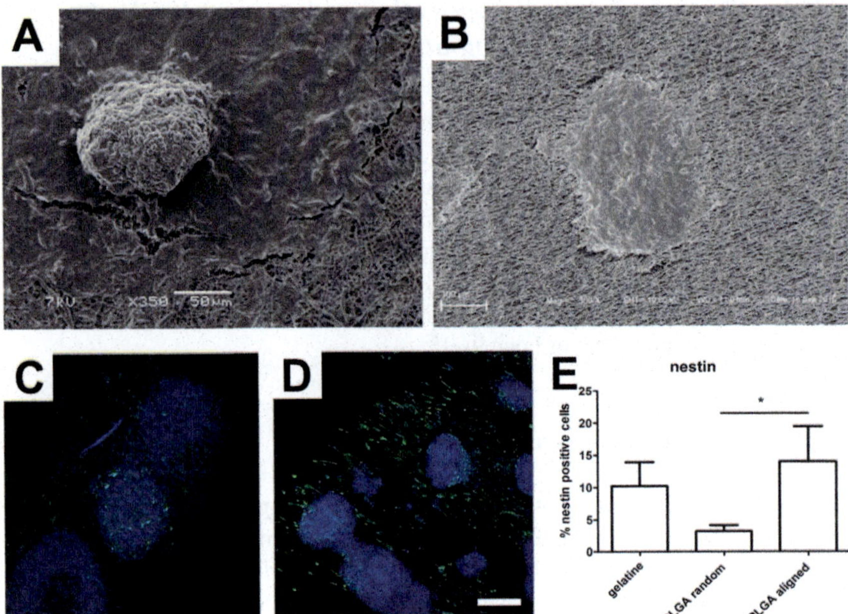

Fig. 13.3 SEM analysis of mESCs growth and differentiation on poly(lactic-co-glycolic acid) (PLGA) fiber scaffolds. (**a**) and (**b**) show differentiated mESCs on random and aligned PLGA fibers respectively. Scale bar (**a**) 50 μm, (**b**) 100 μm. Representative confocal fluorescent images of the neural marker, nestin shown in green, on random (**c**) and aligned (**d**) PLGA fibers. The cell nuclei were counterstained with DAPI (*blue*). Scale bar 100 μm. (**e**) Percentage of positive cells for nestin determined by flow cytometry analysis of mESC differentiation on PLGA nanofibers. * indicates a p values <0.05. (adapted from Sperling et al. 2017)

they are being explored as platforms to induce electric stimulation for electroresponsive cells such as neurons (Ghasemi-Mobarakeh et al. 2011). Polypyrrole is one example of CP demonstrated to differentiate neural cells (Stewart et al. 2015; Jin and Li 2014). In a study by Zhang et al. (2010) it was shown that by controlling the dopant in the polypyrrole scaffold, the cell fate could be directed. Polypyrrole doped with the polymer polystyrene sulfonate (PSS) induced hESCs differentiation after 20 days, whereas polypyrrole doped with the peptide laminin enhanced cell attachment and spreading. This study shows that by adjusting the type of dopant in the electroactive polymer, cell fate and development could be controlled.

13.2.3.2 Myogenic Lineage

Human iPSCs are being applied in cardiac regeneration due to their ability to differentiate into cardiomyocytes. While iPSCs have been shown to improve cardiac function following injection into an infarcted heart (Masumoto et al. 2014), a major limitation is the significant decrease in their viability following engraftment.

A biomaterial-based approach is being sought to resolve this issue. In contrast to 2D, 3D polymeric constructs have been shown to display many features of the native myocardium, such as cell-cell interaction, spontaneous beating activity and increased cardiac specific protein expression (Thavandiran et al. 2013). Design criteria of a biomaterial tailored for the heart include: (i) *mechanical durability under continuous strain*; (ii) *support of contractile functionality of the cardiomyocytes*; (iii) *appropriate stiffness;* and (iv) *conductivity,* to help transduce the bioelectric signal of cardiomyocytes. <u>Elastomeric polymers</u> are being used in cardiac regeneration because of their long-term mechanical stability under exposure to repeated strain cycles. Examples include poly(glycerol sebacate) (PGS) (Ravichandran et al. 2013), poly(urethane) (Alperin et al. 2005) and its biodegradable analogue poly(ester urethane) (Nieponice et al. 2010), PLA and PLGA (Chen et al. 2015). However, these polymers degrade into small acidic compounds that can trigger an immune response in the heart. To accommodate the contractile functionality of the cardiomyocytes, the stiffness of the scaffold could be modulated. Seeding cardiomyocytes in 3D PGS scaffolds of varying stiffness (2.35–5.99 kPa), it was found that the contractile function of the cardiac constructs correlate positively with low stiffness (Marsano et al. 2010). However, this study investigated the role of matrix mechanical properties on the function of differentiated cardiomyocytes. To examine whether elasticity plays a role in directing cardiac differentiation, tissue culture plates were coated with polydimethylsiloxane (PDMS), an inert synthetic polymer of tuneable mechanical properties and embryonic stem cells (ES) seeded on top were monitored for their development (Arshi et al. 2013). The authors reported that the differentiation of pluripotent ES cells into functional cardiomyocytes was better supported by the rigid PDMS substrate (~1000 kPa) in comparison to their softer counterpart (~10 kPa).

The interplay between the material stiffness and differentiation into cardiac lineage is not straightforward and requires consideration of not only the material properties but the polymer chemical composition, the type of stem cells used and the dimensionality of the construct (2D versus 3D). In a similar work by Battista et al. (2005) investigating the effect of elasticity of collagen scaffolds on embryonic stem cells, it was found that cardiac differentiation was significantly inhibited as the collagen scaffold stiffness was increased from 16 to 34 Pa. Also the study investigated the effect of biochemical cues by introducing laminin and fibronectin in the scaffolds. Laminin was found to promote differentiation into beating cardiomyocytes, whereas fibronectin stimulated endothelial cell differentiation and vascularization. These studies point out to the complexity of reproducing the extracellular environment in the laboratory. Not only the polymer chemistry needs to be considered, but its processing into a scaffold with tuneable mechanical closely matching that of the tissue is a key factor to achieve the required objective. Also, biochemical cues either added as soluble factors in the scaffold or anchored covalently on the polymer backbone are essential to mediate cell adhesion and direct differentiation of cardiac progenitor cells.

To date, contractile activity of cardiomyocytes has been investigated in: (i) *thin layers* (Shim et al. 2012; Feinberg et al. 2007), (ii) *3D cardiac microtissues* generated in PDMS molds (Boudou et al. 2012) or non-adherent plates (Ravenscroft et al. 2016); or (iii) *fully bioprinted system* (Lind et al. 2016). Recently, bioprinted vascularized cardiac tissue have been generated by plating iPSC-cardiomyocytes on top of bioprinted vascular network of endothelial cells (Zhang et al. 2016). Further maturation of iPSC-cardiomyocytes into a more adult phenotype has been investigated via electrical, for instance either with electrical impedence spectroscopy (Burgel et al. 2016) or using electrically conductive silicon nanowires (e-SiNWs) (Tan et al. 2017).

13.2.4 Genetic Approaches to Engineer the Stem Cell Niche

Reprogramming of stem cells in 3D cultures via *CRISPR* or *TALEN* technology has the potential to unveil mechanisms regulating the stem cell niche (Sun and Ding 2017; Yin et al. 2016; Gonzalez 2016). Reprogramming of somatic cells into stem cells is dependent on donor age, a factor that limits the use of stem cell-based therapies in humans (Lo Sardo et al. 2017). However, Schwank et al. (2013) have recently demonstrated how the use of CRISPR genome editing in 3D cultures generated from patients with cystic fibrosis (CF) could advance tissue repair and functionality. In this study, targeted genome editing of intestinal spheroid cultures from two CF patients with CRISPR-Cas9-mediated homology-directed repair corrected the mutation (deletion of phenylalanine at position 508) of the CF trans-membrane conductor receptor (CFTR), the primary cause of the disease, supporting the feasibility an autologous gene therapy strategy using 3D cultures in patients with hereditary diseases (Yui et al. 2012). Similarly, gene editing in 3D cultures could be utilized as in vitro disease models to identify novel molecular targets for future therapies, as recently demonstrated in genome-edited human intestinal epithelial organoids (Matano et al. 2015).

13.2.5 Synthetic Biology and the Stem Cell Niche

Synthetic biology allows the integration of highly dynamic and transient multiple stem cell niches at the same time during embryogenesis, which generated promising results in regulating the stem cell niche in vitro with direct application for future stem cell-based therapies (Purcell and Lu 2014). Thanks to the integration between the synthetic and biological components, tissue homeostasis or diverse functions of the stem cell niche can now be controlled by transcriptional activators and/or repressors, and other novel mechanisms, such as: (i) *switches*; (ii) *memory elements*, (iii) *cascades*; (iv) *time-delayed circuits*; (v) *oscillators;* (vi) *logic gates;* (vii) *artificial gene circuits* (Cheng and Lu 2012; Lohmueller et al. 2012; Purcell and Lu 2014; Siuti et al. 2013). For instance, *biosensors* generated using synthetic biology can record the history of cellular exposure to either individual or a sequence of environmental signals, creating a memory within the bioengineered tissue (Siuti et al. 2013). This memory may then be used for the delivery of exogenous signals and apply them in sequence rather than simultaneously, which may help to overcome several limitations occurring with stem cells, such as limited survival and engraftment upon transplantation.

13.3   Conclusions and Future Perspectives

Engineering of three-dimensional niche with defined features demonstrated to improve survival and architectural structure of stem cells, leading to a more physiological microenvironment and behavior. Due to the remaining hurdles in recapitulating in one in vitro model all the biological, morphological and physiological features typical of the in vivo stem cell niche features (Hunsberger et al. 2015), future studies aiming at integrating all the different physicochemical and genetic cues typical of the 3D niche described in this chapter should be considered before engineering stem cell-based therapeutics for humans. Furthermore, engineering of in vitro systems including bioprinted organs-on-a-chip for drug discovery and toxicity testing using stem cells may benefit from the integration of the abovementioned features. This information will become undoubtedly even more relevant with the upcoming interest in "precision medicine" and bio-banking of patient-specific 3D mini-organs (Bredenoord et al. 2017).

References

Adriani G, Ma D, Pavesi A, Kamm RD, Goh EL (2016) A 3D neurovascular microfluidic model consisting of neurons, astrocytes and cerebral endothelial cells as a blood-brain barrier. Lab Chip 17(3):448–459

Alperin C, Zandstra PW, Woodhouse KA (2005) Polyurethane films seeded with embryonic stem cell-derived cardiomyocytes for use in cardiac tissue engineering applications. Biomaterials 26:7377–7386

Arshi A, Nakashima Y, Nakano H, Eaimkhong S, Evseenko D, Reed J, Stieg AZ, Gimzewski JK, Nakano A (2013) Rigid microenvironments promote cardiac differentiation of mouse and human embryonic stem cells. Sci Technol Adv Mater 14(2):025003

Baraniak PR, Cooke MT, Saeed R, Kinney MA, Fridley KM, Mcdevitt TC (2012) Stiffening of human mesenchymal stem cell spheroid microenvironments induced by incorporation of gelatin microparticles. J Mech Behav Biomed Mater 11:63–71

Battista S, Guarnieri D, Borselli C, Zeppetelli S, Borzacchiello A, Mayol L, Gerbasio D, Keene DR, Ambrosio L, Netti PA (2005) The effect of matrix composition of 3D constructs on embryonic stem cell differentiation. Biomaterials 26:6194–6207

Benam KH, Novak R, Nawroth J, Hirano-Kobayashi M, Ferrante TC, Choe Y, Prantil-Baun R, Weaver JC, Bahinski A, Parker KK, Ingber DE (2016) Matched-comparative modeling of normal and diseased human airway responses using a microengineered breathing lung chip. Cell Syst 3:456–466 e4

Bhatia SN, Ingber DE (2014) Microfluidic organs-on-chips. Nat Biotechnol 32:760–772

Bini T, Gao S, Wang S, Ramakrishna S (2006) Poly(l-lactide-co-glycolide) biodegrad-able micro-fibers and electrospun nanofibers for nerve tissue engineering: an in vitro study. J Mater Sci 41:6453

Boudou T, Legant WR, Mu A, Borochin MA, Thavandiran N, Radisic M, Zandstra PW, Epstein JA, Margulies KB, Chen CS (2012) A microfabricated platform to measure and manipulate the mechanics of engineered cardiac microtissues. Tissue Eng Part A 18:910–919

Brafman DA, Chang CW, Fernandez A, Willert K, Varghese S, Chien S (2010) Long-term human pluripotent stem cell self-renewal on synthetic polymer surfaces. Biomaterials 31:9135–9144

Bratt-Leal AM, Carpenedo RL, Mcdevitt TC (2009) Engineering the embryoid body microenvironment to direct embryonic stem cell differentiation. Biotechnol Prog 25:43–51

Bredenoord AL, Clevers H, Knoblich JA (2017) Human tissues in a dish: the research and ethical implications of organoid technology. Science 355:6322. pii: eaaf9414

Burgel SC, Diener L, Frey O, Kim JY, Hierlemann A (2016) Automated, multiplexed electrical impedance spectroscopy platform for continuous monitoring of microtissue spheroids. Anal Chem 88:10876–10883

Carpenedo RL, Bratt-Leal AM, Marklein RA, Seaman SA, Bowen NJ, Mcdonald JF, Mcdevitt TC (2009) Homogeneous and organized differentiation within embryoid bodies induced by microsphere-mediated delivery of small molecules. Biomaterials 30:2507–2515

Chen Y, Wang J, Shen B, Chan CW, Wang C, Zhao Y, Chan HN, Tian Q, Chen Y, Yao C, Hsing IM, Li RA, Wu H (2015) Engineering a freestanding biomimetic cardiac patch using biodegradable poly(lactic-co-glycolic acid) (PLGA) and human embryonic stem cell-derived ventricular cardiomyocytes (hESC-VCMs). Macromol Biosci 15:426–436

Cheng AA, Lu TK (2012) Synthetic biology: an emerging engineering discipline. Annu Rev Biomed Eng 14:155–178

Chowdhury F, Li Y, Poh YC, Yokohama-Tamaki T, Wang N, Tanaka TS (2010) Soft substrates promote homogeneous self-renewal of embryonic stem cells via downregulating cell-matrix tractions. PLoS One 5:e15655

Christoffersson J, Bergstrom G, Schwanke K, Kempf H, Zweigerdt R, Mandenius CF (2016) A microfluidic bioreactor for toxicity testing of stem cell derived 3D cardiac bodies. Methods Mol Biol 1502:159–168

Cosson S, Otte EA, Hezaveh H, Cooper-White JJ (2015) Concise review: tailoring bioengineered scaffolds for stem cell applications in tissue engineering and regenerative medicine. Stem Cells Transl Med 4:156–164

Cozzolino AM, Noce V, Battistelli C, Marchetti A, Grassi G, Cicchini C, Tripodi M, Amicone L (2016) Modulating the substrate stiffness to manipulate differentiation of resident liver stem cells and to improve the differentiation state of hepatocytes. Stem Cells Int 2016:5481493

Dalby MJ, Gadegaard N, Oreffo RO (2014) Harnessing nanotopography and integrin-matrix interactions to influence stem cell fate. Nat Mater 13:558–569

Dee KC, Puleo DA, Bixios R (2003) Protein–surface interactions. An introduction to tissue-biomaterial interactions. Wiley, New York

Dennis SG, Trusk T, Richards D, Jia J, Tan Y, Mei Y, Fann S, Markwald R, Yost M (2015) Viability of bioprinted cellular constructs using a three dispenser cartesian printer. J Vis Exp 103:53156.

Doetschman TC, Eistetter H, Katz M, Schmidt W, Kemler R (1985) The in vitro development of blastocyst-derived embryonic stem cell lines: formation of visceral yolk sac, blood islands and myocardium. J Embryol Exp Morphol 87:27–45

Engler AJ, Sen S, Sweeney HL, Discher DE (2006) Matrix elasticity directs stem cell lineage specification. Cell 126:677–689

Feinberg AW, Feigel A, Shevkoplyas SS, Sheehy S, Whitesides GM, Parker KK (2007) Muscular thin films for building actuators and powering devices. Science 317:1366–1370

Figtree GA, Bubb KJ, Tang O, Kizana E, Gentile C Vascularized cardiac spheroids as novel 3D in vitro models to study cardiac fibrosis. Cells Tissues Organs 204:3

Fleming PA, Argraves WS, Gentile C, Neagu A, Forgacs G, Drake CJ (2010) Fusion of uniluminal vascular spheroids: a model for assembly of blood vessels. Dev Dyn 239:398–406

Frey O, Misun PM, Fluri DA, Hengstler JG, Hierlemann A (2014) Reconfigurable microfluidic hanging drop network for multi-tissue interaction and analysis. Nat Commun 5:4250

Gentile C (2016) Filling the gaps between the in vivo and in vitro microenvironment: engineering of spheroids for stem cell technology. Curr Stem Cell Res Ther 11:652–665

Ghasemi-Mobarakeh L, Prabhakaran MP, Morshed M, Nasr-Esfahani MH, Baharvand H, Kiani S, Al-Deyab SS, Ramakrishna S (2011) Application of conductive polymers, scaffolds and electrical stimulation for nerve tissue engineering. J Tissue Eng Regen Med 5:e17–e35

Gonzalez F (2016) CRISPR/Cas9 genome editing in human pluripotent stem cells: harnessing human genetics in a dish. Dev Dyn 245:788–806

Guidi N, Geiger H (2017) Rejuvenation of aged hematopoietic stem cells. Semin Hematol 54:51–55

Gunter J, Wolint P, Bopp A, Steiger J, Cambria E, Hoerstrup SP, Emmert MY (2016) Microtissues in cardiovascular medicine: regenerative potential based on a 3D microenvironment. Stem Cells Int 2016:9098523

He Y, Lu F (2016) Development of synthetic and natural materials for tissue engineering applications using adipose stem cells. Stem Cells Int 2016:5786257

Hsieh FY, Lin HH, Hsu SH (2015) 3D bioprinting of neural stem cell-laden thermoresponsive biodegradable polyurethane hydrogel and potential in central nervous system repair. Biomaterials 71:48–57

Huang F, Shen Q, Zhao J (2013) Growth and differentiation of neural stem cells in a three-dimensional collagen gel scaffold. Neural Regen Res 8:313–319

Huh D, Matthews BD, Mammoto A, Montoya-Zavala M, Hsin HY, Ingber DE (2010) Reconstituting organ-level lung functions on a chip. Science 328:1662–1668

Huh D, Kim HJ, Fraser JP, Shea DE, Khan M, Bahinski A, Hamilton GA, Ingber DE (2013) Microfabrication of human organs-on-chips. Nat Protoc 8:2135–2157

Hui EE, Bhatia SN (2007) Micromechanical control of cell-cell interactions. Proc Natl Acad Sci U S A 104:5722–5726

Hunsberger J, Harrysson O, Shirwaiker R, Starly B, Wysk R, Cohen P, Allickson J, Yoo J, Atala A (2015) Manufacturing road map for tissue engineering and regenerative medicine technologies. Stem Cells Transl Med 4:130–135

Itskovitz-Eldor J, Schuldiner M, Karsenti D, Eden A, Yanuka O, Amit M, Soreq H, Benvenisty N (2000) Differentiation of human embryonic stem cells into embryoid bodies compromising the three embryonic germ layers. Mol Med 6:88–95

Jaggy M, Zhang P, Greiner AM, Autenrieth TJ, Nedashkivska V, Efremov AN, Blattner C, Bastmeyer M, Levkin PA (2015) Hierarchical micro-nano surface topography promotes long-term maintenance of undifferentiated mouse embryonic stem cells. Nano Lett 15:7146–7154

Jakab K, Norotte C, Marga F, Murphy K, Vunjak-Novakovic G, Forgacs G (2010) Tissue engineering by self-assembly and bio-printing of living cells. Biofabrication 2:022001

Jakobsson A, Ottosson M, Zalis MC, O'carroll D, Johansson UE, Johansson F (2017) Three-dimensional functional human neuronal networks in uncompressed low-density electrospun fiber scaffolds. Nanomedicine 13(4):1563–1573

Jin G, Li K (2014) The electrically conductive scaffold as the skeleton of stem cell niche in regenerative medicine. Mater Sci Eng C Mater Biol Appl 45:671–681

Kamble H, Barton MJ, Jun M, Park S, Nguyen NT (2016) Cell stretching devices as research tools: engineering and biological considerations. Lab Chip 16:3193–3203

Kim JD, Choi JS, Kim BS, Choi YC, Cho YW (2010) Piezoelectric inkjet printing of polymers: stem cell patterning on polymer substrates. Polymer 51:2147–2154

Kohen NT, Little LE, Healy KE (2009) Characterization of Matrigel interfaces during defined human embryonic stem cell culture. Biointerphases 4:69–79

Kramer M, Chaudhuri JB, Ellis MJ (2011) Promotion of neurite outgrowth in corporation poly-l-lysine into aligned PLGA nanofiber scaffolds. Eur Cell Mater 22:53

Kumar D, Dale TP, Yang Y, Forsyth NR (2015) Self-renewal of human embryonic stem cells on defined synthetic electrospun nanofibers. Biomed Mater 10:065017

Lanphier E, Urnov F, Haecker SE, Werner M, Smolenski J (2015) Don't edit the human germ line. Nature 519:410–411

Lee ST, Yun JI, Jo YS, Mochizuki M, Van der Vlies AJ, Kontos S, Ihm JE, Lim JM, Hubbell JA (2010) Engineering integrin signaling for promoting embryonic stem cell self-renewal in a precisely defined niche. Biomaterials 31:1219–1226

Li YJ, Chung EH, Rodriguez RT, Firpo MT, Healy KE (2006) Hydrogels as artificial matrices for human embryonic stem cell self-renewal. J Biomed Mater Res A 79:1–5

Liang Y, Walczak P, Bulte JW (2013) The survival of engrafted neural stem cells within hyaluronic acid hydrogels. Biomaterials 34:5521–5529

Lind JU, Busbee TA, Valentine AD, Pasqualini FS, Yuan H, Yadid M, Park SJ, Kotikian A, Nesmith AP, Campbell PH, Vlassak JJ, Lewis JA, Parker KK (2016) Instrumented cardiac microphysiological devices via multimaterial three-dimensional printing. Nat Mater 16(3):303–308

Lippmann ES, Azarin SM, Kay JE, Nessler RA, Wilson HK, Al-Ahmad A, Palecek SP, Shusta EV (2012) Derivation of blood-brain barrier endothelial cells from human pluripotent stem cells. Nat Biotechnol 30:783–791

Lohmueller JJ, Armel TZ, Silver PA (2012) A tunable zinc finger-based framework for Boolean logic computation in mammalian cells. Nucleic Acids Res 40:5180–5187

Marban E, Cingolani E (2012) Heart to heart: cardiospheres for myocardial regeneration. Heart Rhythm 9:1727–1731

Marsano A, Maidhof R, Wan LQ, Wang Y, Gao J, Tandon N, Vunjak-Novakovic G (2010) Scaffold stiffness affects the contractile function of three-dimensional engineered cardiac constructs. Biotechnol Prog 26:1382–1390

Masumoto H, Ikuno T, Takeda M, Fukushima H, Marui A, Katayama S, Shimizu T, Ikeda T, Okano T, Sakata R, Yamashita JK (2014) Human iPS cell-engineered cardiac tissue sheets with cardiomyocytes and vascular cells for cardiac regeneration. Sci Rep 4:6716

Matano M, Date S, Shimokawa M, Takano A, Fujii M, Ohta Y, Watanabe T, Kanai T, Sato T (2015) Modeling colorectal cancer using CRISPR-Cas9-mediated engineering of human intestinal organoids. Nat Med 21:256–262

Mawad D, Martens PJ, Odell RA, Poole-Warren LA (2007) The effect of redox polymerisation on degradation and cell responses to poly (vinyl alcohol) hydrogels. Biomaterials 28:947–955

Mawad D, Boughton EA, Boughton P, Lauto A (2012) Advances in hydrogels applied to degenerative diseases. Curr Pharm Des 18:2558–2575

Mckinnon DD, Kloxin AM, Anseth KS (2013) Synthetic hydrogel platform for three-dimensional culture of embryonic stem cell-derived motor neurons. Biomater Sci 1:460–469

Mcmurray RJ, Gadegaard N, Tsimbouri PM, Burgess KV, Mcnamara LE, Tare R, Murawski K, Kingham E, Oreffo RO, Dalby MJ (2011) Nanoscale surfaces for the long-term maintenance of mesenchymal stem cell phenotype and multipotency. Nat Mater 10:637–644

Messina E, De Angelis L, Frati G, Morrone S, Chimenti S, Fiordaliso F, Salio M, Battaglia M, Latronico MV, Coletta M, Vivarelli E, Frati L, Cossu G, Giacomello A (2004) Isolation and expansion of adult cardiac stem cells from human and murine heart. Circ Res 95:911–921

Mironov V, Boland T, Trusk T, Forgacs G, Markwald RR (2003) Organ printing: computer-aided jet-based 3D tissue engineering. Trends Biotechnol 21:157–161

Moshayedi P, Carmichael ST (2013) Hyaluronan, neural stem cells and tissue reconstruction after acute ischemic stroke. Biomatter 3(1):e23863

Murphy SV, Atala A (2014) 3D bioprinting of tissues and organs. Nat Biotechnol 32:773–785

Murphy KC, Fang SY, Leach JK (2014a) Human mesenchymal stem cell spheroids in fibrin hydrogels exhibit improved cell survival and potential for bone healing. Cell Tissue Res 357:91–99

Murphy WL, Mcdevitt TC, Engler AJ (2014b) Materials as stem cell regulators. Nat Mater 13:547–557

Nie Z, Kumacheva E (2008) Patterning surfaces with functional polymers. Nat Mater 7:277–290

Nieponice A, Soletti L, Guan J, Hong Y, Gharaibeh B, Maul TM, Huard J, Wagner WR, Vorp DA (2010) In vivo assessment of a tissue-engineered vascular graft combining a biodegradable elastomeric scaffold and muscle-derived stem cells in a rat model. Tissue Eng Part A 16:1215–1223

Oleaga C, Bernabini C, Smith AS, Srinivasan B, Jackson M, Mclamb W, Platt V, Bridges R, Cai Y, Santhanam N, Berry B, Najjar S, Akanda N, Guo X, Martin C, Ekman G, Esch MB, Langer J, Ouedraogo G, Cotovio J, Breton L, Shuler ML, Hickman JJ (2016) Multi-organ toxicity demonstration in a functional human in vitro system composed of four organs. Sci Rep 6:20030

Passier R, Orlova V, Mummery C (2016) Complex tissue and disease modeling using hiPSCs. Cell Stem Cell 18:309–321

Potapova IA, Gaudette GR, Brink PR, Robinson RB, Rosen MR, Cohen IS, Doronin SV (2007) Mesenchymal stem cells support migration, extracellular matrix invasion, proliferation, and survival of endothelial cells in vitro. Stem Cells 25:1761–1768

Polonchuk L, Chabria M, Badi L, Hoflack J-C, Figtree G, Davies MJ, Gentile C (2017) Cardiac spheroids as promising in vitro models to study the human heart microenvironment. Sci Rep 7(1):7005

Preston M, Sherman LS (2011) Neural stem cell niches: roles for the hyaluronan-based extracellular matrix. Front Biosci (Schol Ed) 3:1165–1179

Purcell O, Lu TK (2014) Synthetic analog and digital circuits for cellular computation and memory. Curr Opin Biotechnol 29:146–155

Ravenscroft SM, Pointon A, Williams AW, Cross MJ, Sidaway JE (2016) Cardiac non-myocyte cells show enhanced pharmacological function suggestive of contractile maturity in stem cell derived cardiomyocyte microtissues. Toxicol Sci 152:99–112

Ravichandran R, Venugopal JR, Sundarrajan S, Mukherjee S, Ramakrishna S (2013) Cardiogenic differentiation of mesenchymal stem cells on elastomeric poly (glycerol sebacate)/collagen core/shell fibers. World J Cardiol 5:28–41

Reynolds BA, Rietze RL (2005) Neural stem cells and neurospheres—re-evaluating the relationship. Nat Methods 2:333–336

Saha K, Keung AJ, Irwin EF, Li Y, Little L, Schaffer DV, Healy KE (2008) Substrate modulus directs neural stem cell behavior. Biophys J 95:4426–4438

Schwank G, Koo BK, Sasselli V, Dekkers JF, Heo I, Demircan T, Sasaki N, Boymans S, Cuppen E, Van der Ent CK, Nieuwenhuis EE, Beekman JM, Clevers H (2013) Functional repair of CFTR by CRISPR/Cas9 in intestinal stem cell organoids of cystic fibrosis patients. Cell Stem Cell 13:653–658

Shim J, Grosberg A, Nawroth JC, Parker KK, Bertoldi K (2012) Modeling of cardiac muscle thin films: pre-stretch, passive and active behavior. J Biomech 45:832–841

Siuti P, Yazbek J, Lu TK (2013) Synthetic circuits integrating logic and memory in living cells. Nat Biotechnol 31:448–452

Soleimani M, Nadri S, Shabani I (2010) Neurogenic differentiation of human conjunctiva mesenchymal stem cells on a nanofibrous scaffold. Int J Dev Biol 54:1295–1300

Sperling LE, Reis KP, Pozzobon LG, Girardi CS, Pranke P (2017) Influence of random and oriented electrospun fibrous poly(lactic-co-glycolic acid) scaffolds on neural differentiation of mouse embryonic stem cells. J Biomed Mater Res A 105(5):1333–1345

Stewart E, Kobayashi NR, Higgins MJ, Quigley AF, Jamali S, Moulton SE, Kapsa RM, Wallace GG, Crook JM (2015) Electrical stimulation using conductive polymer polypyrrole promotes differentiation of human neural stem cells: a biocompatible platform for translational neural tissue engineering. Tissue Eng Part C Methods 21:385–393

Sardo VL, Ferguson W, Erikson GA, Topol EJ, Baldwin KK, Torkamani A (2016) Influence of donor age on induced pluripotent stem cells. Nat Biotechnol 35(1):69–74

Sun Y, Ding Q (2017) Genome engineering of stem cell organoids for disease modeling. Protein Cell 8(5):315–327

Tan Y, Richards D, Coyle RC, Yao J, Xu R, Gou W, Wang H, Menick DR, Tian B, Mei Y (2017) Cell number per spheroid and electrical conductivity of nanowires influence the function of silicon nanowired human cardiac spheroids. Acta Biomater 51:495–504

Thavandiran N, Dubois N, Mikryukov A, Masse S, Beca B, Simmons CA, Deshpande VS, Mcgarry JP, Chen CS, Nanthakumar K, Keller GM, Radisic M, Zandstra PW (2013) Design and formulation of functional pluripotent stem cell-derived cardiac microtissues. Proc Natl Acad Sci U S A 110:E4698–E4707

Torisawa YS, Spina CS, Mammoto T, Mammoto A, Weaver JC, Tat T, Collins JJ, Ingber DE (2014) Bone marrow-on-a-chip replicates hematopoietic niche physiology in vitro. Nat Methods 11:663–669

Tsou YH, Khoneisser J, Huang PC, Xu X (2016) Hydrogel as a bioactive material to regulate stem cell fate. Bioactive Mater 1:39–55

Vallier L, Pedersen RA (2005) Human embryonic stem cells: an in vitro model to study mechanisms controlling pluripotency in early mammalian development. Stem Cell Rev 1:119–130

Van der Helm MW, Van der Meer AD, Eijkel JC, Van den Berg A, Segerink LI (2016) Microfluidic organ-on-chip technology for blood-brain barrier research. Tissue Barriers 4:e1142493

Van der Meer AD, Van den Berg A (2012) Organs-on-chips: breaking the in vitro impasse. Integr Biol (Camb) 4:461–470

Villa-Diaz LG, Nandivada H, Ding J, Nogueira-de-Souza NC, Krebsbach PH, O'shea KS, Lahann J, Smith GD (2010) Synthetic polymer coatings for long-term growth of human embryonic stem cells. Nat Biotechnol 28:581–583

Visconti RP, Kasyanov V, Gentile C, Zhang J, Markwald RR, Mironov V (2010) Towards organ printing: engineering an intra-organ branched vascular tree. Expert Opin Biol Ther 10:409–420

Willerth SM, Sakiyama-Elbert SE (2008) Combining stem cells and biomaterial scaffolds for constructing tissues and cell delivery. StemBook, Cambridge, MA

Yin X, Mead BE, Safaee H, Langer R, Karp JM, Levy O (2016) Engineering stem cell organoids. Cell Stem Cell 18:25–38

Yuan N, Tian W, Sun L, Yuan R, Tao J, Chen D (2014) Neural stem cell transplantation in a double-layer collagen membrane with unequal pore sizes for spinal cord injury repair. Neural Regen Res 9:1014–1019

Yui S, Nakamura T, Sato T, Nemoto Y, Mizutani T, Zheng X, Ichinose S, Nagaishi T, Okamoto R, Tsuchiya K, Clevers H, Watanabe M (2012) Functional engraftment of colon epithelium expanded in vitro from a single adult Lgr5(+) stem cell. Nat Med 18:618–623

Zhang C, Zhao Z, Abdul Rahim NA, Van Noort D, Yu H (2009) Towards a human-on-chip: culturing multiple cell types on a chip with compartmentalized microenvironments. Lab Chip 9:3185–3192

Zhang L, Stauffer WR, Jane EP, Sammak PJ, Cui XT (2010) Enhanced differentiation of embryonic and neural stem cells to neuronal fates on laminin peptides doped polypyrrole. Macromol Biosci 10:1456–1464

Zhang YS, Arneri A, Bersini S, Shin SR, Zhu K, Goli-Malekabadi Z, Aleman J, Colosi C, Busignani F, Dell'erba V, Bishop C, Shupe T, Demarchi D, Moretti M, Rasponi M, Dokmeci MR, Atala A, Khademhosseini A (2016) Bioprinting 3D microfibrous scaffolds for engineering endothelialized myocardium and heart-on-a-chip. Biomaterials 110:45–59

Index

© Springer International Publishing AG 2017
A. Birbrair (ed.), *Stem Cell Microenvironments and Beyond*,
Advances in Experimental Medicine and Biology 1041,
DOI 10.1007/978-3-319-69194-7

Printed by Printforce, the Netherlands